Praise for *Power*

Richard Heinberg offers a powerful new way of understanding the historic rise and probable fall of our species. It is an impressive, sweeping, and thought-provoking narrative.

— Dennis Meadows, co-author, *The Limits to Growth*

A profound, rigorous, convincing, and actionable lesson on how to understand power less as the control one has over others and more as the collective capacity we have to share with one another. A rich, moving, and necessary treatise from our most accomplished, coherent, and compassionate thinker on sustainable futures.

— Douglas Rushkoff, author, *Present Shock* and *Team Human*, founder, Laboratory for Digital Humanism

Richard Heinberg is a writer of unfailing interest and this book sums up much of his life's thinking. Understanding our dilemma in terms of power is, well, a powerful way for getting at the predicaments and possibilities of this fraught moment in our evolving history as a species.

— Bill McKibben, author, *The End of Nature*

Power is an extraordinary tour de force. It is a comprehensive compendium of how it has emerged, despite our self-proclamation to be sentient beings, that we now find ourselves scrambling on the edge of a cliff. Ironically this perilous rock-face is one that we ourselves have created. As a species, spurred on by the power of our migrant curiosity, we have exploited the immediate opportunities of the natural world while blindly discounting the future. But our planet keeps score and fortunately so does Richard Heinberg. *Power* is a must read and a call to action for those seeking a sustainable, balanced, human future in harmony with the Earth. No guarantees, of course, but harnessing the power of sentient action certainly beats the alternative; of continuing our blind stumble only soon to be swept aside, as have many creatures before us.

— Peter C. Whybrow, author, *The Well-Tuned Brain*

Heinberg goes to the very heart of the issue. Using his immense knowledge of biology, science, history, psychology, and the politics of energy, he shows that the environmental and social crises we face today have in their origin the insatiable human pursuit, and often abuse, of power, in all its forms. In showing us the path forward, Heinberg guides us to achieve power-limiting behavior so that we cannot just survive but thrive on a healthy planet and in healthy balance with one another.

— Maude Barlow, author, activist, and co-founder, The Blue Planet Project

Power is sweeping in scope and a powerful presentation. Richard Heinberg is willing to face the harsh reality of multiple, cascading social and ecological crises without flinching, and he has written a comprehensive book offering readers a framework for moving forward that isn't based on wishful thinking. Drawing on his decades of activism and research, Heinberg explains why power and energy are central concepts for understanding the human predicament and shaping our future. Equal parts science and philosophy, history and contemporary analysis, *Power* is more engaging than a scholarly tome and more thoughtful than journalism. Heinberg's book is a model of public scholarship about life-or-death challenges to human societies.

—Robert Jensen, author, *The Restless and Relentless Mind of Wes Jackson*

Richard Heinberg's panoramic review of known forms of power is both sobering and inspiring. Given our species' habitual methods for getting its way, be these methods physical, mental, or social, the outlook for our future is bleak indeed. Yet, Heinberg allows for the slim but real possibility of exercising restraint. If we are so persuaded, by wisdom or love for beauty, the future even now remains open. Indeed, such restraint returns us to ancient, almost forgotten appetites and capacities.

— Joanna Macy, author, *World As Lover, World As Self*

Power serves as a Rosetta Stone to decipher how our species went from one of many to apex predator in a very short time. A necessary book to fully understand the imperative that our species returns to "right relation" in this critical time.

— Peter Buffett, composer and philanthropist

For three decades, Richard Heinberg has been foretelling a day when humanity will be compelled to make a fateful choice: either turn away from our path of headlong growth or follow that path into a dark, dystopian future. Now, in 2021, that day has come. As with previous reversals of growth in societies throughout history, Heinberg concludes, humanity's ability to successfully navigate the coming worldwide decline will depend on how we handle power. We must, he says, finally reject vertical social power—the ability to get others to do something—and embrace our collective horizontal power—the ability of a group to self-organize to accomplish something. *Power* is Heinberg's masterwork and it could not be more timely, arriving just as our window for action threatens to slam shut. Ignore this book at your peril.

— Stan Cox, author, *The Green New Deal and Beyond*

Heinberg's *Power* is a searing, unflinching revelation of what has driven us to our current existential crisis: humanity's quest for power. Impeccably researched and masterfully written, this book explains how and why humanity is driving itself off the cliff. If there is any hope for us to continue, Heinberg shows why it must come from efforts to limit our own power.

— Dahr Jamail, author, *The End of Ice*

Power is Richard Heinberg at his synthesizing best. In this sweeping volume, he deftly links raw energy—essential for anything to happen in the physical world—to the exercise of political power in the cultural domain. If the productive use of energy is the ultimate key to evolutionary success, then humanity has no equal on Earth. But energy is also the source of society's addiction to economic growth, and the international power politics that are destroying the planet. In *Power*, Richard Heinberg asks whether we can avoid catastrophe. Will competing nations' primal lust for power give way to high intelligence, mutual trust, and unreserved cooperation in the quest to salvage civilization? Not a trivial question, as only success will grant humanity the chance to scramble yet another rung up the evolutionary ladder.

— William E. Rees, Phd, FRSC, Professor Emeritus,
UBC/SCARP, Faculty of Applied Science,
co-author, *Our Ecological Footprint*

Power reminds us that Richard Heinberg is one of the most important public intellectuals in the conversation about society's future. Eminently readable and engaging, *Power* is breathtaking in its scope and insight. Heinberg persuasively argues that we have reached evolutionary limits to concentrated social power and that empathy and beauty are key to averting ecological and social catastrophe.

> — Chuck Collins, Institute for Policy Studies,
> author, *The Wealth Hoarders*

I turn to Richard Heinberg whenever I need to understand something about energy; he's the "go-to" source. And now this! *Power*, with seamless fluency in paleo-history, economics, psychology, and politics, is the "must-read" for anyone wondering how we can make it through the 21st century. This book is more than informative. It is enlightening. It is essential. It is powerful!

> — Suzanne Moser, climate researcher and consultant

A sobering and timely book just as many governments appear to acknowledge—after decades of inaction—the dangers of climate change.

> — EEnergy Informer

It may be a moral idea that hard work pays off but if we need proof that it counts, this latest from Richard Heinberg carries all the evidence we need. His encyclopedic treatment of power is brilliant. It is sure to pop up in courses and living rooms like toast.

> — Wes Jackson, founder, The Land Institute

POWER

LIMITS
AND
PROSPECTS FOR
HUMAN SURVIVAL

RICHARD HEINBERG

new society
PUBLISHERS

Cover design by Diane McIntosh.

Printed in Canada. First printing September, 2021.

Inquiries regarding requests to reprint all or part of *Power* should be addressed to New Society Publishers at the address below. To order directly from the publishers, please call toll-free (North America) 1-800-567-6772, or order online at www.newsociety.com

Any other inquiries can be directed by mail to:

New Society Publishers
P.O. Box 189, Gabriola Island, BC V0R 1X0, Canada
(250) 247-9737

LIBRARY AND ARCHIVES CANADA CATALOGUING IN PUBLICATION

Title: Power : limits and prospects for human survival / Richard Heinberg.

Names: Heinberg, Richard, author.

Description: Includes index.

Identifiers: Canadiana (print) 20210260416 |
Canadiana (ebook) 20210260521 | ISBN 9780865719675
(softcover) | ISBN 9781550927610 (PDF) | ISBN 9781771423571 (EPUB)

Subjects: LCSH: Human ecology. | LCSH: Nature—Effect of human beings on.
| LCSH: Social history. |

LCSH: Human evolution. | LCSH: Power (Social sciences)

Classification: LCC GF41 .H45 2021 | DDC 304.2—dc23

New Society Publishers' mission is to publish books that contribute in fundamental ways to building an ecologically sustainable and just society, and to do so with the least possible impact on the environment, in a manner that models this vision.

for the survivors

Contents

List of Figures xii

List of Sidebars xiii

Acknowledgments xv

Introduction 1

1. Power in Nature: From Mitochondria to Emotion
    and Deception 15
    The Basis of Life's Power 19
    Power and Bodies 34
    Power and Behaviors 40
    Proto-Human Powers 49

2. Power in the Pleistocene: On Spears, Fires, Furs, Words,
    and Flutes—And Why Men Are Such Power-Hogs . 57
    Hands and Stone 66
    The Fire Ape 69
    Skins 73
    From Grunts to Sentences 75
    Gender Power 83
    The Power of Art 91

3. Power in the Holocene: The Rise of Social Inequality . 97
    Gerdening, Big Men, and Chiefs:
       Power from Food Production 99
    Plow and Plunder: Kings and the First States 105
    Herding Cattle, Flogging Slaves:
       Power from Domestication 114

Stories of Our Ancestors: Religion and Power 121
Tools for Wording: Communication Technologies 131
Numbers on Money 139
Pathologies of Power 147

4. **Power in the Anthropocene: The Wonderful World**
   of Fossil Fuels 159
   It's All Energy 162
   The Coal Train 170
   Oil, Cars, Airplanes, and the New Middle Class 177
   Oil-Age Wars and Weapons 185
   Electrifying! 193
   The Human Superorganism 198

5. **Overpowered: The Fine Mess**
   We've Gotten Ourselves Into 203
   Climate Chaos and Its Remedies 206
   Disappearance of Wild Nature 216
   Resource Depletion 222
   Soaring Economic Inequality 228
   Pollution 234
   Overpopulation and Overconsumption 238
   Global Debt Bubble 242
   Weapons of Mass Destruction 248

6. **Optimum Power: Sustaining Our Power Over Time** . . 257
   Involuntary Power Limits: Death, Extinction, Collapse . . . 260
   Self-Limitation in Natural and
      Human-Engineered Systems 266
   Taboos, Souls, and Enlightenment 273
   Taxes, Regulations, Activism, and Rationing:
      Power Restraint in the Modern World 279
   Games, Disarmament, and Degrowth 288
   Denial, Optimism Bias, and Irrational Exuberance 292

7. The Future of Power: Learning to Live
   Happily Within Limits   303
   All Against All   305
   Trade-Offs Along the Path of Self-Restraint   312
   The Fate of the Superorganism   319
   Questioning Technology   325
   Learning to Live with Less Energy and Stuff   328
   Lessening Inequality   334
   Population: Lowering It and Keeping It Steady   337
   Fighting Power with Power   341
   Long-Term Power Through Beauty, Spirituality,
   and Happiness   350

Notes   361

Index   389

About the Author   399

About New Society Publishers   400

Figures

1.1 Proton pumping in a bacterium   24
1.2 Exponential growth   30
1.3 The energy pyramid in nature   33
1.4 Visualization of Kleiber's law   35
2.1 Global temperatures during the Pleistocene
        and Holocene   63
2.2 Key language, motor, and auditory areas of the brain . . .   76
3.1 Placement of people on a slave ship   119
5.1 Global wildlife decline   219
5.2 Global extraction/production of resources in
        1875, 1945, and 2015 (metric tons)   227
5.3 Power Center import vulnerability ratings of
        nonrenewable resources   227
5.4 United States top one percent income share (pre-tax) . . .   234
5.5 Growth in total global debt by sector, 1999–2019   247
5.6 Gun ownership and violent deaths   255
6.1 Predator/prey dynamics   262
6.2 The adaptive cycle   264

Sidebars

1. Defining Power 11
2. Powers in Math 18
3. Measures of Physical Power 21
4. How Much Power? 47
5. Ominous Power from Space: Asteroids, Comets, and Climate Change 62
6. A Watery Theory of Human Origins 64
7. Male Violence 86
8. Human Aesthetic Decadence 93
9. Measures of Social Power 98
10. Pandemics and the Evolution of States and Empires . . . 112
11. Slavery and Power 116
12. DNA Evidence for Steppe Invasions 124
13. Justifying Colonialism 130
14. The Original Sins of Mainstream Economists 145
15. Institutional Power 150
16. Genocide as Ultimate Exclusionary Social Power 152
17. Key Writers on Power 155
18. An Arrested Industrial Revolution in China 173
19. Geopolitics: Global Power 188
20. Personal Carbon Output 208
21. Rising Risk of Disease and Pandemic 217
22. Inequality in Economic and Political Power in the US, and Its Consequences 229
23. Guns: The Power to Kill Cheaply and Easily at a Distance . 253
24. The 2,000-Watt Society 269
25. Viktor Frankl and the Will to Meaning 317

26. Dethroning GDP: Key to Limiting the Power
    of the Superorganism 320
27. Energy and Human Values. 323
28. Advice to Young People in the 21st Century. 332
29. Power Analysis and Organizing for Activists 343

Acknowledgments

First off, I acknowledge the staggering level of privilege that made this book possible. I have been fortunate to be able to spend much of my life reading, thinking, and writing; and I've traveled to roughly 20 countries, rich and poor, where I've had the opportunity to learn both from trained observers of nature and society, and from groups of people self-organizing to solve environmental and social problems. Further, pursuing my career has entailed damage to ecosystems (as a result of transportation and other resource consumption) that will impact other living beings and future generations of humans. Relatively few individuals have been able to enjoy such advantages. I can only hope that this book represents a worthwhile return on society's and nature's investment—if I may be permitted to frame the situation this way.

I am deeply grateful to my colleagues at Post Carbon Institute, and to those who have donated to our organization, for materially supporting me as I worked uninterrupted for many months on the manuscript. Thanks especially to Asher Miller, the Institute's Executive Director; Rob Dietz, our Program Director; and Daniel Lerch, our Publications Director, for being thoughtful and knowledgeable sounding boards for drafts of the book as it developed. Daniel also expertly shepherded the process of finding a publisher, putting together the illustrations, and handling planning and correspondence related to the publication process. Thanks also to Amy Buringrud for spearheading our marketing efforts, and Desiree Cesarini for managing events in support of this project. I again feel privileged—in this case, to work with such competent and conscientious colleagues.

Thanks to Kristin Anton, who did the extensive, revelatory research and calculations for the sidebar, "How Much Power?" My appreciation also goes out also to Craig Collins, who made key suggestions regarding several chapters, especially the last. Special thanks to Joni Praded, who read versions of the Introduction and the early chapters and helped me make these key parts of the book more easily understandable to a wider range of readers.

I'd like to publicly acknowledge my gratitude to the team at New Society Publishers—notably Acquisitions Editor Rob West, who somehow understood the usefulness of this project at a time when "big picture" books about the human condition have fallen out of fashion. Kudos also to Julie Rayddish, publisher; Sue Custance, publishing director; EJ Hurst, sales director; Sara Reeves, marketing director; and the rest of the staff, who all contributed to a successful release. Murray Reiss deserves credit for meticulous copy-editing.

Finally, I must once again acknowledge the support of my wife Janet Barocco, whose love of nature, art, music, and good food fills our lives with beauty, thereby providing a stable foundation from which I've been able to launch my unsettling investigations and critiques of the unsustainable underpinnings of modern industrial society.

INTRODUCTION

Many people are searching for a magic formula to save the world from the converging crises of the 21st century. Climate change, economic inequality, air and water pollution, resource depletion, and the catastrophic disappearance of wildlife threaten to upend society while destabilizing our planet to such a degree that it may be impossible for future generations of humans to persist. What if we could solve all these problems with one simple trick?

Don't hold your breath. A single solution doesn't exist: it's not socialism or capitalism, it's not renewable energy or nuclear power, it's not religion or atheism, and it's not hemp. However, I believe there is a single causative agent in back of most of our troubles, the understanding of which could indeed help us emerge from the hole we're rapidly digging for ourselves.

That causative agent is power—our pursuit of it, our overuse of it, and our abuse of it. In this book, I argue that all the problems mentioned above, and others as well, are problems of power. We humans are nature's supreme power addicts. Power—the ability to do something, the ability to get someone else to do something, or the ability to *prevent* someone else from doing something—is everywhere in the human world. We obsess over power in its various forms, from wealth to governmental authority to weaponry to the concentrated energy sources that make modern industrial societies run. We seek power in many ways. But doing so often gets us into trouble. And it may be our downfall as a species.

1

Seeing the converging crises of this century as problems of power doesn't change much, in that we're still left fighting a host of individual battles. After all, recognizing that climate change is a problem of power, as I argue in Chapter 5, doesn't make it easier for nations to reduce their carbon emissions. Yet it also changes everything. It reveals how our current existential crises are related, and suggests a common meta-strategy for dealing with them.

◆   ◆   ◆

One might think that everything that could possibly be written on the subject of power already has been. At least, that's what I thought when I began the journey that led to this book. There are thousands of tomes that discuss subjects related to power in one or another of its many manifestations, and hundreds with the word *power* in their titles. But no book that I'm aware of has systematically examined the sundry forms of power, and investigated how they are related, how they arose, and what they mean for us today.

Perhaps the reason no author has addressed power so broadly is that it is a topic that's both huge and apparently nebulous. How to make sense of something so incomprehensibly vast and varied? Why even try?

When I started the research that would culminate in this book, I wasn't compelled by a burning interest in power *per se*; rather, I was driven to better understand the problems that imbalances and abuses of power have caused. I was determined to find answers to three survival-level questions:

1. How has *Homo sapiens*, just one species out of millions, become so powerful as to bring the planet to the brink of climate chaos and a mass extinction event?
2. Why have we developed so many ways of oppressing and exploiting one another?
3. Is it possible to change our relationship with power so as to avert ecological catastrophe, while also dramatically reducing social inequality and the likelihood of political-economic collapse?

In their essence, these questions had dogged me my entire adult life, though it's only in the last few years that I've been able to distil them down to these few words. As I pondered these questions, it became increasingly clear that reliable answers required a clearer understanding of power in and of itself, since it's the thread tying together our critical human problems and their potential solutions.

What *is* power? I decided to do a literature search. Not only was I dismayed to find no existing comprehensive investigation of the nature and workings of power, but I began to notice that, in books that discuss it, power is often poorly defined, if it's defined at all. I wondered if that was because no one had thought to trace the story of power back to its beginnings.

Physicists define power as the rate of energy transfer. That, at least, is a precise definition, and one that enables power to be measured quantitatively. Does it provide a good place to start in better understanding the power of, say, great wealth or high political office? That seemed doubtful at first.

Nevertheless, I already knew the importance of energy in recent history, having written several books about fossil fuels and renewable energy alternatives. Further, one of the most important lessons I had learned during my years of examining these subjects was that, if you want to understand any ecosystem or human society, a good rule of thumb is to *follow the energy*. I wondered if, by starting with the process of energy transfer and tracing its development through biological evolution and human history up to the present, it might be possible to better grasp what power is and how it works—and, in the bargain, to get a better idea of how to deal with our converging power problems. My goal would not be to reduce the complex world of social power to energy (so that political influence, for example, could be measured in watts), but simply to better grasp how our many forms of power arose and how they relate to one another, and thereby discover if there are indeed solutions to our impending survival dilemmas.

This focus on energy turned out to be a way not just of making power more comprehensible, but also of tying together a wide range of disparate phenomena in fields from cell biology to ecology to psychology to geopolitics. Most importantly, it threw new light on my three questions, leading me to surprising ideas for changing power dynamics, changing our personal behavior, changing our communities, and changing the world.

The third of my motivating questions, the most crucial one, is of course still open. But in the pages that follow I test the widespread belief that the pursuit of power is irrepressible, that bullies will forever be bullies, that the high and mighty will ultimately triumph, and that people in wealthy countries will never willingly give up comforts and conveniences in order to forestall global environmental catastrophe.

Boiled down to its basics, this belief holds that the will to power overwhelms all other human motives. There is evidence to support that belief. As I discuss in Chapter 1, biologists tend to agree that evolution has been driven by the *maximum power principle*—according to which, among directly competing systems, the one that harnesses available energy most effectively will prevail. Human beings' pursuit of power is rooted in nature: evolutionary precursors of it can be seen in competition between species for territory and food, and between members of the same species for mating opportunities. Nature is a power struggle.

However, it's also clear there is more going on, both in nature and in human societies. Evolution has found ways of preventing species from attaining so much power that they overrun environmental limits, and human societies have evolved ways of reining in tyrants, sharing and conserving resources, and limiting inequality. In Chapter 6, I propose a new bio-social principle in evolution—the *optimum power principle*—to describe these pathways.

Strategies to avert the concentration of too much power, whether in nature or human affairs, are partial and imperfect. They can't prevent occasional excesses. A case in point: evolution has no precedent for the immense power that humanity has recently derived from fossil fuels. Coal, oil, and natural gas have enabled humans to increase their total energy usage forty-fold in the span of just three human lifetimes—a rate of increase that's likely far greater than any previous power shift since the dawn of life on Earth. But fossil fuels are finite and depleting resources, and burning them destabilizes the global climate. So, we are left in a precarious spot: we will have to adapt at an unprecedented pace to limit this excess power, or risk societal and ecosystem collapse.

Help may come from an unexpected source. Beauty, compassion, and inspiration can influence or motivate human behavior. In a sense, then, these are powers too—though of a kind fundamentally different from the political, military, and economic powers that run our world. Yet, as we will see, beauty has helped drive biological evolution, and transcendent qualities of human character have shaped history. If we are to survive this century, we may need to rely on and develop these powers as never before.

Again, there is no silver bullet here. Even though I suggest specific ways of limiting power that could address the main crises of our time, I can't promise that these suggestions are politically realistic or likely to be implemented. What's the point, then? By better understanding power, I believe we can gain a clearer view of the human condition, reaping not just knowledge, but perspective and perhaps even wisdom. Practical applications may follow.

◆　◆　◆

The story of humanity's fascination with power begins with four key advantages that we have exploited to a far greater degree than any other species on Earth: the ability to make and manipulate *tools*, which we were able to develop because of our opposable thumbs and big brains; *language*, which enabled us to coordinate our behavior over time and space; *social complexity*, which makes human societies more cooperative and formidable, though usually at the expense of increased inequality; and our *ability to harness energy* beyond that which is supplied by food and exerted through muscle. *Power* chronicles how these advantages have propelled us on a trajectory from hunter-gatherer life in small, wandering bands to modern existence in huge cities, surrounded by machines and able to summon highly desirable foods and manufactured goods with a keystroke.

At the same time, power inequality (including the power differential between women and men) has tended to grow in human societies—though in fits and starts, and with occasional reversals. In the story of social power, starring roles have been played by three kinds of tools: *money*, which can best be thought of as quantifiable, storable, and transferrable social power, and as a token for the ability to command energy; *weapons*, which have enabled ever more sophisticated and deadly forms of warfare, while also helping drive cultural evolution; and *communication technologies* (from writing to social media), which have given some people the ability to influence the minds of many others.

Money, weapons, and communication technologies—the key tools of social power—have enabled relatively few human beings to wield extraordinary influence. Today just a few extremely rich individuals claim as much wealth as the poorer half of humanity, and we find ourselves asking whether such extreme levels of inequality are sustainable. New cross-disciplinary analysis of hundreds of societies from the past 5,000 years offers unprecedented insight into how and why social and economic inequality arises, and the trajectory on which it tends to propel societies. As we'll see, worsening inequality tends to lead to cycles of societal expansion followed by ages of discord.

Military power is perhaps the rawest form of social power. We'll trace how it has evolved from hand-to-hand combat to the point where a single soldier can obliterate hundreds, even millions of combatants and noncombatants without ever seeing their faces or hearing their anguished cries. In the past century, some of our weapons have become so awesome that our only prospect for collective survival lies in never using them. We'll also see how warfare contributed to the origin of early city-states with full-time division of labor, the creation of empires, and the modern trend toward economic globalization.

The powers to communicate quickly over great distances, to heal injuries and cure diseases, to travel safely and quickly halfway around the world in a day, and to explore other planets seem benign by comparison. Yet all these powers—whether used to dominate or enable—are connected and follow some of the same basic principles.

We'll explore in Chapter 1 how living creatures evolved fascinating and ingenious ways of deriving and exerting power. Power enables not just individual survival, but the production of breathtaking variety and beauty throughout nature. But it is possible to have too much of a good thing; and that, I argue, is the essence of the human predicament in the 21st century. It is also possible, of course, to abuse power; and, as we will see, the problem of the abuse of power is often closely related to that of the overaccumulation of power (as historians who study the careers of dictators continually remind us). Indeed, the overaccumulation of power makes the abuse of power increasingly likely.

Altogether, in response to the three questions I posed earlier, this book will leave the reader with six takeaways:

1. Power is everywhere. We can't understand nature or human society without investigating the workings of power.
2. Our human ability to overwhelm nature and our tendency toward extreme inequality have both evolved in discrete stages. That is, in neither case was evolution a steady process. It's possible to pinpoint key moments in biological evolution and social evolution when everything changed due to a dramatic power shift.
3. There is a fundamental correlation between physical power and social power. Social scientists sometimes tend to downplay this point. But throughout history, dramatic increases in physical power, derived from new technologies and from harnessing new energy sources, have often tended to lead to more *vertical social power* (that's a phrase we'll unpack as we go along; it basically means a few people having more wealth than everybody else, or being able to tell lots of other people what to do).
4. Our problems with power result not just from abuse. We're rightly outraged by abuses of power in the forms of slavery, despotism, corruption, racism, sexism, and so on. But sometimes the accumulation of too much power within a system is problematic regardless of the benign or sinister intent of system managers.
5. The "will to power," about which German philosopher Friedrich Nietzsche wrote, is real—but it isn't everything. We humans have

other instincts that counteract our relentless pursuit of power. Efforts to limit power are deeply rooted in nature's cycles and balancing mechanisms, and have been expressed in countless social movements over many centuries, including movements to curb the power of rulers, to abolish slavery, and to grant women political rights equal to those enjoyed by men.

6. The power of beauty has driven biological evolution as surely as has the pursuit of dominance, and the power of inspiring example has shaped human social evolution as much as the quests for wealth and superior weaponry. These aren't just feel-good sentiments; they're research-based observations.

◆   ◆   ◆

Finally, an overview of the book's structure.

Chapter 1 explores power in nature. When discussing power, it's tempting to jump directly into an examination of social power in human groups; however, social power is based in biology, and it's necessary to investigate power's evolutionary roots if we're to understand its manifestations in the modern world. This chapter is a whirlwind tour of the biology of power. It addresses the ways power moves through the living world—starting with the cell's ability to capture, store, and controllably dissipate energy. We'll explore the manifold powers of living things, including motion, perception, communication, reproduction, emotion, and deception. As we'll see, power comes in many forms, and specializing in any one kind of power tends to result in a trade-off with others.

Chapter 2 focuses on how humans developed extraordinary and unique powers during the Pleistocene epoch, starting roughly 2.5 million years ago. Using recent findings in archaeology and anthropology, we will trace the earliest human uses of stone tools, fire, and clothing, revealing how they enabled our ancestors to expand their range and their competitive advantages over other large-bodied mammals and other human species. As we'll see, tools and fire changed us as much as we changed the world by using them. We'll explore how language supercharged our other powers by enabling us to explain processes (like the manufacturing of ever-more complicated tools), tell stories, and ask questions. Finally, we'll trace the trajectory of power relations between women and men in prehistory, and explore *Homo sapiens'* long-standing fascination with the power of beauty.

Chapter 3 outlines the evolution of vertical social power—how some people gained influence over others. Most of the key milestones in this process occurred in the Holocene epoch—that is, during the past 11,000 years. As our early ancestors began domesticating animals and plants and growing crops, they also created a "wealth pump" that continually generated economic inequality. We'll see how the adoption of a symbolic medium of exchange and basis for the creation and payment of debt (i.e., money) led to a near-universal, self-regulating system of wealth and poverty. We'll see how early agriculturalists applied the skills they developed in domesticating animals to the project of controlling other people. We'll also investigate the power conferred by communication tools—from writing to the printing press to social media. Finally, in a section titled "Pathologies of Power," we'll examine the ways in which vertical social power often makes us literally crazy.

Chapter 4 focuses on the period of time I'll be calling the Great Acceleration. It's during this brief historical moment—roughly the past two hundred years—that trends like climate change and rapid population growth have really taken off. This chapter reveals how the fossil fuel revolution enabled us dramatically to increase resource extraction, manufacturing, and consumption in a historic eye-blink—causing the middle class to balloon and consumerism to blossom. We'll also see how the worst human impacts on our environment—and the most lethal wars ever—have occurred due to the powers that coal, oil, and natural gas have conferred on us.

Chapter 5 explores the consequences of unleashing the energy of millions of years' worth of ancient sunlight, stored as fossil fuels, into a world of finely-tuned natural checks and balances. This chapter is an unflinching look at what's going wrong in the 21st century.

10

Climate change, resource depletion, and species extinctions all result from humans exercising enormous and growing power over our environment. Meanwhile, our expanding energetic powers have sent the wealth pump common to all complex societies into overdrive, so that economic inequality is growing to absurd extremes. People in past societies amassed power in similar ways, but on a smaller scale, and it never ended well.

Fortunately, it is possible to rein in excessive power. In Chapter 6 we'll see how evolution has provided for the limitation of power in other species, and how, throughout history, we humans have found ways to check our own powers, both over nature and over one another. We'll see how our inner powers of empathy and self-control have led to moderation and peace in previous historical moments, and could do so again. In the course of pursuing power over nature and one another, we have also created art, music, literature, spirituality, and science—our most sublime achievements. And we are capable of remarkable acts of creativity, compassion, mercy, and self-sacrifice.

Chapter 7 is about the future of power. It examines the specific ways we must alter our relationship with power in order to prevent a premature end to the human experiment. We'll look at the worst- and best-case scenarios for our deep future, and the power tradeoffs that will determine our actual trajectory. As we will see, limiting our dominance of nature and of other people while maximizing self-control could make life not only more secure, but also far more beautiful for ourselves and our descendants.

This book argues that *in principle* we can indeed overcome our current crises of over-empowerment and power imbalance. That doesn't mean we *will*; as I point out in the last section of Chapter 6, humanity is in a unique situation now, wherein (because of fossil fuels) we've succeeded in overcoming certain limits—on energy, food, population, debt, and scale of social organization—that kept previous societies within critical bounds. Our very success is blinding us to the fact that limits nevertheless still exist, and are in fact looming. So, the odds are that we *won't* escape some form of societal collapse (fast or slow; complete or partial) during the current century.

My reason for writing this book is that I believe it is vital that as many people as possible understand the following point: Whatever degree of resilience or sustainability we *can* achieve prior to, during, or after collapse *must* come from a return to self-limiting behaviors. The call for power-limiting behavior is implicit in a great deal of existing environmental, social justice, and spiritual literature. This book makes that call explicit; grounds it in physics, biology, anthropology, and history; brings it up to date; and underscores what is at stake.

In the years ahead, human survival will depend on our ability to reckon with power in all its forms. If this book can help even marginally in that process, it will have done its job.

SIDEBAR 1

Defining Power

Power is a familiar word, but it has several distinct meanings. I explore its meanings throughout this book, and specify which meaning I am using in any given context. To avoid confusion, let's consider these definitions at the outset.

In physics, power is defined as *the rate of transfer of energy*. Using a unit of measurement such as the watt, it is possible to precisely specify the power of an engine, the power of an asteroid crashing into a planet, or the power of human arm and leg muscles used in digging a ditch or carrying a heavy weight. (See sidebar 4, "How Much Power?" in Chapter 1.)

Physical power can be used to accomplish goals. We often think of physical power simply as *the ability to do something*. We speak of "the power of speech" or "the power of flight," but these are just two of an astonishing number of things that can be done by harnessing energy in particular ways. Chapter 1 explores how living organisms gradually evolved to use energy to gain an ever-expanding repertoire of abilities. Chapter 2 zeroes in on the evolution of key human abilities—especially language, toolmaking, and control of fire—that have given us dramatic advantages over other creatures.

In normal conversation, most of us tend to use the word *power* to refer to social power, which can be defined as *the ability to get other people to do something*. A political leader or a billionaire has this kind of power. Chapter 3 explores the evolution of social power since the emergence of kings and money roughly 4,000 years ago.

Not all social power is the same. Horizontal social power is *the ability of a group to self-organize to accomplish something*. The implicit message is, "We can do this together." This is the power at work in a self-organizing social movement or a mob. Societies characterized by horizontal power include hunter-gatherer bands and, to a degree, successful modern democracies.

Vertical social power is *the ability to get others to do something through a threat or incentive*. The implicit message is, "You must do this or else," or "If you do this, I will give you that." This is the power wielded, for example, by a judge or an employer. Societies highly characterized by vertical power include early agrarian kingdoms and modern totalitarian dictatorships. The line separating vertical from horizontal power can be somewhat blurry: social movements have leaders, dictators often appeal to mob instincts, and modern democracies have police and prisons.

Still another form of social power is the ability to exclude others from having access to something. The implicit message is, "This is mine, and I will prevent you from having it." Like many other forms of power, exclusionary power developed in nature long before the appearance of humans (as we will see in Chapter 1). However, modern humans have developed exclusionary power to a high degree through laws defining property rights, and through wars over land and other resources.

Some forms of social power are hard to define or measure. We say that ideas have power, but ideas must have human hosts to continually give them life. Creative artists have the power to move their audiences emotionally and aesthetically. Many spiritual and political leaders gain followers through the power of inspiration and example, or force of personality. Finally, sexual attraction exerts an undeniable power that shapes nature and human affairs.

These definitions and types of power may seem disparate and unrelated. However, as we will see, one kind of power tends to lead to another. By tracing the relationships between varied manifestations of power, starting with energy transfer, we begin to see patterns and connections that would otherwise be invisible. History and current events emerge with greater clarity when viewed through the lens of power.

POWER IN NATURE

From Mitochondria to Emotion and Deception

The way in which mitochondria generate energy is one of the most bizarre mechanisms in biology. Its discovery has been compared with those of Darwin and Einstein. Mitochondria pump protons across a membrane to generate an electric charge with the power, over a few nanometers, of a bolt of lightning. This proton power is harnessed by the elementary particles of life—mushroom-shaped proteins in the membrane—to generate energy in the form of ATP. This radical mechanism is as fundamental to life as DNA itself, and gives an insight into the origin of life on Earth.

— NICK LANE, *Power, Sex, Suicide*

The maximum power principle can be stated: During self-organization, system designs develop and prevail that maximize power intake, energy transformation, and those uses that reinforce production and efficiency.

— HOWARD T. ODUM, *Maximum Power*

Everything in the world is about sex except sex. Sex is about power.

— OSCAR WILDE

POWER IS EVERYWHERE AND IS THE BASIS OF, WELL, EVERY-thing. Without it, literally nothing can happen. Exploring the origins and evolution of power helps us better grasp how it shapes the human world today—and why our recently developed abilities to dramatically magnify and concentrate human power now threaten both humanity itself and the natural world on which we all depend.

While we have come to dominate other species and to transform our planet, and some of us have grown far wealthier and more influential than others, our powers are puny in universal terms. The universe is shaped by cosmic forces—gravity, which is nearly undetectable as an attractive force between two human-sized masses, but which shapes galaxies and the orbits of planets; nuclear fusion, which occurs due to forces in atomic nuclei and causes stars to emit enormous amounts of energy; and the electromagnetic force, which is responsible for the intense magnetic fields in rapidly rotating, charged black holes that can accelerate particles to spectacular speeds and energy levels. One particular star, our Sun, is the ultimate source of most power on our planet—whether it's the physical power of a hurricane, or the social power of a successful political movement (after all, the people who form that movement have to eat, and the energy in their food comes from the Sun).

The Sun radiates energy, largely in the form of light, far and wide throughout space, but only a tiny fraction of the Sun's total output falls on Earth. Even so, this minuscule portion is enough to heat the planet's surface so as to keep most ocean water in a liquid state, and to drive the weather that stirs our atmosphere.

Sunlight has also powered the most amazing development in the entire solar system—the evolution of living things. The process by which biological evolution got started is still the subject of research and speculation (we'll explore it more in a moment), but the results—after over four billion years—are all around us in the forms of millions of species of plants, animals, microbes, and fungi, and of complex ecosystems, each containing many species, each species adapted to others, and all adapted to particular regimes of moisture, temperature, and climate.

Every organism is able to capture some of the Sun's energy as that energy flows through Earth's systems.[1] And each organism has found a way to dissipate that captured energy in a controlled way. In doing so, every living thing wields powers of its own.

Indeed, evolution favors those organisms, and systems of organisms, that use power more effectively than others do. Early natural scientists and philosophers, including Gottfried Leibniz and Vito Volterra, deduced somewhat intuitively that evolution works this way, but the process wasn't described in detail or given a name until the 20th century. Ecologist Howard Odum, who worked on the problem in the 1960s by building on the earlier efforts of biophysicist Alfred Lotka, called this evolutionary tendency the *maximum power principle*. It's a key concept for understanding power anywhere and everywhere in the living world.[2] One way to think of this principle is that the species that exploits a given resource most effectively will tend to crowd out competing species.

But if evolution favors power maximization, then why didn't a single powerful organism emerge early in Earth's history and dominate the planet from then on? The diversity of life results from the fact that there are many ways to exert power, and many different environments in which to do so. As we'll see during the course of this book, one species *has* recently taken charge of virtually the entire planet as a result of its ability to maximize power in a host of ways— and we, of course, are members of that species.

However, many fundamental powers began to evolve in organisms long before humans appeared. While this book is mainly about the evolution of power in human societies, especially in recent decades, it's much easier to grasp the nature and sources of human power if we ground our exploration of the subject in the wider evolution of power throughout natural systems. Doing so also reminds us of some biological principles that we'll refer back to as we consider natural limits to the seemingly endless extension of human powers.

In this chapter, we'll take a look at some of the powers that arose in living things long before humans emerged. Then in subsequent chapters we'll see how humans have amplified these already existing potentials. We'll see how the drive for power makes us both cooperative and competitive; how all organisms have learned to limit their powers in order to develop and diversify; and how evolution turned higher animals like us into worshippers of beauty that are often willing to sacrifice some of our other powers for purely aesthetic purposes.

SIDEBAR 2

Powers in Math

Exponents, or *powers*, are a way of showing that a number is to be multiplied by itself repeatedly. In the expression 2^5, 2 is called the *base* and 5 is called the *exponent*, or *power*. 2^5 is shorthand for "multiply five twos together": $2^5 = 2 \times 2 \times 2 \times 2 \times 2 = 32$.

Powers of ten are often used to express really big and very small numbers. This is called *scientific notation*. For example, the total power of modern industrial civilization can be expressed as 4×10^{13} watts. In ordinary decimal numeric notation, that would be 40,000,000,000,000 watts.

When using scientific notation, increasing the exponent by one (i.e., multiplying the number by ten) is often described as increasing its *order of magnitude* by one. A quantity that is ten times greater than another is therefore said to be larger by an order of magnitude: 2.5×10^4 is an order of magnitude larger than 2.5×10^3.

To express very small numbers, negative exponents can be used: $10^{-3} = (1/10)^3 = 1 / (10 \times 10 \times 10) = 1/1000$. The number 10^{-4} is *an order of magnitude smaller* than 10^{-3}.

Scientific notation is useful in fields as varied as astro-physics, particle physics, biology, and engineering. It enables us not only to understand the vast variance in scale between the ultra-tiny and the incomprehensibly vast, but to operate across scales with technologies ranging from telescopes and microscopes to microprocessors, chemicals, spacecraft, and medicines. Powers of ten enable us to peer into the nucleus of an atom or estimate the size of the universe. They broaden our understanding while also greatly enhancing our ability to influ-ence and control our environment.

The Basis of Life's Power

In the final analysis, life is all about power. Every living organism wields the powers of growth, metabolism, and reproduction. Whether a bacterium, a blade of grass, or a towering giraffe, all life forms cap-ture energy from their surroundings and dissipate that energy in a controlled way to maintain steady conditions within the boundary that separates them from their environment. Even the simplest or-ganism is a complex system able to control what's going on inside it-self, and able to manipulate its environment in some way to get what it needs to sustain itself.

But how did life get started? Biologists have been puzzling over the question for at least the last couple of centuries. The process must have geared up remarkably early in Earth history. In 2017, researchers in Quebec found tiny filaments, knobs, and tubes—effectively, fossils of single-celled creatures—in rocks thought to be up to 4.28 billion years old. The planet itself is believed to have formed 4.6 billion years ago, so evolution apparently wasted comparatively little time in get-ting going.[3] These minute structures formed around hydrothermal vents at the bottom of ancient oceans; we'll see the significance of that in just a moment.

Until recently, some researchers thought the evolution of life must have begun with viruses, since they're much simpler than bacteria—they consist merely of bits of DNA or RNA wrapped in a protective coat of protein. But that theory has fallen out of favor. The problem is that viruses have no energy source: they depend on the energy of a host cell they've hijacked in order to reproduce themselves. Prior to the time such host cells existed, there would have been no way for viruses to have perpetuated themselves. Viruses must have evolved later. They likely started out as fully developed cells that gradually streamlined themselves by giving up their vital functions so they could more effectively piggyback on fully developed organisms.

The most intriguing and convincing current theory of the origin of living things centers precisely on energy and power. Billions of years ago, volcanic seepage sites spewed iron and sulfur into ancient oceans. Those ocean waters were acidic close to the surface, but alkaline further down close to the seafloor. Tiny iron-sulfur bubbles formed around the volcanic vents, with membranes often just a single molecule in thickness (similar bubbles can be observed today around deep-water volcanic vents). The outside of the bubble was more acidic than the inside. The membranes were also electrically conductive, and simple electrochemical processes would have shunted electrons to the inside of the bubbles, and protons to the outside. This inside-outside difference in acidity and charge could have generated a difference in electrical potential across the bubble membrane of several hundred millivolts—roughly the same level of charge that powers bacteria to this day. The process just outlined would supply not only a source of energy, but also the basis for the synthesis of biochemicals, including the proteins necessary for life—amino acids, peptides, and RNA.[4]

This story of life's origin is convincing because it describes a process—separating electrons and protons—that still generates energy in every living cell throughout all of nature. Everything that every cell does requires energy. And each of the three basic energy-generating processes of life—respiration, fermentation, and photosynthesis—involves proton pumping. The proton pump that started life's evolution over four billion years ago still powers all of life on Earth today.[5]

Just how powerful is life? Astoundingly, on a gram-for-gram basis, the average organism is 10,000 times as powerful as the Sun.[6] How could this be? The Sun generates an enormous amount of power, but it is also very massive. The math is simple: dividing luminosity by mass yields a mere 0.0002 milliwatts of power per gram for the Sun. A human, eating an average diet and converting that food energy into heat and work, averages two milliwatts per gram. Now, as we've seen, just about all the energy on Earth originated from the Sun, so life's power is derivative—it comes from somewhere else; in contrast, the Sun generates its own power. Nevertheless, the effectiveness of living things at appropriating and using energy is astonishing. Life is powerful indeed.

SIDEBAR 3

Measures of Physical Power

Leaving aside its social manifestations, power is, at its simplest, the rate of energy transfer. Think of energy as water flowing from a firehose. In terms of this metaphor, power is the rate at which water is emerging from the hose—which is determined by two factors: the diameter of the hose and the water pressure. If the hose is large but water pressure is low, there's not a powerful flow; likewise, if the pressure is great but the hose is small. If the hose is large and the pressure is high, you'll have a powerful flow on your hands. The total amount of energy (or, in the metaphor, water) that is released is determined by rate of flow (power) over time: a rate of flow of one liter per second would, naturally, yield sixty liters in a minute.

The unit of power most commonly used by scientists is the *watt*, named after the Scottish inventor James Watt (1736–1819). The watt is defined as the transfer of one joule of energy per second. How much is that? Using a hand crank, you can apply roughly eight watts of mechanical power. In terms of electrical power, a watt can run a single LED. *Energy* can be measured in watt-hours (power over time), as well as joules, calories, British Thermal Units (BTUs), therms, and other units.

Very small amounts of power can be expressed in terms of a nanowatt (one billionth, or 10^{-9} watt), a unit that can be useful in discussions about power in cell biology or the gravitational pull of a human-sized object. Large amounts of power can be measured in terms of a kilowatt (a thousand, or 10^3 watts), a megawatt (a million, or 10^6 watts), a gigawatt (a billion, or 10^9 watts), a terawatt (a trillion or 10^{12} watts), or a petawatt (one quadrillion, or 10^{15} watts). A typical modern middle-income American household can operate all its appliances using approximately one to three kilowatts of electrical power.

Sometimes power is discussed in terms of *horsepower* (hp). However, there are several different standards of horsepower. Two common ones are the *mechanical* (or *imperial*) horsepower, which is about 745.7 watts, and the *metric horsepower*, which is approximately 735.5 watts.

One horsepower is only vaguely equivalent to the power of an average horse—and therein lies a story. When he was developing the steam engine, James Watt was visited by a brewer, who requested an engine that could match the power output of his biggest and strongest horse. Watt accepted the challenge and built a machine that exceeded the power of the brewer's horse; it was the output of that machine that became the basis for horsepower.

So, how much power can an actual horse apply? A 1993 paper in *Nature* by R. D. Stevenson and R. J. Wassersug cited measurements made at the 1926 Iowa State Fair, where one horse was recorded at peak power over a few seconds at 14.9 hp (11.1 kW). However, they observed that for sustained activity, a work rate of about 1 hp (0.75 kW) per horse is indeed realistic.

A healthy human can produce about 1.2 hp (0.89 kW) briefly and sustain about 0.1 hp (0.075 kW) for longer periods; trained athletes can manage up to about 2.5 hp (1.9 kW) briefly and 0.35 hp (0.26 kW) over several hours.

Three Big, Powerful Families

The earliest cells must have been simpler than today's single-celled organisms, which are the beneficiaries of billions of years of evolution: even though they're tiny, they are highly organized internally and have diverse and sophisticated ways of making their way in the world. Today there are three broad kinds of cells, and therefore three branches of the tree of life: bacteria, archaea, and eukaryotes. Bacteria are almost certainly the oldest of the three.

Bacteria and archaea exist only as single-celled organisms (unless you count a bacterial colony as an organism: in some colonies of Myxobacteria, individual bacteria specialize). They are also similar in that they lack nuclei. For a long time, biologists assumed that archaea *were* bacteria, but as more about them was discovered, it became clear that these are two fundamentally different kinds of life forms.

Most bacteria are shaped like rods, spheres, or spirals. Most are also tiny, 0.5 to 5 micrometers in length, though a very few, such as *Thiomargarita namibiensis*, are up to half a millimeter long—big enough to see with the naked eye.

An individual bacterium is like a battery. The cell continually pumps protons out across its membrane into the periplasmic space between the membrane and the outer cell wall, thereby creating an electrical gradient. It does this by gathering electrons from its environment, passing them along a chemical chain, and using their combined negative charge to pump the protons outward. The creation of an electrical gradient around the bacterium can be compared to blowing air into a balloon: pressurized air can later be used to do work, like turning a pinwheel; the electrical potential around the bacteria can be used as well, for work such as chemical synthesis and reproduction. Bacteria can even use their proton-pumping force to power locomotion.

In order to maintain their electrical gradient, bacteria must have a source of extra electrons, and they have evolved several strategies for getting them. Some bacteria derive their electron energy from light

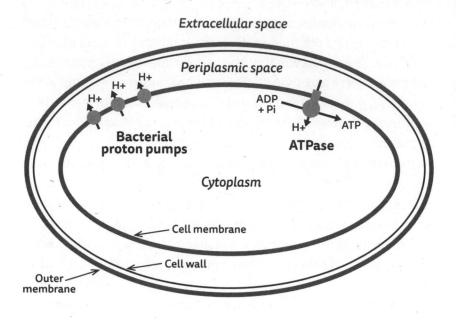

Figure 1.1. Proton pumping in a bacterium (highly simplified).

Source: Adapted from an illustration by Steve Mack, MadSci Network.

through photosynthesis. Others, called chemotrophs, use chemical compounds as a source of electrons (see "Power of Eating," below). Chemotrophs are further divided by the types of compounds they use. Ones that use inorganic compounds like hydrogen, carbon monoxide, or ammonia are called *lithotrophs*, while those that use organic compounds are called *organotrophs*. Within bacteria, a different compound must receive the electrons from "food" in a chemical reaction in order for the cell to obtain and use energy; aerobic bacteria use oxygen as the electron acceptor, while anaerobic bacteria use compounds such as nitrate, sulfate, or carbon dioxide.

Even though bacteria generally lack the ability to form complex, multicelled individuals, they have evolved the ability to live in extremely varied environments (hot or cold; with air or without air; acidic or alkaline). They are also so good at producing such a huge range of chemicals that industrial biochemists are constantly finding new ways of harnessing them for commercial applications.[7]

Although archaea were at first thought of as bacteria, their genes are significantly different from those of bacteria, as are their cell walls and other cell structures, and so is their energy metabolism. They were "discovered" only in 1977, and are extremely difficult to culture and study in the laboratory. Only 250 species have been identified (compared to some 30,000 kinds of bacteria).[8] Genetic differences suggest that archaea and bacteria probably split from a common ancestor 3.7 billion years ago. Though they are less numerous in the gut than bacteria, archaea are still plentiful in your digestive tract and on your skin.[9]

Eukaryotes, which have cell nuclei, appeared more recently, about 2.7 billion years ago. Unlike bacteria and archaea, they were able to form complex multicelled individuals: all plants, animals, and fungi are eukaryotes (you're a eukaryote and so am I, glad to meet you!). Even single-celled eukaryotes can be vastly larger and more complicated than typical bacteria or archaea. They're able to grow so big because they found a way to delegate their energy production activities to mitochondria—vital internal cellular structures we'll explore in more detail in a moment.

ATP Power

We have seen how bacterial cells generate an electrical potential by pumping protons into the space between their cell membrane and their outer cell wall. But how do they actually use that power? In any energy-using system—such as an organism, a household, or a city—it helps to have a medium for storing and transferring energy. In a typical medieval household, firewood served that function; in modern cities, gasoline is one of our primary media for storing and transferring energy. Adenosine triphosphate, or ATP, is the firewood or gasoline of the cell.

The primary energy-yielding activity of the eukaryote cell is respiration, which, at the cellular level, is the breaking down of glucose (food) with oxygen. (There's also an anaerobic pathway for ATP creation—fermentation, which yeast are really good at...Care for a beer?—but let's stick with respiration, because it's more relevant for understanding the energy pathways in most plants and animals.) As a result of various chemical transformations, one molecule of glucose will yield up to 36 molecules of ATP, which can be stored or circulated until needed. The ATP molecule features a chain of three phosphate atoms; when the third of those atoms is broken off through a chemical process, ATP becomes ADP (adenosine diphosphate) and energy is released and available for immediate use. The ADP can then be recycled by adding back another phosphate atom (which, of course, requires energy). ATP is actually better than firewood or gasoline, in that it can be endlessly recycled!

In bacteria, energy generation can take many routes, but all have one process in common: the electrical potential that develops across the inner membrane, which generates a proton-motive force that drives protons through a molecular structure known as the ATP synthase complex, generating ATP from ADP by adding a phosphate atom to the chain. (By the way, if this sounds complicated, it is—there are many textbooks devoted to explaining it, and researchers are still learning more about the energy pathways of super-tiny individual cells.)

Cells use the energy from ATP for just about everything they do, including synthesizing DNA and RNA. Bacteria and archaea make ATP in ATP synthase complexes on the cell membrane; in eukaryotes, those structures are located in mitochondria. Now's the time to talk about these remarkable structures.

Mitochondrial Power

Eukaryotes differ from bacteria and archaea not just because the former have nuclei, but also because they have organelles, which are like cells within the cell. The most important of the organelles are mitochondria, of which there may be hundreds in each eukaryote cell.[10]

Mitochondria generate virtually all of a eukaryote cell's ATP. They look like bacteria, and that's because they once were free-living bacteria that took up residence inside larger cells (probably a type of archaea) some two billion years ago. They have their own genomes, separate from the cell's genome that's housed in the cell nucleus.

Eukaryotes have a cell membrane but, unlike bacteria, no rigid cell wall. That means they cannot use the space between membrane and wall as a battery, the way bacteria do. The ancestors of eukaryotes had cell walls, but once they acquired mitochondria it was possible to tear down that wall and delegate energy generation to their mitochondria.[11] Ditching the cell wall meant that eukaryotes could change shape and become true predators—like the amoeba, which engulfs its prey (see "Power from Eating," below).

Within mitochondria, proton pumping occurs across a highly folded inner membrane. The cell delivers electron source material to the mitochondria; the latter then store the energy from the source material as ATP, and deliver the ATP to the cell for all its various energy-using functions. This neat trade-off made complexity possible.

The Power of Complexity

For bacteria, not having mitochondria is a limit on size. Double the size of a bacterial cell and it can produce only half the ATP per unit of volume; meanwhile the larger cell needs *more* energy for all its cellular functions. That's why bacteria, on the whole, have stayed extremely small.

Not so the eukaryotes. Single-celled eukaryotes can be hundreds, even thousands of times the size of a typical bacterium. Need more energy? Just make more mitochondria! Once energy production was delegated to mitochondria, cells could grow larger, and also join each other in coordinated groups in which they each took on specialized tasks.

Mitochondria made complexity possible; natural selection made complexity attractive. Bigger, more specialized, and more complex creatures could more efficiently exploit resources or escape predators. Natural selection acted like a ratchet, turning random variation into a trajectory. And that trajectory has led us up the ramp of complexity.[12]

If eukaryotes with mitochondria hadn't evolved, the story of life on Earth would have been all about single-celled bacteria and archaea.

Animals and plants would never have seen the light of day if it hadn't been for that incident, two billion years ago, when a primitive bacterium took up residence inside an archaeon and they found a mutually beneficial way to live and work together. This may have been a fluke, unlikely to be duplicated in bacterial colonies on other planets throughout the universe. If that's the case, complex life may be unique to Earth.

Gene Power

Consider the lowly *Escherichia coli*, usually known by its nickname *E. coli*, one of the many species of bacteria found in the human gut. (You are probably host to millions of them right now.) An organism that can live with or without air, *E. coli* is harmless in most of its strains—but some forms, if ingested, will make you very sick. These tiny rod-shaped critters derive their energy from semi-digested food in the lower intestines of animals, from feces, or from prepared cultures in Petri dishes (*E. coli* are by far the most popular subjects for lab experiments, so scientists know a lot about them).

Though it is one of the simplest organisms and its size is measured in microns, each individual *E. coli* has a cell wall composed of cellulose, plus a plasma membrane and all the various proton-pumping and ATP-generating equipment discussed above. Internal systems regulate its functions and transport materials through its cell wall to spots where metabolism occurs. Altogether, the *E. coli* cell is like a tiny factory, with control centers, energy production sites, and chemical synthesis chambers. It also contains a string of DNA organized as genes. Without this DNA string, the *E. coli* cell would not be able to organize and reproduce its other structures.

All living organisms are made up of cells (in many cases, like *E. coli*, only a single cell) that contain genetic material. In bacteria and archaea, DNA is just a single strand floating in the cell's cytoplasm. In eukaryotes, DNA is stored in the cell's nucleus. Wherever you find it, DNA acts as a blueprint for the manufacture of proteins, which the cell uses for everything it does.

Genes conserve the organism's basic structure and functions through generations of reproduction. As the building blocks of heredity, genes also confer the power of mutation, enabling successive generations to adapt to environmental change. Mutations, however, can pose a risk: many actually reduce the cell's viability and the organism's survival potential; that's one of the reasons why nuclear radiation, which triggers higher rates of random genetic mutations, is generally to be avoided.

The Power of Reproduction

While single-celled organisms like *E. coli* reproduce through cell division (mitosis), most multicelled plants and animals have adopted sexual reproduction. Either way, the power of reproduction is one of the keys to life's success. Reproduction enables any given type of organism to grow in number, taking advantage of more resources and occupying more territory—hence maximizing its collective power.

The reproduction of any cell requires energy, some of which is used for copying the DNA blueprint. Mistakes in copying can be expensive or even deadly.

The vast potential of reproduction is illustrated in the concept of "doubling time"—the amount of time in which a species can double its population size. Imagine a pond with a patch of algae on its surface. If the algal population grows at ten percent per hour, the total number of algae will double every seven hours. If the algae occupied one-twentieth of the pond surface when our observations started, they would take up one-tenth of the surface after 7 hours, one-fifth after 14 hours, two-fifths after 21 hours, and four-fifths—nearly the entire surface—after a mere 28 hours. The compounded, or exponential, growth of any physical system accelerates until it hits a limit—such as the boundaries of the pond's surface, in our example.

Bacteria can reproduce much faster than that. They can divide every 20 minutes, through an energy-intensive process that requires making a complete copy of the entire DNA blueprint. Starting with a single bacterium, if doubling continued for just two days, the resulting colony would weigh 2,664 times the weight of the Earth.[13] Obviously, this never actually happens. That's because bacteria are normally half-starved, and when they run out of food they pass into a nearly inert state in which their rate of mitosis slows dramatically.

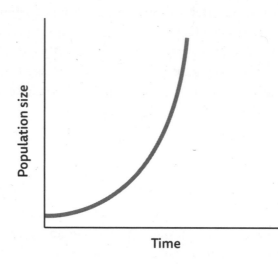

Figure 1.2. Exponential growth.

It's worth noting that even though humans reproduce far more slowly than bacteria or algae, the doubling-time principle still holds. At one percent annual population growth rate (which is lower than the current global average), the total human population doubles in about 70 years. In reality, one billion humans—our global population level in 1820—became two billion in 1927, four billion in 1974, and we are roughly eight billion today. While population growth confers the advantage of greater collective power, it runs the risk of overwhelming limited resources. Hence some human societies (such as China and Cuba in recent decades) have sought to moderate population growth, partly as a way of staving off resource scarcity, while other societies (including ancient states in Mesopotamia and China, and European-Americans during the 18th and 19th centuries) have deliberately sought to increase their population growth rate as a way of competing with other societies or maximizing their power in other ways.

The Power of Self-Limitation

Complexity confers benefits, as discussed above. But it also entails costs and trade-offs.

To live in a multicelled organism, each cell must submit to the severest of limits—death. If cells didn't die, the multicelled organism couldn't live. That's because individual cells eventually wear out or become damaged; if the body had no way of eliminating and reabsorbing them, whole subsystems would become dysfunctional fairly quickly, imperiling the entire individual. This does happen occasionally anyway, and the result is cancer. To prevent this, cells are programmed to commit suicide when instructed to do so by a molecular "police force." It turns out that mitochondria and power play key roles: the loss of electrical potential across the folded membranes within the mitochondria of a cell is a trigger that invariably leads to cell suicide.

This self-subordination of the individual to the needs of the whole is seen also in groups of highly cooperative organisms, such as ants. Among exploding ants (*Colobopsis saundersi*, found in Malaysia and Brunei), workers produce a toxic fluid in their abdomens; when the colony is attacked, some of the workers sacrifice themselves by exploding, releasing the toxin and killing the invaders. Self-limitation and self-sacrifice might seem to contradict the maximum power principle. They don't, because groups of ultra-social creatures act as superorganisms; in these cases, evolution is acting at the group level, rather than the individual level. We'll explore that concept further, especially as it relates to humans, in chapters 3, 6, and 7.

Power from Photosynthesis

Photosynthesis, which powers nearly all multicelled life on Earth, got its start in cyanobacteria, which appeared around 2.7 billion years ago. The chemistry of photosynthesis essentially runs respiration backward. Respiration uses glucose and oxygen to make water, CO_2, and energy; photosynthesis uses energy (from sunlight) plus water and CO_2 to make glucose and oxygen. Plants use some of the glucose they make to fuel respiration and release energy; the rest is incorporated into their tissues and becomes leaves, stems, and roots.

Within plant cells, photosynthesis takes place in organelles called chloroplasts. These appear to have originated as cyanobacteria (i.e., bacteria able to perform photosynthesis), which, roughly 500 million years ago, found a cozy living arrangement within some eukaryote cells, much the way mitochondria had already done.

Green plants have the ability to use the energy of sunlight directly to build body tissues from little more than air and water. (Plants also need nitrogen, phosphorus, potassium, and trace elements, but these do not make up the bulk of plant tissues.) In doing so, they store energy in the chemical bonds between the atoms of carbon, hydrogen, and oxygen that make up the carbohydrates composing a plant's body. The power of photosynthesis has enabled green plants to spread across the surface of the planet, wherever there are sufficient water sources and nutrients; plants (most often in the forms of phytoplankton and kelp) are also dispersed throughout suitable areas of oceans, streams, and lakes.

It is largely through photosynthesis that the Sun's energy is able to power multicelled organisms throughout the biosphere. Biologists call green plants "producers," because they essentially make their own food—as opposed to "consumers" (a category that includes all animals), which have to eat other organisms or products of organisms to obtain their energy.

Power from Eating

By eating other organisms, consumers obtain the energy that is the source of their powers. At the same time, eating requires the exertion of energy; on a net basis, the energy derived from food must exceed the energy required to obtain and eat food if the organism is to survive.

Bacteria are able to transport food through their cell wall by secreting enzymes onto the food to break it down into simpler forms that can be transported through the wall. Eukaryote cells, which don't have rigid walls, are able to transport nutrients directly through the thinner cell membrane. The first true predators were eukaryotic cells that were able to engulf their prey (mostly bacteria, algae, and particles of formerly living cells) by flowing around it and enclosing it as a food vacuole, within which it could be digested. This process, known as *phagocytosis*, is how the amoeba and many other protozoa still make their living.

Eating is easier to understand intuitively in big, multicelled animals, because it's an activity we ourselves engage in frequently—sometimes with great pleasure. Herbivores eat plants. Carnivores eat other animals; secondary carnivores, like many snakes and fish, eat other carnivores. Decomposers, such as fungi, eat the dead bodies of other organisms, gradually turning them into soil—which helps support the growth of producers.

When a whitetail deer (an herbivore) stops along its journey through a meadow to eat a dandelion (a producer), only about a tenth of the dandelion's total energy stored as carbohydrate can be absorbed and used to power the animal. Similarly, when a mountain lion (a carnivore) stalks and kills that deer, only about a tenth of the total energy is transferred. Biologists often describe the overall results in terms of a food (or energy) pyramid by which energy moves through the ecosystem. At each stage, most energy and materials are lost as heat and waste rather than being converted into work or bodily tissues. Thus, a typical terrestrial ecosystem can support only one carnivore to ten or more herbivores of similar body mass, one secondary carnivore to every ten or more primary carnivores, and so on.[14]

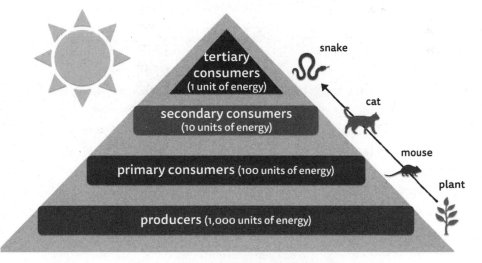

Figure 1.3. The energy pyramid in nature. Credit: Post Carbon Institute.

Think of the energy pyramid as a visualization of power getting distributed through the biological world: it enters at the bottom and works its way up, at each level becoming more "concentrated" (Howard Odum termed this "concentration" of energy *transformity*). Yet organisms at the base of the pyramid, as a result of being far more numerous and having far greater total biomass, are collectively just as powerful, in their way, as the top predators at the pyramid's apex (though humanity's use of fossil fuels has temporarily upset that balance, as we'll see later on).

Power and Bodies

The Power of Size

We've already discussed some of the advantages of bigness, but there's a lot more to say on this score. On average, larger animals have a slower rate of metabolism and enjoy a longer life than smaller ones. The important thing to note is that their individual cells actually need fewer nutrients: as biologist Nick Lane colorfully puts it, "an elephant-sized pile of mice would consume 20 times more food and oxygen every minute than the elephant itself does."[15] A mouse has to eat half its body weight every day to fend off starvation, while a human survives eating only two percent of body weight daily. It turns out that, as animals get bigger, their metabolic rate slows by a factor that roughly corresponds to the ratio of surface area to mass. This is known as Kleiber's law, one of the power laws of biology.

There are two ways of looking at this situation. First, the cheery way: being big offers the advantage of efficiency and economy of scale. The less cheery way: this increase of efficiency with growth in size is at least partly a result of the burgeoning bulk of the supply network required to deliver food and oxygen to each cell; as the network grows, it forces constraints upon the animal's individual cells and their mitochondria, so these cells have to economize on energy and they're not able to work as hard. Therefore, the ratio of the animal's maximum power to its weight falls with size.

When it comes to lifespan, size seems to confer a less ambiguous advantage. The mouse lives two years on average, while the elephant

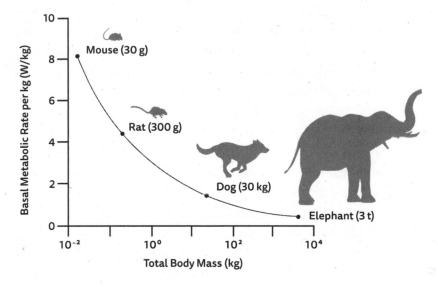
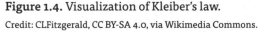

Figure 1.4. Visualization of Kleiber's law.

Credit: CLFitzgerald, CC BY-SA 4.0, via Wikimedia Commons.

may live a hundred years. Altogether, the mouse is working harder and burning out quicker. However, there are a few anomalies in the size-lifespan continuum. Birds, bats, and humans live longer than they should, based on their size: a pigeon may live 35 years, while a similar-sized rat lives only two. Many humans live longer than they otherwise would due to modern medicine, but it appears that the long lives of birds and bats are due to a genetic adaptation having to do with the production and accumulation of free radicals within cells. Scientists are trying to better understand what helps birds live so long (some seabirds appear entirely immune to the diseases of old age) and to see whether humans could benefit from that knowledge.

Some additional advantages of bigness: being of larger size aids in retaining heat; offers greater strength; reduces problems having to do with water surface tension (insects have a hard time taking a drink); and tends to optimize the functioning of organs such as eyes, which are composed of cells of fixed size (more cells, better vision— or that's the theory, anyway; I had to wear glasses when I was a kid despite having eyes bigger than a hawk's).

But there are also some additional disadvantages to bigness that are worth mentioning: larger birds have to work harder to fly, and larger mammals struggle to make their way through thick vegetation or to walk on boggy ground. For a tiny animal, falling down is no big deal; for an elderly human, it can be fatal.

Muscle Power

There are lots of kinds of specialized cells within larger multicelled organisms. Let's look at just two kinds, muscles and neurons, which often work together and are particularly important for developing many of the unique powers enjoyed by animals such as ourselves.

First, muscles. They are the source of animals' powers of lifting and moving (see "Powers of Motion," below). But muscles do more than this: they are essential for breathing and for moving blood through veins and arteries (the heart is a muscle) and food through the digestive system.

Muscles are made of long, thin cells grouped into bundles, or fibers. They're powered by ATP (of course), and controlled by neurons. When a muscle fiber gets a signal from its nerve, proteins and chemicals release energy to either contract the muscle or relax it.

Muscles make up about 40 percent of the total weight of a typical human, and perform many specialized tasks including moving the eyes and protecting the inner ear from loud noises.

Neuron Power

If muscles supply the brawn for animals, neurons are literally the brains—as well as providing the signaling channels for brains to convey their messages. Outside the brain, there are three types of neurons: sensory, motor, and interneurons (which connect spinal motor and sensory neurons). But within the brain, there are so many kinds of neurons that they are hard to neatly classify.

Neurons get their messages across to other neurons using neurotransmitters. These are chemicals that can excite, inhibit, or modulate a neuron's activities. There are dozens of neurotransmitters, including dopamine. The brain has many dopamine pathways, and this particular neurotransmitter is involved in motor control, behavioral reward and reinforcement, and motivation. We'll discuss dopamine more in Chapter 6.

Most neurons have a cell body, an axon, and dendrites. The axon extends from the cell body and often branches several times before ending at nerve terminals. Dendrites receive messages from other neurons, and are covered with synapses connecting them to axons from other neurons. The nervous system is a communications marvel.

The human brain contains, on average, 100 billion neurons, and operates at about 23 watts of power during wakefulness. Compared with electronic computers, this is an amazing level of power efficiency. While it is impossible to calculate precisely, the human brain's computational power has been estimated at a billion billion calculations per second. The world's fastest supercomputer, using several megawatts of power, takes over half an hour to do what the human brain does each second.[16]

The Power of Sex

Asexual reproduction by way of mitosis has worked fine for bacteria and other single-celled creatures for billions of years. Why wasn't it maintained in all bigger, multicelled creatures via self-cloning?[17] After all, the alternative—sexual reproduction—reduces reproductive opportunities: with two sexes, reproduction is only possible with, at best, half the members of your species. Finding a mate can take time and effort, and there's no guarantee of success. Therefore, there must be some advantage that outweighs that cost.

The advantage is genetic diversity. Among animals, males contribute sperm while females contribute eggs. Both pass along genes to the offspring—but only the mother passes along mitochondria (and their genes) to the next generation.[18]

Sexual reproduction generates unique and often beneficial combinations of genes that in most cases have already proven not to endanger survival. Greater genetic variety helps, for example, by increasing the chance that some of a population of bees, soybeans, or blackbirds will survive if a deadly infection occurs.

Of course, sex introduces the possibility for all kinds of social power intrigues, imbalances, and abuses, some of which we'll explore further in later chapters. But in addition, sexual reproduction opens the door to sexual selection and beauty, which we'll turn to in a moment.

Powers of Perception

Sense organs give organisms powers to adapt to their environments, find food, locate potential mates, and avoid predators. Seeing, hearing, feeling, tasting, and smelling are only the most familiar of the possible senses, which also include the abilities to detect the Earth's magnetic field and to judge spatial orientation and degrees of moisture and acidity.

Among animals, champions of perception include dogs, whose noses are about 40 times more acute than our own; birds, which have an ultraviolet color receptor in their eyes and can therefore see a whole "dimension" of color that's invisible to us; and bats, whose hearing can distinguish high-pitched echoes accurately enough and quickly enough to enable them to locate and catch flying insects.

Plants don't have neuronal networks the way higher animals do. Nevertheless, plants have perception. Nearly all exhibit photosensitivity and the ability to sense orientation in space (i.e., they perceive and actively respond to gravity). Most are also sensitive to an array of chemicals in the soil or air, which are sometimes interpreted either as warnings of inhospitable conditions or as invitations to bloom and reproduce. Some plants are even able to detect the amount of soil around them and adapt their growth accordingly, regardless of the availability of nutrients.

Powers of Motion

Because animals eat other organisms, they often have to travel to find their next meal.[19] They may also have to travel in order to find a mate. Animals ooze, creep, jet, swim, slither, crawl, walk, run, jump, fly, and reach. If you've ever watched swallows catching flying insects, you probably share my admiration for these avian aerial acrobats. Even the lowly single-celled amoeba (which isn't really either an animal or a plant) has the power to extend and retract pseudopods—arm-like projections of cytoplasm—in order to move and to reach for food.

Most plants, able to take advantage of ambient sunlight, stay in one place; however, plants do have limited ability to move, as can be observed by watching a sunflower bend its head toward the changing relative position of the Sun throughout the day. Plants also use air currents, or other organisms, to move their seeds and pollen far and wide. And communities of plants, such as forests, can slowly shift their range in response to changing climate conditions—as many North American trees did at the end of the last Ice Age, moving their range by up to 30 miles per century (a speed that may not be fast enough to enable many tree species to adapt to current human-driven climate change). Some plants move parts of themselves quite rapidly: the Venus flytrap closes its trap in about a hundred milliseconds, while the white mulberry tree can catapult pollen from its flower stamens at a velocity of over half the speed of sound.

The Power of Warm Blood

Warm-blooded animals (biologists call them *endotherms*), such as mammals and the lineage of dinosaurs now restricted to birds, have a power advantage over cold-blooded animals (*ectotherms*), which include reptiles, amphibians, and many aquatic vertebrate and invertebrate lineages. When animals gained the ability to closely regulate their own temperature, they could stay active in spite of daily and seasonal temperature variations that send ectotherms into hiding or inactivity. This gave endotherms a higher baseline demand for energy but also the ability to forage for energy much more of the time. Therefore, in many environments, the average power output of an endotherm is much higher than the average power output of an ectotherm; this enables endotherms to outcompete ectotherms for resources. An evolutionary example of the maximum power principle at work!

But this advantage comes at a cost. If a reptile and a mammal maintain the same temperature, the mammal needs to burn six to ten times as much fuel. If the outside temperature falls, the difference grows: in cold weather, the reptile uses less energy than in warm weather, while the mammal uses more. On average, a mammal uses 30 times as much energy as a same-sized reptile. This translates to more meals eaten more often—and more work in obtaining food. That's why birds and mammals tend to have much more stamina than reptiles (i.e., they can maintain a higher pace of work for longer). For this reason, mammalian and bird muscle cells tend to have far more mitochondria than muscle cells in snakes.

Thus, it turns out that there are advantages to both endothermy and ectothermy—which is why there are still reptiles, even though there are also birds and mammals.

Power and Behaviors

Powers of Emotion

We know that emotions can be powerful. They can cause us to flee danger, fight an attacker, or risk humiliation and rejection in pursuit of a mate. Emotions are also key to social life: through them, we communicate urgent motives and maintain group cohesion. As such, emotions have played an essential (though still largely unexplored) role in evolution.

New research is showing that emotions are everywhere in nature. Primatologist Frans de Waal, who has spent decades studying the behavior of chimps and bonobos, has concluded that there may be no uniquely human emotions. Similarly, observing my backyard flock of chickens on a daily basis, I see clear expressions of rage, joy, contentment, jealousy, fear, curiosity, and affection; they can hold grudges or choose to forgive past insults. These birds, which are highly social, are also very emotional creatures.

Do plants have emotions? A couple of decades ago, only a tiny minority of scientists would have said that they do, but that is changing. Plants respond to opportunity and threat—even threats to their neighbors. In one experiment, deliberate damage to a cucumber plant caused measurable chemical responses (fear?) in nearby chili peppers and lima beans.[20]

But plants don't have brains or neurons; how, then, could they possibly feel emotion? The answer may have to do with *convergent evolution*, in which organisms not closely related evolve similar traits through different pathways in order to solve similar problems. Moths and geese fly, but they have found anatomically different ways of doing so. In the same way, both animals and plants appear to exhibit what might be considered emotional responses, transmitted on one hand by neurons, on the other by volatile chemicals and what researcher Kat McGowan has called "electrical pulses and voltage-based signaling that is easily reminiscent of the animal nervous system."[21]

Powers of Intelligence

The abilities to discern cause and effect, and to remember, imagine, compare, and plan ahead offer higher animals with lots of neurons the advantage of being able to foresee the behavior of prey, predators, or competing members of the same species, and to devise strategic responses.

For decades, most scientists assumed that only humans exhibit intelligence. In recent years, however, that attitude has shifted dramatically. Research has shown that all living things have at least some vestiges of cognition.[22]

Most scientists have tended to assume that the animals most like ourselves (i.e., other primates) must be the most intelligent. Yet elephants, porpoises, and many birds (such as corvids, parrots, and starlings) are champs at remembering, counting, calculating costs and benefits, and doing lots of other things we associate with being smart.

Further, intelligence is certainly not limited to vertebrates. In her book *The Soul of an Octopus*, Sy Montgomery shares her alternately eerie and heartwarming experiences with cephalopods—"alien" intelligences who easily recognize and remember particular people, display unique personalities, and, if kept in captivity, can devise ingenious and detailed plans of escape.

As with emotion, it was long the assumption of scientists that, without brains, plants are incapable of the basic functions of cognition. Yet new research shows that plants nevertheless exhibit clear signs of intelligence—of which the ability to plan is generally agreed to be a key. Anthony Trewavas, writing in *Trends in Plant Science*, notes that

> Branch and leaf polarity [i.e., relative growth patterns] in canopy gaps have been observed eventually to align with the primary orientation of diffuse light, thus optimizing future resource capture. The internal decisions that resulted in the growth of some branches rather than others were found to be based on the speculatively expected return of future food resources rather than an assessment of present environmental conditions.[23]

Once again, convergent evolution seems to have provided plants and animals with different pathways to the same power.

Powers of Deception

The ability to deceive predators, prey, potential mates, or rivals of the same species can sometimes confer a power advantage. Both animals and plants have learned to deceive by mimicking other objects or organisms.

The *Kallima* butterfly in Sumatra has wings that closely match the color and shape of dead leaves, thus tricking potential predators to ignore it. Other edible butterflies mimic the color and behavior of toxic insects in order to avoid being eaten. The power of deception can also be deployed on an offensive basis: the bolas spider hunts by releasing an odor that exactly matches the one given off by female moths ready for mating.

Plants are just as good at deception as animals. Stinging nettle, as the name implies, delivers a nasty surprise to those who touch it. A plant of the mint family, known as dead nettle, likes to grow nearby; it looks just like stinging nettle and, though it isn't painful to the touch, is avoided by browsers and thereby benefits from the association. Other plants (such as the bee orchid and tongue orchid) use color, shape, and smell as sexual lures to attract pollinators, which believe they are approaching a potential mate.

Deception can even be used to get others to do the work of parenting for you. The cuckoo is a notorious trickster, laying its eggs in the nests of other bird species, which rear the cuckoo chicks as their own. Cuckoos are even able to change the color, shape, and size of their eggs to match the appearance of the host species' own eggs.

Powers of Communication

Communication is the basis of social power—which existed long before humans. Cooperation requires communication, and extreme cooperation (as among ants) requires an almost constant exchange of information. Communication is also essential for deception, mating, hunting, and avoidance of predators. It can occur through sounds, visual displays, scents (i.e., chemicals), heat, electromagnetic fields, or vibration; and it can be intentional or unintentional. When it results in a behavior change in the receiver, researchers call it a signal.

Ants use body language, scent, touch, and sound to navigate their hypersocial world. Each colony has its own cocktail of pheromones, so any two ants can instantly smell whether they are relatives and teammates, or from rival colonies. If an explorer ant finds some interesting food on her travels, she will lay down a scent trail as she hightails it back to her colony to inform the other workers. She even describes (by antenna touching and body language) what to expect at the end of the trail. If a part of the underground colony collapses, the trapped ants will scrape their legs on a washboard-like section of their bodies to make a sound that alerts the rest of the colony.

Birds are also enthusiastic communicators, but only recently have researchers begun to appreciate the range of possible avian signals. Nearly everyone is familiar with the sound of crows, but that insistent "Caw! Caw!" conveys far more useful information to other crows than it does to nearby humans. Not only does each crow recognize the sound of other individual crows, but these birds are able to share remarkably detailed and specific information about their surroundings. In crow-speak, the same call can mean different things depending on tone, timing, the space between repetitions, and speed of utterance.

In the wild, parrots use body postures, their eyes, and their feathers to communicate, along with a wide variety of vocalizations—which vary from region to region, like human dialects. In captivity, parrots are easily able to mimic the sound of human speech. More impressively, where researchers have taken the time to teach them the names of objects and actions, parrots have learned to form meaningful sentences, commenting on what they see and how they feel, and asking for what they want.

Similarly, chimps and gorillas have been taught to use sign language. Washoe, the first chimpanzee to learn American Sign Language, could communicate with approximately 350 signs; Koko, a gorilla born at the San Francisco Zoo, understood about 2,000 words of spoken English, in addition to 1,000 signs.

Researchers are still only beginning to understand the communication abilities of cetaceans—dolphins and whales. These sea mammals rely primarily on sound for communication (a dolphin's hearing range is six times broader than a human's), as well as for echolocation. Each pod of dolphins has its own dialect, and each dolphin uses a unique signature whistle—imparted by its mother—by which it is known. Groups communicate in order to hunt more effectively, and the pod is able to stay together because individuals keep within hearing range of one another.

Plants communicate using electromagnetic and chemical signals that seem to mimic the functions of neural circuitry in animals. For example, wounded tomato plants give off the volatile odor methyl jasmonate, which researchers believe acts as an alarm signal for neighboring plants. The neighbors can then prepare for attack by producing chemicals that either defend against consumer insects or attract insect predators.[24]

Fighting Power

Animals sometimes use jaws, beaks, teeth, hooves, horns, and claws as weapons. Carnivores use them as offensive weapons, or as defenses against other carnivores; herbivores use them to ward off carnivores or in competing with other members of their species for territory or mates.

When members of the same bird or mammal species fight, it is rarely to the death (this is not always the case with spiders and insects: female praying mantises devour their mates). Toucans, for example, use their giant beaks for sword-play, dueling with rivals; but one bird hardly ever seriously injures another.

When members of different species fight, the outcome is more often lethal. Predators, after all, depend on killing for their survival. They typically choose prey that are incapable of putting up much of a defense; but sometimes, if the prey animal is large enough (for example, a zebra being hunted by a lion), a battle ensues. If the prey is dispatched, predators of the same or different species may then fight over the carcass.

Some predator species (including wolves, lions, dolphins, and hyenas) often hunt in packs, attacking many prey (such as a herd of impala or a school of anchovies) at once. But warfare—in which many members of one group systematically attack another group of the same species—is rare; humans, ants, and chimpanzees provide the main observed instances. Some ants even raid colonies of other ant species to enslave their captives.

Plants fight, too, though usually not as photogenically as animals. There are over 500 species of carnivorous plants that use a variety of "trap" strategies to snare their prey (mostly insects). The main categories of such plants are: pitfall traps, which catch prey in a rolled leaf holding a pool of digestive enzymes; bladder traps, which create an internal vacuum to suck in prey; flypaper traps, which use glue to snare the feet of prey; snap traps, which use quick-moving leaves; and "lobster-pot traps," which force prey to move down a tunnel with inward-pointing hairs toward a digestive organ.

In addition (and more commonly), plants can give off chemicals that discourage insects and other predators, or outright poison them. Almond trees and sorghum stalks can release hydrogen cyanide as a defense; a cow munching on sorghum can be sickened as a result.

Many plants are able to chemically sense the presence of insect saliva, which causes the plant to quickly manufacture defensive chemicals.

Exclusionary and Territorial Power

Exclusionary power—the ability to prevent others from accessing resources—is the essence of competition, and it's common throughout nature.[25] It's seen in animals competing for foraging areas, mates, and nesting sites, and even in baby birds vying for their parents' attention during feeding. Exclusionary jostling occurs both among different species and among members of the same species.

Territoriality—the behavior of an animal in defining and defending its space for feeding or breeding—is a version of exclusionary behavior. I was recently reminded of it by some unusual activity around my bird feeder. Finches, nuthatches, and sparrows were gathering to feast, as usual, only to be chased away by an aggressive mockingbird—who didn't stay to eat. After a minute or two, a few of the smaller birds would return, only to be rousted again by the mocker. A little research revealed that my mockingbird's behavior, while unusual, is hardly unheard of: the little bully has somehow decided that the feeder is part of its territory, and, even though he's not much of a seed aficionado, he spends considerable time and effort making sure other birds stay away.

Territorial animals typically mark their territory using smells, calls, or visual signs. Defense of territory usually begins with the territorial defender confronting the interloper with ritualized aggression. Actual fighting, which could injure either or both animals, is typically a last resort.

Territoriality is seen in only a minority of species. And in the vast majority of those, exclusionary competition results in no injury (ants and chimpanzees, which sometimes fight to the death, provide exceptions to this rule). Territoriality and exclusionary power will take on a special significance in later chapters of this book, when we explore the origins of human behaviors related to private property and war—which have resulted in a great many injuries and deaths.

How Much Power?

It's interesting and helpful to have a sense of scale with regard to physical power. Here are some examples:

Nonhuman animals:

- **Mouse:** 0.8 watts, based on the daily energy expenditure of an average lab mouse.[26]
- **Horse:** 798.6 watts, based on the daily energy expenditure of a mid-sized horse.[27]
- **Elephant:** 2,292.4 watts, based on the daily energy expenditure of an average Asian elephant.[28]

Man-made systems:

- **A space shuttle taking off:** 1.2×10^{10} watts or 12 gigawatts, based on the combined power of solid rocket boosters and main engines.[29]
- **A wind turbine:** 2.43×10^6 watts or 2.43 megawatts, based on the average rated capacity of new turbines.[30]
- **A solar panel:** 320 watts, based on the average new solar panel.[31]

Natural systems:

- **The Mississippi River at New Orleans:** 8.6×10^7 watts or 86 megawatts, based on a calculation of kinetic energy using known average flow rate and an estimate of cross-sectional area made with known width and depth.[32]
- **A rainstorm over San Francisco:** 1 inch of rain over 3 hours: 6.4×10^{11} watts or 640 gigawatts; 1 inch of rain over 1 hour: 1.93×10^{12} watts or 1.93 terawatts; 2.5 inches of rain over 1 hour (as in a January 2017 rainstorm): 4.8×10^{12} watts or 4.8 terawatts, based on calculations of energy released through condensation using latent heat of condensation.[33]
- **A hurricane (category 3):** Total amount of energy released through condensation (cloud/rain) formation: 6×10^{14} watts. Energy dissipated by winds: 1.5×10^{12} watts.[34]

Humans:

- A theoretical primate the size of a human, without fire or other human ways of gathering/amplifying power (just food, muscle): 109.6 watts, based on average daily calorie expenditure of a hunter-gatherer.[35]
- Hunter-gatherer, with fire and some tools: 476 watts, based on average daily calorie expenditure of a hunter-gatherer and fire for cooking, heating, and tool-making. This number is highly variable depending on climate (hunter-gatherers in cold climates burn more wood for heating). This range would be approximately 272 watts in warm climates to watts to 1,625 watts in very cold climates.[36]
- Early farmer with an ox: 718 watts, based on average daily calorie expenditure of humans, oxen, and fire for cooking, heating, and tool-making. This too depends on the need for firewood as well as the number of draft animals per person—which is widely variable, but a 466 watt to 2,109 watt range can be approximated.[37]
- An ancient king or pharaoh: Up to a million watts, depending on the size of workforce commanded. If only personal direct food and fuel consumption is taken into account, the number would not be much higher than the upper end of early farmer range (718+ watts). However, if the human labor a ruler directs is included, the number is much higher. Take the example of Khufu, the Pharaoh believed to have commissioned the Great Pyramid of Giza. Archaeologists estimate that 4,000 laborers worked directly on the construction of the pyramid, and another 16,000–20,000 laborers provided supporting functions such as tool-making and food-preparation. Roughly 5,000 were likely permanent workers while the rest were temporary workers. We can estimate the energy output of the workforce on any given day. If there were 22,000 laborers burning an average of 1,000 kilocalories during working hours (caloric

expenditure was likely more for those doing construction and less for those providing supportive functions), this would amount to around 1 million watts.[38]

- **Modern industrial human, poor country (India):** 319 watts, based on per capita GDP and energy intensity of GDP; 812 watts, based on per capita energy consumption. The latter is likely more accurate. In India, a significant portion of the population uses wood cooking stoves, and this household use of biomass isn't accounted for in GDP.[39]
- **Modern industrial human, rich country (US):** 9,925 watts, based on per capita GDP and energy intensity of GDP; 9,024 watts based on per capita energy consumption. In this case, the GDP figure more accurately reflects US consumption. Since the US imports many energy-intensive products, counting only the energy used within the nation's borders can underestimate energy consumption.[40]
- **The human Superorganism (at its current size):** 1.84×10^{13} watts, calculated using 2018 primary energy consumption.[41]

Proto-Human Powers

The Power of Beauty and Attraction

Why do many animals—especially birds—put so much effort into breathtakingly beautiful displays of sound and color? Are the virtuosic songs and striking plumage of the males of many avian species simply signals to potential mates of their overall fitness? Or is more going on here? Charles Darwin, the father of evolutionary theory, believed the latter. The hunger for beauty, he argued, is an evolutionary force separate from natural selection, and sometimes on par with it in terms of its influence on the development of species.

Darwin's first book, *On the Origin of Species by Means of Natural Selection*, became a bible for evolutionary biologists. But his second book, *The Descent of Man, and Selection in Relation to Sex*, confounded many of his followers. In it, he described instance after instance where animals invest extraordinary effort in display activities. He wrote, "The sight of a feather in a peacock's tail, whenever I gaze at it, makes me sick!" The feather is unquestionably beautiful, but its evolution is nearly impossible to explain in terms of fitness and natural selection. Why is nature so filled with apparently useless beauty?

Darwin's solution to the conundrum was the principle of sexual selection. In species that reproduce through sex, the successful transmission of an individual's genes to the next generation depends not just on that individual's relative vigor, size, or strength (qualities we intuitively, though often mistakenly, associate with fitness), but also on its ability to attract a mate. Females of the species often choose males with whom to get it on (more rarely it's the other way around), and the criteria for choosing or being chosen sometimes appear bizarre.

Most of his followers, rejecting this aspect of Darwin's legacy, have attempted to explain these displays by hypothesizing natural selection benefits. But ornithologist Richard O. Prum, in *The Evolution of Beauty*, makes a convincing case that Darwin got it right in concluding that such displays often aren't tied to objective measures of competitive fitness. "Individual organisms," he writes, "wield the potential to evolve arbitrary and useless beauty completely independent of (and sometimes in opposition to) the forces of natural selection."[42]

Consider the bowerbird of Australia. The adult male builds a bower (which is an elaborate structure that's not a nest and has no other use) to attract a female. Various bowerbird species build differently sized and designed bowers; some carefully arrange colored objects—pebbles, petals, feathers, insects, bottlecaps—to decorate the structure. Why do these birds go to so much trouble? Meticulous research has shown that artistic effort on the part of the male, and selective preference on the part of the female, have coevolved in a self-reinforcing feedback process. In the male bowerbird, and many other creatures, the power to attract a mate has become inextricably tied to the activity of producing expressions of beauty that have no other practical value and are not signs of overall male fitness. Beauty has become a powerful end in itself. (In Chapter 2 we will explore how

sexual selection and the power of beauty and attraction may have impacted human evolution as well.)

Evolutionary investments in display can proceed to such extremes
that they lead to "aesthetic decadence," contributing to a species' decline and even extinction. When the males and females of a given species come to agree that only a particularly extravagant display—one
whose costs impair the species' survival abilities—are worthy of mate
choice, then attraction can truly become fatal. The fossil record probably holds plenty of examples, though teasing out the exact cause
of extinction in any given case is often difficult (the Irish elk, with
its impossibly bulky antlers, is probably a good example-candidate).
It's perhaps easier to show instances of aesthetic decadence among
creatures still living, such as the club-winged manakin, a small bird
that lives in the Amazon rain forest. The male courts its potential
mate by clapping his wings together at over 100 times per second,
much faster than the flapping of a hummingbird's wings, producing
an oboe-like tone. Females adore the sound and choose their mate
based on the excellence of his wing-clapping performance. Unfortunately, however, in order to effectively make their unique sound,
club-winged manakins need solid wing bones—which they have duly
evolved. As a result, their flight is slow and clumsy, putting them at a
distinct disadvantage compared to other birds.

Powers That Derive from Being a Specialist or a Generalist

Specialist species get better and better at exploiting a certain type
of food within a specific environment. Perhaps the most famous example: when Darwin arrived on Galapagos Island in 1836, he found 15
species of finches, each of which had evolved a different form of beak
for opening a particular kind of seed available on the island.

Generalists do many things, but are typically champions at none.
Humans, crows, cockroaches, rats, and racoons are all generalists,
and are able to eat a wide variety of animal and plant foods.[43]

Specialists have the advantage of being able to out-compete most
other creatures that would otherwise take advantage of a targeted
food source. The risk of specialization is that if the preferred food

source is suddenly less abundant, the organism may have difficulty adapting. Generalists tend not to use any particular resource very efficiently. But, because they can harness a wider range of resources, generalists may have a power advantage in a varied, unpredictable, or rapidly changing environment.

There are fewer generalist than specialist species, but some generalists (such as crows and rats) are abundant and wide-ranging. One generalist species, *Homo sapiens*, has become especially abundant and wide-ranging, and hence collectively powerful, for reasons we will explore in the next three chapters.

The Power of Cooperation

Many of the powers we've been discussing are useful in competition—whether between one individual and another, or one species and another. However, evolution works not just by fierce competition, but also—perhaps even more so—by cooperation. Ironically, organisms that develop ways to cooperate often thereby derive a competitive advantage: working together, many individuals can become more powerful than they would be if working in isolation.

Nature offers innumerable examples of cooperative behavior; indeed, without it, there would be no multicellular organisms. We've already seen how eukaryotes are the result of an ancient symbiosis between two very different kinds of cells. Green plants—containing both mitochondria and chloroplasts—are double symbionts.

At the macro scale, cooperation among members of the same species is seen in ants, honey bees, fish that "school" (such as herring), prairie dogs, members of the canine family, many bird species, and all primates (this is only a partial list).

Cooperation between members of different species is also common. Many flowering plant species have evolved cooperative relationships with specific pollinators, in which both participants benefit. Sometimes smaller animals work out cooperative relationships with larger ones—such as cleaner fish (including cleaner wrasses, neon gobies, and some catfish) that feed on dead skin and parasites on larger predatory fish that could easily eat the cleaners, but don't.

You needn't go far to observe interspecies cooperation; indeed, each of us is host to trillions of microbes, including about a thousand different bacterial species (like our old friend *E. coli*), that live in and on the human body. Their total number is typically greater than the number of strictly human cells in our bodies. These various kinds of microorganisms generally complement each other and their host, fulfilling functions essential to life, including synthesizing vitamins and neurochemicals, aiding digestion, and strengthening the immune system.

Extreme cooperation can lead to extreme power. Ants, the most cooperative organisms on the planet until civilized humans came along, are one of the most successful groups in evolutionary history and account for about a third of all insect biomass. Humans, because they are extraordinarily cooperative, are able *individually* to specialize to an astonishing degree (especially so since the adoption of agriculture, as we will see in Chapter 3), thereby enabling *Homo sapiens* to take advantage of both generalist and specialist strategies to an extent unmatched elsewhere in nature.

Tool Power

Animals and plants obtain and exert many of their powers by developing parts of themselves for a particular purpose—such as a wing ideal for soaring on wind currents, or a tail perfect for swinging from tree branches. One way of understanding tools is to think of them as detachable organs. Since tools are not part of the organism and can be replaced or modified at will, an organism can use different tools for different purposes.

Decades ago, most scientists thought that the ability to make and use tools was uniquely human, but this attitude has gradually changed. Researchers were initially curious to see if other primates used tools. Chimps, for example, were observed to have specialized tool kits for hunting ants. Gorillas use tools less frequently than chimps and bonobos (perhaps because gorillas live in environments with an abundance of food plants), but they have been seen using sticks to gauge the depth of water.

However, nonhuman primates may not be the best examples of animal tool users. Crows have been observed modifying twigs, leaves, and their own feathers to serve as tools. In carefully designed experiments, crows have even figured out how to drop stones into pitchers to raise the level of water inside.

Other avian tool users include finches and woodpeckers, which have been seen to insert twigs into trees in order to catch or impale larvae. Several parrot species have been observed using tools to wedge open nuts they're trying to crack, or to scratch the backs of their heads. One parrot species, the palm cockatoo, occasionally fashions a drumstick from a tree branch, then strikes it repeatedly against a hollow tree trunk to make a distinctive sound that can be heard for miles. Biologists are still unsure of the purpose of this remarkable behavior.

Bottlenose dolphins in Shark Bay, Australia, carry marine sponges in their beaks to stir the sand on the ocean floor to uncover prey. It's said that they spend more time hunting with tools than any other nonhuman animal.

One could argue that a spider's web is a tool. The spider uses its spinneret glands to produce silk, which it then weaves into a web for catching unwary insects. The process by which spiders make their webs, and the designs they use, are instinctual and genetically coded. Yet each spider must make many decisions to creatively solve unique problems. And among spiders of the same species, some are clearly better than others at building and managing their webs. The net-casting spider weaves a small net, attaches it to its front legs, then waits for potential prey. When the victim arrives, the spider lunges forward to wrap the net around it, then bites and paralyzes its prey. Observed in any mammal, similar behavior would almost certainly be described as the making and use of a tool.

Prior to the appearance of humans and close human ancestors and relatives, toolmaking and tool use already existed in nature. But tools clearly represented an enormous opportunity for the further harnessing and leveraging of power, and also helped open the way to a different and much faster kind of evolution—cultural evolution—

which we'll discuss at more length when we get around to exploring language (in Chapter 3).[44]

Fire Power

The controlled use of fire constitutes a special instance of tool usage. Until recently, it was commonly thought that only humans had harnessed fire, but this is not strictly true. At least three Australian raptor species (the black kite, the whistling kite, and the brown falcon) have been observed to pick up burning twigs or branches from wildfires and then drop them at a distance, deliberately starting fires elsewhere in order to flush prey from undergrowth.

Still, the ability to harness fire is so rare in the nonhuman world that it might hardly be worth mentioning, except that it serves to put into perspective the vastly greater power ramifications that fire has had as result of human beings' ability to control it. We will unpack some of those implications in chapters 2, 4, and 5.

Trade-Offs of Specialization

As we've seen, organisms have developed amazing abilities. Why isn't every organism good at everything? Because getting really good at one thing tends to hamper your ability to do something else. To put it more formally, the extreme expression of any given trait is likely to reduce fitness in some way. The giraffe's long neck enables it to browse on the leaves of trees, but makes it highly vulnerable to predators when it needs to bow low to drink water. A bacterial mutation that confers resistance to an antibiotic may weaken the bacteria if that antibiotic isn't present. The bat can hear certain frequencies with extraordinary acuity, but if it simultaneously had the eyesight of a hawk, the ability to sense infrared radiation the way a pit viper does, and the ability to see ultraviolet colors the way many birds do, its brain would have to be very large and would likely be overloaded with stimuli.

Generally, if the benefits of an extreme specialization outweigh the costs and the organism is able to survive and reproduce, the extreme ability will be preserved; if the costs outweigh the benefits, evolution will de-emphasize the specialization or the organism will fade away and evolution will move on.

All of this suggests there are probably downsides to humans' specialized powers (extreme intelligence, a highly developed ability to communicate, and proficient tool use). We humans tend to emphasize the advantages of these traits, but it's always important to look for hidden costs. We'll be discussing these evolutionary costs at some length in later chapters.

✦   ✦   ✦

In short, prior to the appearance of humans, the natural world was already a complex system of powers and power balances. Energy originating in our nearest star was being cycled and recycled in intricate and beautiful ways via biochemical processes in individual cells—processes that enabled the flourishing of both single-celled and multicelled creatures, and ecosystems of unfathomable complexity.

Into this milieu came a handy animal, a generalist *and* a specialist, that was able to use many different powers in increasingly effective ways—and able, as we will see, to develop many powers (motion, perception, cognition, deception, communication, exclusion, cooperation, fighting, toolmaking, and control of fire) to extraordinary degrees.

POWER IN THE PLEISTOCENE

On Spears, Fires, Furs, Words, and Flutes— And Why Men Are Such Power-Hogs

A system of obedience depends on punishment. Within families or small groups, the mechanism of punishment can be emotional manipulation or physical beatings, but in the politics of large-scale groups, a proactive coalition provides the power. An order given to a subordinate is essentially a threat to use aggression unless the order is obeyed. If the threat depended merely on the fighting power of the leader, it would rarely be convincing. No leader could risk repeated fights. Even an alpha chimpanzee tries to avoid fighting when possible. However, a human leader does not have to fight personally; a coalition of supporters guarantees the value of the leader's threat, and the likely danger from an aggressive encounter is low for the supporting coalition because their overwhelming power can be brought to bear on the subordinate.

— RICHARD WRANGHAM, *The Goodness Paradox*

Right, as the world goes, is only in question between equals in power, while the strong do what they can and the weak suffer what they must.

— THUCYDIDES

THE STORY OF HUMANITY'S DEVELOPMENT OF ITS AWESOME and unique array of powers likely began, ironically, with climate change. About 13 million years ago, several centuries of drought ravaged the forests of east Africa. As a result, many tree-dwelling primates were forced to adapt to living in expanding savannahs. The primates that stayed in the shrinking forests were the ancestors of today's nonhuman great apes—gorillas, chimps, and bonobos. The primates that came down from the trees evolved into several lineages of *hominins* (i.e., proto-humans). One of those lineages ultimately led to us.

It's still unclear whether the savannah dwellers developed an upright stance, walking on two legs instead of four (a process that must have taken millions of years), before, during, or after leaving the forest. Nevertheless, living in grasslands rewarded the trait of bipedalism. It was now advantageous to stand tall, so as to peer over high grass to see predators and potential food. Crucially, this increasingly upright posture conferred power in several other ways:

- Bipedal hominins burned fewer calories as they foraged, compared to knuckle-walking forest primates (walking on two legs isn't calorically superior to the quadrupedal locomotion of deer or horses, at least over short distances—but our ancestors weren't directly competing with those kinds of animals; instead, they were competing with other apes).
- Hominins and their descendants could walk and run long distances at a constant speed. This enabled them to chase down prey such as deer that could sprint much faster, but tired quickly.
- Hominins' (and later humans') ability to walk long distances also helped them to spread out geographically—ultimately hiking,

boating, and sledding their way throughout all the Earth's continents except Antarctica.

- Crucially, bipedalism freed hominins' front limbs for tasks other than locomotion.

Meanwhile, life on the savannah required more emphasis on long-distance vision. As their sense of sight increasingly predominated over other senses, proto-humans began to coordinate their eyesight with small, controlled movements of their freed-up hands. Even though the hands of other primates are anatomically similar to those of humans, new neural pathways in hominins helped in the development of the muscle control necessary for grasping, manipulating, and throwing—using what paleoanthropologists appropriately call the "power grip."

But walking upright imposed costs and required adaptations. Notably, it created the need for a thicker pelvis; and this, in turn, resulted in changes to the process of childbirth. Babies were born smaller, so they required more lengthy parental care before they could fend for themselves. And that period of extended care promoted extended families—which offered more opportunity for social interaction and for the development and transmission of culture.

Increased social contact contributed to one more crucial change in hominins and their descendants: a substantial growth in brain size. Bigger brains co-evolved with walking; with finer, more controlled hand movements; and with increased social interaction. Each of these developments compounded the others. Big brains, for example, also reinforced changes in childbirth, already underway as a result of bipedalism: because of their larger crania, babies had to be born in a less mature state, and were even more dependent on long-term familial care. Meanwhile, more social interaction required more memory and thought—hence bigger brains. Walking, the development of better hand-brain coordination, bigger brains, and increased social interaction, in effect, together constituted a self-reinforcing evolutionary feedback loop. Over nearly seven million years our brains tripled in size, with most of that growth occurring in the past two million years. And, during that same period, humans became some of the most social creatures on the planet.

Crucially, it's important to remember that these changes were occurring among several pre-human and human species. Even after the appearance of true humans (genus *Homo*), perhaps three million years ago, several distinct species persisted.

- The earliest known of the true human species was *Homo habilis*, which must have emerged nearly three million years ago; all known signs of this species are restricted to Africa.
- Signs of *Homo erectus* date from two million years ago; this human species emerged in Africa but then spread to Eurasia. It may have survived in Java as late as 70,000 years ago.
- *Homo denisova*, which probably appeared in Africa at least 400,000 years ago, migrated northeast to Siberia, and southeast to Indonesia and New Guinea. Denisovans persisted longest in their southeastern range; genetic evidence suggests they may have interbred with *Homo sapiens* in New Guinea as recently as 15,000 years ago.
- *Homo neanderthalis*, with which we share about 99.8 percent of our genetic code, emerged in Africa at least 400,000 years ago and spread north and northwest to the Middle East and Europe; and later eastward, where Neanderthals probably interbred with Denisovans. Neanderthals had similar-sized brains (perhaps even slightly bigger), on average, to those of *sapiens*. But the formation of Neanderthal skulls suggests that there may have been differences in how this large brain was organized and used, compared to ours. Neanderthals died out by about 40,000 years ago.
- There were other human species as well. *Homo ergaster*, which may have been a variant of *Homo erectus* or a separate species (experts disagree on this point), lived in southern Africa between 1.9 and 1.4 million years ago. Two species, *Homo soloensis* and *Homo floresiensis*, have been identified just in Indonesia, where human fossils seem to have survived particularly well. *Homo naledi* lived in part of southern Africa 300,000 years ago. *Homo heidelbergensis* preceded Neanderthals in Europe, and persisted alongside them for

a time (it may have been the common ancestor of *Homo neanderthalis* and *Homo sapiens*).

All the species mentioned are simply ones whose remains happened to be preserved and were later discovered by modern researchers; there may have been more, evidence of which hasn't been found. What's crucial to our story is the clear fact that none of them survived to the present—except one.

Homo sapiens first appeared in Africa roughly 300,000 years ago, then spread out from that continent in at least three waves, the first of which occurred over 250,000 years ago, and another about 100,000 years ago. *Sapiens* individuals who were part of those early migratory waves interbred with Neanderthals and probably with *Homo erectus* as well. But these *sapiens* evidently died out. A later wave of migration began about 70,000 years ago. This time *sapiens* survived and flourished nearly everywhere it went. All humans outside Africa who are alive today—Asians, Pacific Islanders, Europeans, and Native Americans—descended from that last migratory wave. Meanwhile, in the process of flourishing, *sapiens* profoundly altered landscapes nearly everywhere it went, wiping out many large mammal and bird species. This was a uniquely powerful kind of primate.

During the last three million years, as these various human species came and went, the environment was in a state of enormous flux. Ice ages lasting tens of thousands of years were interrupted with shorter warming periods during which the ice retreated, opening up space for grasslands and forests. Then the ice would return. Some of these shifts were fairly abrupt, punctuated by sudden, cataclysmic floods and extreme weather events. Neurophysiologist William H. Calvin has theorized that these repeated bouts of climate change acted as an evolutionary pump, forcing humans to adapt repeatedly to dramatically altered circumstances, thereby selecting for larger brain size.[1] The notion is hard to prove or disprove, and it raises the question whether there were equivalent evolutionary impacts on other animals. However, it also offers food for thought about our future human response to the climate change that we ourselves are now causing. But—back to our story.

Ominous Power from Space: Asteroids, Comets, and Climate Change

In their 1980 book, *The Cosmic Winter*, astrophysicist Victor Clube and astronomer Bill Napier argued that Earth was bombarded 13,000 years ago by the fragments of a giant comet; this resulted in a cosmic winter that changed the global climate for centuries. Ancient peoples' later worship of vengeful sky gods, according to the authors, may have been partly a response to periodic rains of fire that persisted, as the orbits of the comet's remnants occasionally crossed Earth's path. In the past four decades new findings have lent increasing support to this notion of cosmic catastrophes as contributing drivers of planetary and cultural change.

Abundant evidence shows that Earth underwent rapid cooling 12,800 years ago. Average temperatures dropped as much as 14 degrees Fahrenheit in just a couple of years in some parts of the Northern Hemisphere, and ice cores from Greenland show that this cool period lasted about 1,400 years. Climate historians call this the Younger Dryas period. It marked the beginning of a decline in ice-age megafauna, such as mammoth and mastodon, eventually culminating in the extinction of more than 35 genera of animals across North America.

For decades, most geologists tied the Younger Dryas to the failure of glacial ice dams holding back huge lakes in central North America. The sudden influx of freshwater into the North Atlantic would have shut down ocean circulation and cooled the temperate regions of the Northern Hemisphere.

However, a competing hypothesis holds that an extraterrestrial impact contributed to this chain of events by igniting massive wildfires that blocked sunlight with their smoke. Ocean, lake, soil, and ice samples from around the world show large peaks in particles associated with burning in the layers associated with the start of the Younger Dryas. So much soot was released so quickly that scientists estimate that as

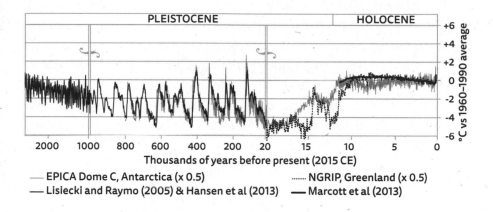

Figure 2.1. Global temperatures during the Pleistocene and Holocene.

Credit: Adapted from an illustration by Glen Fergus, CC BY-SA 3.0, via Wikimedia Commons.

much as ten percent of the world's forests and grasslands may have burned. The Younger Dryas boundary layer is also rich in impact-related materials such as silica-rich tiny magnetic spheres, nanodiamonds, melt-glass, nickel, osmium, iridium, and platinum.

How much power can an asteroid or comet impact deliver? It's estimated that the Tunguska detonation of 1908, in which a comet fragment exploded over Siberia, released 20–30 megatons of energy, while the Chicxulub impact in Central America that likely killed off the dinosaurs 60 million years ago is estimated to have had the explosive power of 100 trillion tons of TNT, or over a billion Hiroshima bombs. The detonation of the object that triggered the Younger Dryas would have had a power somewhere between those two values.

All human species walked upright. All used their opposable thumbs to make and use tools. Why did the last great migration of *Homo sapiens* out of Africa succeed, when earlier ones didn't? It would appear that *sapiens* had acquired some crucial power advantage in the meantime that enabled it to outcompete all other human species, either deliberately or inadvertently driving them to extinction.

The prime candidates for this hypothetical new source of power, according to most paleoanthropologists, are the fabrication and use of more sophisticated tools, control of fire, and language-based social skills.[2] Spoiler alert: the question of how and why other human species died out is still open and the evidence is inconclusive. Nevertheless, it's worth exploring these three advantages one by one, as they profoundly shaped who we are today. Whether or not they enabled *sapiens* to outcompete other sorts of humans, these were unquestionably the sources of power that ultimately led our kind to dominate the rest of the biosphere.

SIDEBAR 6

A Watery Theory of Human Origins

Initially proposed by marine biologist Alister Hardy in 1960, and developed by Elaine Morgan in her books *The Aquatic Ape Hypothesis* (1982) and *The Scars of Evolution* (1990), the aquatic ape theory is a heterodox alternative to the widely accepted notion that humans evolved mostly in the African savannah. The theory argues instead that early *Homo sapiens* became adapted to swimming and diving in freshwater lakes, and that various unusual human anatomical features resulted from adaptation to watery surroundings.

Geological evidence shows that deep freshwater lakes appeared and disappeared all along the East Africa Rift Valley between 2.6 million and 1 million years ago—a key period in human evolutionary history. The Valley was a highly variable environment that forced our primate ancestors to accommodate themselves to alternating aquatic ecosystems and relatively dry ecosystems, and that also forced them to migrate.

Plentiful remains of *Homo sapiens*, about 200,000 years old, have been found in the Rift Valley and along the African coast. These coastal humans gathered, opened, and cooked shellfish, and made spears—possibly for fishing. By 90,000 years ago, modern humans had begun making tools that were unambiguously designed for fishing. They lived in caves in at least a dozen sites along the coast. And they were the last hominins to leave Africa, around 60,000 years ago.

Proponents of the aquatic ape theory suggest that humans' upright posture developed for wading in shallow water and swimming, not seeing over tall grass; that early stone tools were used more often to break open shellfish than to kill and butcher land animals; and that we lost our body hair and replaced it with fat for positive buoyancy and warmth in the water—as did seals and manatees. They point out that we are the only primates that can float, and that we are the only primates born with a breath-holding instinct. The crab-eating macaque is the only other primate that spends much time in the water and that can hold its breath for any length of time (about 30 seconds), and it lives in coastal mangrove swamps. Compared to our closest primate relatives, the chimps and bonobos, humans are far more attracted to water and are much better swimmers and divers.

The human speech apparatus allows the nasal cavity to be either connected or disconnected from the other supralaryngeal air passages. It's not clear how or why this feature evolved; however, one benefit from this arrangement is that it prevents water entering the nasal cavity from reaching the lungs. This anatomical feature allows us to control our breathing more than other primates and allows us to make a wider variety of sounds, while also enabling us to swim and dive. Thus, according to the aquatic ape theory, our watery early environment may have played a key role in the human development of speech and language.

The aquatic ape theory is discounted by most anthropologists, who offer alternative explanations for the development of several key human anatomical features. Nevertheless, this neglected theory may hold clues to the origins of at least some uniquely human adaptations and powers.

Hands and Stone

Tools leverage power. A stone axe, for example, can focus the force of a swinging arm onto a tiny area, delivering a blow that can break open a hard nut or bone, or crush the skull of an animal. Something similar can be said for most of the vast range of other tools that have followed upon the simplest and earliest ones used by our ancestors, from the wheel to the smartphone: they offer means of transferring energy in ways that increase useful power.

While chimpanzees use sticks and simple stone tools, they don't do much to alter their stones, which they employ mostly just for breaking open nuts. The earliest proto-humans, when they first appeared roughly three million years ago, did a little more than that: they began to reshape the stones they used. Mainly this consisted of striking one stone against another to knock off a flake, thereby creating a cutting edge. But this was a crucial difference: making and altering tools was what set genus *Homo* on the path toward ever-growing infatuation with and reliance on technology. Still, as archaeologists survey strata from the past three million years, they find surviving stone tools that remained relatively plain and simple for a very long time—until just the past few tens of thousands of years.

There is evidence that early humans also made tools out of wood, leaves, reeds, skin, bone, sinew, and other animal and plant materials. However, most of that evidence is indirect: these kinds of tools were highly perishable (unlike stone tools, which by their nature are more apt to survive). Some of them, such as boats made from tree trunks or hides and poles, and sleds made of poles and split wood—the earliest transportation tools—would be essential in the process of human dispersal around the world. Stone tools were used in manufacturing nearly all of these ephemeral technologies.

Many early tools were weapons—axes, knives, and projectiles that expanded humans' power over other animals and other humans. Weapons were a leveling force in human social relations: without them, the individual with greater size and physical strength typically prevailed in any contest. With weapons as part of the equation, strategy took on greater importance. A smaller but cleverer individual could realistically hope to overcome a bigger opponent. Through the millennia, weapons would become pivotal tools, with technological advances contributing to the beginnings of warfare, the formation of early states, and the rise of empires (these social developments would, of course, come much later; we'll discuss them in the next chapter).

Innovations in designing and crafting stone tools probably occurred redundantly as well as repeatedly: individuals in more than one place would come up with a new way of making or using a tool, then forget it, then think of it again as the need arose. Gradually and fitfully, tools became more specialized and sophisticated.

Homo habilis mastered the simplest human tool set, which consisted of flaked stones with suitable characteristics. These stones, intended to have a cutting edge, were probably mostly used to butcher and skin animals. With tools like these, *Homo habilis* had the power to thrive in environments previously inhospitable to primates; that's because it had access to new food sources and could make clothing (more on that below).

Roughly at the same time as the appearance of *Homo erectus* came tools that were more thoroughly worked. Stone tools were more carefully shaped, and their edges were retouched with softer hammers of wood or bone to produce knives and axes that were finely chipped all over. Such tools could be used to slice, something previous tools couldn't effectively do.

Still another characteristic tool set was associated with *Homo neanderthalis* and early *Homo sapiens*; it included smaller, carefully worked flakes that would have required more hand strength and flexibility for their manufacture and manipulation.

A much more diversified and refined tool set is found in *sapiens* sites dating between 50,000 and 10,000 years ago. The people who made these tools seem to have relied increasingly on finely worked stone blades for hunting, and there is clear evidence that the blades were mounted on shafts, using string or sinew and glue, to reduce breakage and improve leverage. There is also evidence that rocks with particularly desirable properties (such as obsidian, which can be flaked to an extremely sharp cutting edge) were sometimes sourced from hundreds of miles away, implying the existence of trading networks.

Early archaeologists ventured some faulty guesses about stone tools—how they were made, how they were used, and what they were used for. Researchers (some of whom were credentialed archaeologists, some of whom were amateurs) took a big step forward in their understanding when they started to make and use stone tools themselves. Today, primitive technology classes are offered in many places around the world; in a two-day or week-long workshop, you can learn the very basics of how to flake and haft stone tools, among other useful skills. As a result of what they've learned this way, scientists now appreciate much better the care and skill required to craft and use stone implements—especially ones made after about 50,000 years ago.

The manufacture of these sophisticated tools required skills developed during hundreds or thousands of hours of practice on the part of the manufacturer, as well as the ability to pass those skills on to others, no doubt using language (which we'll discuss shortly). These tools took a variety of forms—from hoes to hammers, bowls to baskets, and needles to knives—and included an increasing variety of projectile weapons (spears and spear throwers, bows and arrows, and slings).

Could it be that these sophisticated tools gave an edge—perhaps quite literally in some cases—to *sapiens* over other human species? That was long the assumption of many theorists. But it's unclear whether this was really the case, or whether improved tools were merely part of a more complicated story. Neanderthals, after all, made and used tools, in some cases ones almost indistinguishable from those crafted by *sapiens* during roughly the same period.

Setting aside once again the problem of what enabled *sapiens* to survive when other human species didn't, let's return to the more basic question of how tools expanded humans' power over the rest of nature. At the time of their earliest tool kits, humans were probably scavengers. Like today's great apes (all of which are endangered), early humans were also prey to big predators. They were capable of killing only small animals for food and probably led a precarious existence.

With spears for killing at a distance, and knives for butchering, all that changed. Humans became formidable hunters. And away went the wooly mammoth, the short-faced bear, the hornless rhino, the giant wombat, the giant ground sloth, the moa, the giant wallaby, the marsupial tapir, the giant platypus, the giant echidna, and many other species of megafauna. Moreover, big predators—including lions, tigers, and bears—now had to fear armed humans as much as humans feared them.

But long before many of these developments, we *sapiens* had set in motion other elements in the feedback process that propelled our primate lineage to the pinnacle of power. Let's turn to what was almost certainly the next element in line, in terms of chronology: controlled combustion.

The Fire Ape

As we saw in Chapter 1, a few bird species in Australia have been observed deliberately spreading fires in order to flush out prey. Our distant ancestors adopted a similar behavior and developed it further by learning to make and manage fire. This gave them significant power in three ways.

First, when fire was used for cooking, it enabled early proto-humans and humans to increase the amount of energy their bodies could derive from food. As anthropologist Richard Wrangham argues in his book *Catching Fire: How Cooking Made Us Human*, cooking "gelatinizes starch, denatures protein, and softens everything." As a result, "cooking substantially increases the amount of energy we obtain from our food."[3]

The benefits of cooking aren't obvious to everyone. Many people extol the advantages of raw food over cooked—including the preservation of essential enzymes in the food itself and the promotion of stronger teeth and jaws in those who eat it (there is good evidence that eating softer foods has led to narrower jaws, malocclusion, and impacted wisdom teeth, among other maladies). However,

Wrangham points out that, whatever the costs, cooking nevertheless makes it much easier for us to digest a range of foods. Evolutionary success is mostly concerned with energy, so, "if cooking causes a loss of vitamins or creates a few long-term toxic compounds, the effect is relatively unimportant compared to the impact of more calories."[4] Cooking, in other words, gives us more nutritional power—and the advantages conveyed by accessing more power, in this case as in so many, overwhelm other concerns.

Second, campfires offered our ancestors warmth and light in the night. They also kept predatory animals away while humans slept— not a small advantage if one happens to have big cats or other fearsome predators as neighbors. Some have argued that campfires helped spur the development of language and religion; after all, what do we ourselves do when we sit around a campfire at night? We tell stories to entertain one another. Maybe ancient humans did the same.

Third, fire enabled early humans to alter woody landscapes to make them more productive for human purposes. This turned out to be a world-changing sort of power. Over a few tens of thousands of years, humans had vast impacts on landscapes on every inhabited continent.

There is no starker example than that of Australia. Humans arrived on this fairly isolated continent roughly 50,000 years ago. As they advanced, they set fire to thickets and dense forests in order to force game into the open, and to clear space for grassland ecosystems that were more amenable to human habitation. Before 45,000 years ago, Australia's forests consisted of trees that are now rare; eucalyptus trees (which are fire-resistant, contrary to popular myth) and grasslands took their place. Australia would never be the same.

Other ecosystems were similarly transformed. When Europeans arrived in the Americas, they observed and described landscapes that likewise had been profoundly altered, and were being deliberately managed by the resident humans—again, largely with fire.

How humans started taming fire is a question that will probably never be answered with direct evidence, but it's pretty easy to infer the process. Research has shown that hunter-gatherers who live in lightning-prone environments tend more commonly to use fire; this supports the widespread speculation that the first uses of fire by humans consisted simply of removing burning limbs from lightning-caused fires, then using those blazing logs to start controlled fires elsewhere. Learning to make fire with friction (using flint or sticks) was a big advance, but it might have occurred more than once among various human groups. (Again, it's instructive to take a class on primitive technology in order to learn first-hand the various ways of starting a fire from scratch, how much work it is, and how much skill it takes.)

Exactly *when* humans started taming fire is another question that can't easily be answered with direct evidence. At a site in the northern Jordan River Valley, dating from almost 800,000 years ago, there are clear signs of people who "had a profound knowledge of fire-making," according to the archaeologist who uncovered the site, Nira Alperson-Afil.[5] But Wrangham invokes two lines of indirect evidence to argue that the first controlled use of fire occurred earlier still.

The first line of evidence: The study of *Homo erectus* skeletons suggests that this human species was able to climb only poorly. Therefore, *erectus* must have slept on the ground—but how could it do this without fire to keep predators away?

The second: Among humans we can observe today, softer foods make for narrower faces and smaller teeth than was the case with their ancestors. Therefore, the adoption of cooking should have left similar signs in ancient skeletons. Wrangham argues that skeletal evidence suggests the adoption of cooking most likely corresponded with the emergence of *Homo erectus*, 1.8 to 1.9 million years ago. If he's right, we have been using fire a long time indeed.

Once the use of fire took off, there were multiple side effects, not the least of which was an increase in the size of the human brain as a result of increased food energy from cooking. Brains require a lot of energy: for humans today, the brain uses about 20 percent of resting metabolic energy, though it accounts for only 2.3 percent of body weight. This proportional difference is greater in humans than in any other mammal. In *Catching Fire*, Wrangham examines the whole process by which brains grew, starting five to seven million years ago when the proto-human lineage diverged from lineages of the other great apes. The first spurts of brain growth can be attributed to eating different foods (roots instead of leaves). But later and more dramatic phases of brain enlargement seem tied to cooking. Further, in Wrangham's words, cooking "surely continued to affect brain evolution long after it was invented, because cooking methods improved."[6]

Cooking also made us more social: we brought much of our food back to the campfire to be cooked, rather than simply eating it where we found it. In addition, keeping a fire going required somebody to stay home to look after it. Thus "home" became a place of greater social importance. Further, humans likely developed more sharing behaviors as a result of cooking.

Today we regard open fire as an inefficient and unsustainable way of heating and cooking. But that's partly because we likely haven't thought very hard about how our ancestors actually used it. Our ancestors didn't just use their campfire for these two obvious purposes. They also used it for illumination, food preservation, water heating, clothes drying, signaling, and protection from predators and insects, among other things.[7] When you add all those services together, fire begins to look like a cheap and ingenious solution to innumerable problems.

Altogether, fire helped fuel the evolutionary self-reinforcing feedback process that also led to tool making, big brains, and greatly increased social interaction. Fire gave us power. But was it fire that led *sapiens* to dominate other human species? It appears not. As we've already seen, *Homo erectus* probably used fire. And there is clear, direct evidence that Neanderthals not only used fires, but knew how to start them.[8]

Nevertheless, if stone tools made us a more powerful animal in relation to other animals, fire made us powerful in relation to whole environments. Crucially, burning stuff made us unique in another important way. It enabled us to derive useful energy from our environment via a source other than food. Gradually we would learn to use fire for ever more projects: to transform materials, to smelt and work metals, and (much later) to run engines. Moreover, as a result of our discovery and use of fuels other than firewood, fire would eventually open the way for technological marvels that have recently changed the world in truly remarkable and ominous ways, as we will see in Chapter 4.

Skins

In Chapter 1, I called tools "detachable organs." The idea goes back at least to William Catton, who, in his marvelous book *Overshoot: The Ecological Basis of Revolutionary Change*, described technologies as prosthetics.[9] We modern humans can, at least metaphorically, strap a Boeing to our waists and fly across a continent with prosthetic wings. In a similar light, clothing can be thought of as prosthetic fur—a detachable layer of insulation that enables us to live in environments that would otherwise be forbidding.

Humans evolved as furless apes. Why we lost our fur is a matter of speculation; it could have been so that we could cool more efficiently during and after a long-distance run. This explanation is supported by the fact that we also evolved dense sweat glands; we are by far the sweatiest of all primates. But after exiting Africa, humans encountered Ice Age conditions. For creatures with no fur, clothing meant the difference between life and death. With time, suitably insulated humans could take over increasingly remote ecosystems. In doing so, they expanded their range, their numbers, and their power.

The first clothes were probably animal skins, but using skins this way requires that they be scraped and dried, and in some cases tanned—processes that entail tools and skill. Later developments included the ability to cut hides to a desired size (this depended on finely sharpened slicing tools), and the ability to sew (which requires needles). We don't know when people made their first shoes or hats, but these could again have been fashioned from fur or other materials. Some early clothing was no doubt made from leaves or grass draped, wrapped, or tied around the body or feet.

During the many thousands of years in which they refined their clothing, people gradually figured out which pelts were warmer, easier to work, and more comfortable. The wolverine, for example, has particularly thick fur, which makes its pelt even today a top choice among the Inuit.

The earliest use of clothing cannot be dated by direct evidence, since furs and fibers don't fossilize well. As a workaround, genetic researchers have been able to sequence the DNA of parasites that are well adapted to living in human clothing, and that mutated to their current form in order to thrive up close to humans.[10] Ralf Kittler, Manfred Kayser, and Mark Stoneking, anthropologists at the Max Planck Institute for Evolutionary Anthropology, were able to trace the crucial mutation in human body lice to around 170,000 years ago. However, a second group of researchers, using similar genetic methods, arrived at a much earlier date of around 540,000 years ago.[11]

Assuming the more recent date, was something happening then to encourage people to adopt clothing? Around this time, several human species were radiating out from Africa into new territories. Meanwhile, the Northern Hemisphere was in the early phase of a glacial period that would last roughly 50,000 years. Time to bundle up.

Sewing needles found in the Denisova Cave in Siberia have been dated to at least 50,000 years ago, while the earliest dyed flax fibers, dated to 36,000 years ago, were found in a cave in the modern-day nation of Georgia. By 25,000 years ago we see carved images offering us pictures of what then-current clothing must have looked like. The "Venus of Lespugue" figurine, found in southern France, appears to wear a cloth or twisted fiber skirt. Other figurines from western Europe sport woven hats, belts, and cloth body straps worn just below shoulder level. Thus, during the period from 50,000 to 25,000 years ago, as tools were becoming more refined and specialized, clothing was becoming more sophisticated as well.

What started as a way to stay warm became part of our quest for beauty, as we will discuss in more depth later in this chapter. Humans may have been using body painting before they adopted clothing, and, starting quite early on, some of the features of the human body likely evolved to look the way they do today partly in response to sexual selection pressures. After the invention of clothing, items of decoration quickly appeared, including shell necklaces, beads, and dyes. Thus, almost from the very beginning, clothes were an aesthetic statement. Clothing would later become a badge of social status, a consumer product, and a sign of conformity or nonconformity.

Clothing made us more powerful, but—like tools and fire—it wasn't the special sauce that led *sapiens* to outcompete other human species. After all, Neanderthals wore clothes, and it's likely that *Homo erectus* did too. It was probably something else that made *sapiens* a lethal threat to other human species. Could it have been language?

From Grunts to Sentences

Language is power. But it's a kind of power that differs from the measurable force exerted by a hammer or axe. Language gives us *social* power—the ability to influence the behavior of other people. And language can thereby magnify all the other powers available to us.

Language is also a tool, though a different kind of tool from a basket or knife. The latter entail the use of our hands, whereas the speaking of words involves our brains, mouths, throats, and lungs. As it happens, the parts of the brain used in speech—Broca's area and Wernicke's area—are very close to those that are activated when we manipulate tools with our hands. That's probably why we often use our hands to gesture while we speak. Think of language as a "soft" technology. If stone tools were our first technological hardware, words and grammar were our initial software.

But language is a very special tool, as we're about to see.

All animals communicate, and plants do as well. Some people speak of the "language" of plants or birds, and there is nothing wrong with using the word this way. But here I will employ the word *language* to refer only to the unique verbal communication used by humans, with its abstract symbolism and rules of grammar. Human language is clearly different from the communicative habits of other species. I don't believe that humans are intrinsically superior to other animals because of language (a judgment our kind all too quickly

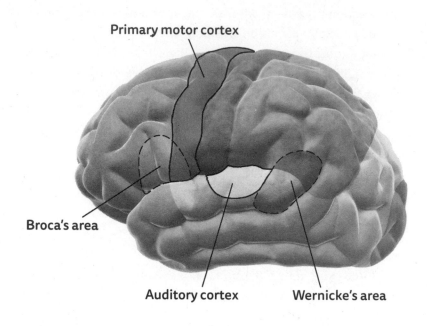

Primary motor cortex

Broca's area

Auditory cortex

Wernicke's area

Figure 2.2. Key language, motor, and auditory areas of the brain.

Credit: Adapted from an illustration by Blausen.com staff in "Medical gallery of Blausen Medical 2014."

makes, in my view). However, if we hope to understand the enormous power that language gives us, we must start by appreciating what sets it apart.

Language consists of a set of essentially arbitrary sounds or vocalizations (or gestures or marks), used as symbols, that can be combined and recombined in endless ways, but always according to a consistent set of rules. Those rules enable us to string together sounds that, in the proper order, signify thoughts; further, we can string verbal thoughts together, and insert thoughts within thoughts; and we can express thoughts about imaginary things, and about other places and times. Those sounded thoughts can be framed as questions or speculations, as well as statements or commands, including deliberately misleading ones. Using language, it's possible to say not just, "Hey, everybody, I just laid an egg!", as the chicken in the backyard coop outside my office window has just announced, but, "The wholesale price of eggs has been rising in recent weeks, posing a financial obstacle to the profitability of the mayonnaise industry." Just about every word in the latter sentence conveys a thought that would be inexpressible by any nonhuman animal, using its native communicative abilities. And this was just a single sentence I made up on the spot, to make a point. Language enables us to convey far more expansive, detailed, and abstract thoughts than that weird little sentence does.

The origin of language is a real problem for theorists, since language seems only to work once the basics of grammar are in place. But how did grammar arise? It's difficult to imagine a step-by-step process by which all the interlocking parts of language could have come into being.

Early theorists reasonably assumed that human speech evolved from primate communication, and therefore studied monkeys and apes for clues as to how the process might have occurred. The difficulty here is that there's little discernible gradual increase in complexity of communication among primates as you get genetically closer to humans. Vervet monkeys are as avid communicators as chimps. And, to further confound the situation, it turns out that nonhuman primates are far from being the best communicators among all animals. Birds generally do much better in this regard. Vervets have about ten "words," but my backyard chickens have roughly 25 unique vocalizations with which they say everything from the near-daily call that roughly translates to, "I just laid an egg!" to one that means, "Watch out for that hawk!" And the starling's extraordinary communicative abilities are just coming into focus.

In his book *The Language Instinct*, evolutionary psychologist Steven Pinker calls the ability to learn and use language an instinct, similar in this regard to spiders' web-weaving or beavers' dam-building behavior. It wasn't just an intensification of what other primates were already doing, and it's not something that each infant has to develop from scratch. It was a whole new package of behaviors that, once acquired, became largely innate.

But that doesn't tell us *how* language evolved. In order to develop language, people would have had to be spending a great deal of time in one another's company, presumably cooperating to ever higher degrees. Such high levels of cooperation are hard to evolve. As biologist Edward O. Wilson has noted, there are only 19 kinds of animals in all of nature that have evolved to be even nearly as cooperative as humans are. What got the ball rolling? Language certainly made us more cooperative—but we already would have had to be highly cooperative to invent language. Therefore, in order to explain the origin of language, we also have to explain how and why people were becoming more cooperative.

Earlier in this chapter we saw how walking upright led to the birthing of babies that needed longer periods of care before they matured. This probably led to greater sociality among early humans, but it almost certainly wasn't the only cause of our cooperativeness, perhaps not even the main one. Anthropologist Richard Wrangham, in *Catching Fire*, suggests that campfires also made us more social and therefore more prone to start talking to one another. But in a more recent book, *The Goodness Paradox*, Wrangham digs deeper, arguing that the beginnings of speech were tied to human "self-domestication" (domestication also figures strongly in Chapter 3 of this book).[12] What does he mean by that?

Domesticated animals have been deliberately and selectively altered by humans, most of them in prehistoric times. The process probably often started with choosing the least aggressive individuals from a group of captive wild animals, and allowing them to mate and have offspring. From those, again the least aggressive and most docile were selected. After 30 to 50 generations of this, the wild animal would have become permanently tame; the fierce wolf became a friendly dog. But domestication has unintended consequences. Across species, artificially reduced aggression seems consistently to correlate with a set of other characteristics: floppy ears, white spots of fur, less gender dimorphism (males and females are closer in size), a smaller or flatter face, an over-bite or under-bite in the jaws, and a smaller brain. The reasons for this well-studied "domestication syndrome" have to do with the multiple roles of genes: the DNA sequences that control emotional reactivity also code for proteins involved in other, seemingly unrelated traits.

Wrangham's point is that humans show many of the signs of domestication, with one partial exception: throughout its evolution the human brain was growing, not shrinking—until it reached a peak size about 15,000 to 10,000 years ago; since then it has indeed, on average, shrunk a bit. Otherwise, we fit the description quite well: compared to our closest primate cousins, we are less aggressive, we have flatter and smaller faces, and we exhibit less gender dimorphism (since we're furless, we don't have white patches).[13] The domestication syndrome appears to be related to another syndrome called *juvenilization*, whereby adults of a domesticated species tend to show physical characteristics and behaviors of the juveniles of related wild species. A human face in profile, for example, looks more like a juvenile chimp's profile than that of an adult chimp.

But if humans are a domesticated species—an observation that was familiar to Darwin, though he ultimately rejected it for lack of a good explanation—then who domesticated us? Unless you happen to believe in prehistoric visits by space aliens, the only realistic candidate is *ourselves*.

Self-domestication *could have* happened this way—though this is only a hypothesis, and we may never know the real story. Suppose primordial groups of humans systematically began killing the most troublesomely aggressive males among them (indeed, capital punishment of extremely troublesome people is a practice observed among all hunter-gatherer groups that have been studied). Highly aggressive men might therefore have less opportunity to mate and leave offspring. Over time, people would have become more docile, thus better able to live together in larger groups and cooperate.

One thing would have been especially useful in the primordial capital punishment of overly aggressive males: the ability to gossip and plan. That means language. Thus, language may have co-evolved with human self-domestication.

Whatever its contributing causes, this gradual shift toward increased docility and cooperation likely brought with it the rest of the domestication syndrome. And, as people became less aggressive and spent more social time together, there was at least the opportunity for language to develop, or to develop much further. With a changing climate, with new environments to explore via migration, and with more time spent making and using tools, there would have been plenty to talk about.

Here's another hypothesis. In a much-discussed 2018 paper, authors Oren Kolodny, a biologist at Stanford University, and Shimon Edelman, a professor of psychology at Cornell University, claimed that the development of sophisticated tools drove language acquisition.[14] Making a stone tool requires a specific sequence of processes, and teaching others how to make the same tools depends on communicating that same sequence. Kolodny and Edelman argue that ancient humans began inadvertently to couple the neural networks required for complex, hierarchical, sequence-dependent tool production with the cognitive processes in their brains' communicative apparatus. Thus, word order and sentence structure began to play key roles in the production of meaning. Gradually, the rudimentary language that had developed in the context of toolmaking and teaching broke free from these immediate contexts, enabling new cognitive pathways to be used for a wide array of other purposes. The result was our modern wide-ranging faculty for language.

There are still huge explanatory gaps in getting from primate grunts to human sentences, but theories about self-domestication and increased time and space for cooperation and tool making provide a good start toward bridging some of those gaps.

When language originated is, again, a thorny dilemma for theorists. There's no direct evidence, since human vocal utterances disappear into the air far more quickly than ancient clothing and wooden tools biodegrade. Anthropologist John Shea, in *Stone Tools in Human Evolution*, argues that the *ways* later human tool manufacture and use differ from early hominin tool use are similar to the ways that animal communication differs from human speech.[15] And, as we have already noted, the parts of the brain used in speech and tool use are connected. For Shea, dramatic changes that occurred in tools roughly 50,000 years ago suggest that this was the time language appeared.

Many scientists believe that the appearance of cave art around

40,000 years ago (which I'll say more about below) indicates that the people who produced it must have been able to think symbolically, and therefore to speak. If so, this would establish the very latest date at which language could have emerged.

But the process may have gotten started much earlier. Richard Wrangham argues that language began to take form as early as 300,000 years ago, since it was implicit in most of the evolutionary and cultural changes that have occurred in *sapiens* since its appearance. Kolodny and Edelman imply the process of language development could have gotten started even earlier, perhaps when tools began to become more sophisticated, roughly 1.5 million years ago. Did language start 40,000 years ago or 1,500,000? That's a huge range.[16]

Altogether, there is startling lack of agreement among experts. It's conceivable that some integrated aspects of language appeared quite early, then other important chunks fell into place later. The process may have been complex, with more than one contributing cause. Much of the story of language's origin is still guesswork.

What language did for us is less controversial. Language served our survival by enabling us to cooperate in domestic tasks and in searching out and harvesting wild foods. It empowered people to associate in larger, more cooperative groups, thus making them more formidable against big animals and other human groups. And it enabled our ancestors to make more effective tools (all of our subsequent development of technology, from wheels to typewriters to nuclear-tipped guided missiles, has occurred via the ever-greater magnification of our tool-making ability through our use of words).

Was the extinction of non-*sapiens* human species a result of our development of language? Neanderthals may have had some ability to speak, but such ability, if it existed, was probably more limited than that of *sapiens*. Neanderthal tools didn't quickly become more refined and diversified 50,000 years ago; they continued to look like ones made tens of thousands of years previously. And no cave paintings can be clearly attributed to Neanderthals. Therefore, language and increased cooperative behavior are probably the best candidates, among those we've surveyed, for the key factor that enabled *sapiens* to out-compete other species of humans.

Whatever the original motives for the development of language, it's safe to say it has had consequences that were far more sweeping than early humans could have conceived. Language has proven to be such a powerful and versatile ability that we have used it for activities that were unimaginable to early humans—from writing books about human evolution to tweeting insults about political rivals. And, in the process, language has reshaped the human mind. Thinking with words has led to a deepened split between the functions of the left and right hemispheres of the human brain (in most people, the left brain is more verbal), and between the conscious mind (containing all that can be verbalized) and the unconscious (what can't).

The ability to ask questions must have led to the posing of questions that no one could factually answer (What happens to us when we die? Why are we here? How did the world come to be?), but to which imaginative answers could be supplied. Hence came religion (as we will discuss more in Chapter 3)—just one byproduct of language, but one that has played an incalculable role in cultural evolution and the history of ideas.

Despite its extraordinary nature and consequences, we take language for granted. How could we do otherwise? We inhabit a mental world of language-mediated thoughts that shape our consciousness as much as, and often far more than, our awareness of our immediate surroundings. When we see someone crossing a street staring at their smartphone but ignoring oncoming traffic, that's a testament to the power of language. Most of us spend virtually our entire waking time talking, listening, reading, or thinking in words; categorizing, commenting, instructing, complaining, criticizing, praising, describing, or explaining. It's difficult for us to imagine consciousness without words, because we are so habituated to using words for thinking.

Biological evolution is slow, and so is the evolution of behavior that's purely gene-coded, such as the complex mating calls and behavioral displays of some birds. Cultural evolution, which occurs primarily via language and technology, can be very fast by comparison.[17] Watch a movie from the 1930s—less than a century ago. If dogs, cats, or chickens appear in it, you will note that they engage in dog, cat, or chicken behavior that is identical to the behavior of present-day dogs, cats, and chickens. But the people in the film are using different figures of speech than we use today, as well as different means of travel and different modes of communication; they're wearing different clothes, and (for the most part) listening to different music. Yet just one human lifespan of time separates us from the somewhat alien cultural world we glimpse in those old movies. Language and tools have enabled humans to dramatically increase the pace and scope of cultural evolution. And that has greatly accelerated our attainment of power.

Gender Power

One of the central problems this book proposes to address is how some humans became so much more powerful than others. It is therefore crucial that we explore perhaps the oldest power imbalance within our species—that between women and men.

I must emphasize that all of the human behavioral and anatomical gender differences discussed below, and elsewhere in this book, are based on statistical averages. In every society and culture, and presumably stretching back into prehistory, the human family has included individuals who expressed extremely polarized gender qualities, as well as individuals who expressed characteristics of both genders—anatomically, behaviorally, and psychologically. This is a fascinating area of discussion on its own, but this is not the place for it. This briefest of disclaimers is merely intended to prevent the misconception that gender differences are always expressed in the same ways or to the same degrees.[18] That said, we can proceed.

Human gender relations are often described as a "battle of the sexes." It's a peculiar battle, though, to which the two sides bring different kinds of power. Males more often use displays of dominance (via signs of physical strength, wealth, or social influence) to impress potential mates, and more often resort to violence or intimidation in order to control women's behavior and reproduction. Men are also more aggressive with one another. Indeed, among primates, males are responsible for nearly all violence (in some evolutionarily distant species, such as hyenas, females are more dominant and violent; but this is not the case for any monkeys or apes—with the partial exception of bonobos, which we will discuss shortly). The historical and intercultural evidence is overwhelming and inescapable: while men aren't always brutes, they all too often are. Women more often use attraction, nurturing, and persuasion to achieve their power. In anthropological literature, women's power is often described as informal or non-hierarchical. Especially in pre-industrial societies, women exerted considerable, if not primary, control over domestic affairs, and most of society was domestically-oriented.[19]

In different cultures, and at different times, the balance of power between the genders has shifted. As we're about to see, evidence suggests that in human evolution female sexual selection via mate choice appears to have contributed substantially to the development of certain uniquely human physical and behavioral characteristics. On the other hand, as we will see in Chapter 3, in early agricultural and animal-herding societies men became more controlling, especially in the public sphere: they regarded women as property, and often exerted sexual violence with impunity. Still more recently, in much of the world during the 20th century, women gained political and economic equality to a degree probably unprecedented since the origin of agrarian societies, and possibly since the dawn of the species.

But let's back up for a moment and look at the subject of gender power through an evolutionary lens. Yes, men are often brutes. But, among primates, human males aren't always the worst offenders. For perspective, it's helpful to compare our species with our closest primate cousins, chimpanzees and bonobos, with which we share about 96 percent of our genetic material. At first glance the bonobo, which used to be called the "pygmy chimp," does look like a smallish chimp, but it's actually a separate species (among other differences: bonobos have a flatter, more open face and a higher forehead than chimpanzees—that is, they show signs of the domestication/juvenilization syndrome). Chimps have been known to Europeans since at least the 17th century. However, the bonobo wasn't discovered by scientists until 1929 and its behavior wasn't studied by university researchers until the 1970s.

Chimps have a strict male dominance hierarchy, and more-dominant males generally engage in more aggression. Even if they aren't at the top of the hierarchy, aggressive males tend to sire more offspring. However, alpha males at the top of the hierarchy aren't bullies; their social role includes stopping fights and distributing food. Adult male chimpanzees travel with other males, grooming one another frequently. They have to stick together to fight off males of neighboring communities, as chimp "wars" are vicious and lethal. The dominant male of the group will typically mate with all the females, but every male in the troop is dominant in relation to every female. Males often charge at females, and may rip out their hair and slap, kick, or beat them. Dominant males sometimes kill the babies of their rivals; therefore, female chimps mate with many males in the troop in order to create uncertainty about the paternity of their offspring. When they are at their most fertile, however, females seek to mate with the males they prefer, presumably so as to increase their chances of producing fit offspring.[20]

Bonobos occupy the other end of the spectrum in terms of primate gender power. Bonobo society is both female-centered and female-dominated. If a male bonobo tries to harass a female, the other females band together to chase him off. Bonobos use sex to reduce tension and aggression. They engage in sex more often than any other primate, and in nearly every partner combination (though less often with close family members). Bonobo rates of reproduction in the wild are about the same as those of chimps, so most of this frequent sex is not aimed at reproduction. Females are almost continuously sexually attractive and active. All of this sexual activity is quick, casual, and relaxed, and appears to be an expected part of social life.[21]

Male Violence

In the modern world, where statistics can be gathered with some precision, it is clear that the great majority of human-on-human violence is attributable to men.

In the United States, during the years in which data have been available, men have been held responsible for roughly 90 percent of homicides. In addition, according to 2011 FBI data, males constituted 98.9 percent of those arrested for forcible rape, 87.9 percent of those arrested for robbery, 85 percent of those arrested for burglary, 79.7 percent of those arrested for offenses against family and children, and 77.8 percent of those arrested for aggravated assault.[22]

In addition, war—the deadliest and most socially valorized form of violence—has traditionally been a male pursuit. There have been exceptions: for example, during World War II, the Soviet Union sent many women into combat, and roughly eight percent of troops were female. In the United States in 2014, approximately 14 percent of active duty Army personnel were female, as well as 23 percent of the Army Reserve and 16 percent of the Army National Guard. While there are and have long been women warriors, as anthropologist Richard Wrangham notes, "...They are women serving in a men's army, fighting men's wars."[23]

Two broad explanations have been proposed to explain men's greater tendency toward violence: genetics and socialization. The evidence for a genetic basis for men's greater proclivity for violence starts with men's greater height and strength, on average. Most surveys show women to be about 50 to 60 percent as strong as men in the upper body, and 60 to 70 percent as strong in the lower body. Over the course of human evolution, this sexual dimorphism probably contributed to men specializing in violent activities.

The argument for socialization as the source of men's scrap-
piness stresses the roles of education, child rearing, and social
expectations in encouraging more violent behavior in boys and
men than in girls and women.

There are no known societies in which the violence of females exceeds that of men. However, the levels of overall violence in societies diverges significantly. For example, homicide
rates in recent years have ranged from 0.27 per 100,000 of population in the nation of Oman, to 43.85 per 100,000 in Jamaica
(the US rate is 4.96 per 100,000 per year).[24] This wide variation
offers hope that, even if males are genetically more prone to
violence than women, human violence overall can be reduced
very substantially. Over the long term, humans could conceivably evolve to become more bonobo-like, with reduced sexual
dimorphism and male violence.

Gender relations among humans have some chimp-like and some
bonobo-like qualities, but have also evolved their own unique patterns. Let's look at some of these patterns as possible evidence for how
power relations between the sexes might have shifted in prehistory.

Along with bipedalism and bigger brains, discussed earlier, modern humans differ from all other great apes, and presumably our
early pre-human ancestors as well, in terms of some gender-specific
physical characteristics:

- Adult women have large breasts compared to other primates, and
  their breast tissue is permanent (in other primates, breast tissue
  develops only during lactation).
- Human females have concealed ovulation: even women themselves may not know when they are fertile, and men have no clue
  whatever (among other primates, estrus is conspicuously signaled by swollen or brightly colored tissue around the genitals).
- Compared to the females of other primate species, women have
  narrow waists, wide hips, and fatty hips and buttocks.

- Men have the largest penis of any primate, and lack the baculum, or penis bone (all other male primates except the spider monkey have one of these).
- As we've already mentioned, there is less sexual dimorphism among humans than among other primate species.
- While humans lack fur, both genders grow pubic hair at puberty, and have scalp hair that grows continually; men begin to grow facial hair at puberty that, like scalp hair, can reach extreme length.

In addition to these physical characteristics, humans have gender-related behaviors that distinguish them from other primates:

- Human males and females are sexually picky (in other primate species, males will mate with any available female, and, among both chimps and bonobos, females mate with all available males).
- Human males and females form pair bonds to a far greater extent than chimps or bonobos.
- As pointed out earlier, men make a significant reproductive investment, which they share with women, via care for their infants and children (among both chimps and bonobos, females have all the child-care responsibilities).

Also, as we've already noted, human males (unlike male chimps) don't routinely kill the offspring of sexual rivals. Male infanticide among humans does indeed occur during wars, but it's not routine. Over all, human males are less aggressive than male chimps.

Now, how did we get this way? There are only a few possibilities. The one that comes first to mind is, of course, natural selection—adapting to the environment in order to promote survival and the ability to leave behind more offspring. But there is also sexual selection, involving patterns of mate choice, which can, over time, produce changes to the gene pool that may not result in increased survival fitness. In addition, there is group selection, in which it is competition not between individuals but groups that is decisive (we haven't explicitly discussed group selection so far, but will do so in more detail in the next chapter). All three of these are processes of biological evolution. Then there is cultural evolution via language, customs, and cultural artifacts, including all forms of technology. Finally, we know that some characteristics are not themselves adaptive, but arise as a result of being genetically tied to adaptive changes (we see this with the domestication syndrome and the related juvenilization syndrome, and with the cascade of changes to human anatomy and behavior that flowed from walking upright).

Ascribing specific uniquely human characteristics to one or another of these pathways is a theoretical problem that calls for conjecture, and scientists differ in their opinions, sometimes strenuously. Some of these questions may never be settled. However, I am persuaded by Richard Prum's argument, in his thought-provoking book *The Evolution of Beauty*, that reduced sexual dimorphism, the large male penis, and reduced male infanticide all suggest evolution by mate choice (i.e., sexual selection) on the part of women. Others of these characteristics (the tendency of women to be large-breasted and small-waisted, compared to other primates) suggest sexual selection by men. Male reproductive investment and pair bonding no doubt flow from our adoption of upright posture and the resulting birth of less-mature and more-dependent infants; however, female sexual selection may have accentuated these behavioral trends. In Prum's view, humans' sexual pickiness is also a product of aesthetic evolution (i.e., sexual selection), perhaps on the part of both genders. The evolutionary cause of concealed ovulation in women is widely disputed; Prum and others link it to female sexual selection aimed at reducing male infanticide (because of concealed ovulation, it's harder for males to tie a particular act of copulation to a resulting pregnancy; thus there's more uncertainty with regard to paternity). Some of our unique characteristics (like men's facial hair) seem pretty arbitrary and may end up being difficult to explain by any of these routes.

The physical evidence is hazy and open to different interpretations. The best window we have into primordial gender relations is probably via hunter-gatherer societies that survived into the 20th century, when anthropologists and ethnographers could visit and study them. Social relations in such societies are universally described as egalitarian. Rarely could anyone compel anyone else to do something against her or his will, and almost everyone had at least some say in decisions affecting them. Modesty and generosity were expected and commonly shown, at least partly because vain, pushy, or greedy people (typically men) were usually dealt with harshly by the rest of the males in the group through banishment or capital punishment.

Gender roles among hunter-gatherers have long been stereotyped in the popular imagination: men hunted, while women gathered wild plant foods and did the cooking. But, in reality, life prior to the agricultural revolution was more complicated. In some societies, women shared in hunting at least the smaller prey, and men did some of the gathering and cooking. Since meat often made up only a relatively small portion of total caloric intake, women were substantial providers, and were respected as such. Baskets served group survival, in their way, as much as spears did. Men had their secret societies in which they engaged in sacred rituals related to imaginative and supernatural powers, but, at least in some societies, so did women. Child care was mostly women's responsibility, but, once boys were old enough, their fathers and other male relatives would spend considerable time with them, teaching them practical skills and cultural traditions. The ability of women to contribute to group decision making varied greatly from one group to another, and in many societies men's voices predominated. Still, at least in some groups, unusually skilled and sociable women gained high social status. In short, while the hunter-gatherer way of life was hardly a feminist ideal, from women's point of view it typically offered substantially greater freedoms and opportunities than the agrarian and animal-herding lifestyles that would appear later.

It's interesting also to inquire briefly about same-sex sexual relations in such societies. Many hunter-gatherer societies, such as the Aché, who inhabit a region in Paraguay, have special categories for individuals who defy binary gender description (this was true of many native North American societies as well). The Aché maintain a category called *panegi* for transgendered androphilic males; *panegi* look, act, and talk like females and engage in female work and social activities. In other societies, such as the Aka of central Africa, same-sex sexual behavior appears to have been rare, even if tolerated.[25] But such examples give us relatively little insight into same-sex sexual behavior in human evolution. Richard Prum hypothesizes different pathways for the development of female same-sex preference as opposed to male same-sex preference, with female same-sex preferences arising from protective female social networks. Male same-sex preference could have resulted from female sexual selection, because it contributed to "male supportive and protective nonsexual relationships with females." But these are really just informed speculations. What we do know is that later societies, including many animal-herding groups and early agrarian states, banned and punished same-sex sexual behavior, often with death, as a few countries still do. The advance of women's power in the modern world has been accompanied by dramatically increased acceptance of variety in sexual expression and gender identity.

Throughout human evolution, women appear to have found ways to assert freedom and agency through mate choice, through provision of the bulk of necessary nutrition for the community, through domestic work, through participation and leadership in cultural activities, through child-rearing, and by cooperating with one another to overcome male intimidation or violence. Environmental or cultural circumstances have sometimes enabled men to (re)assert chimp-like patriarchal dominance. Shifts in gender power have continued right up to the present day, and will likely persist as long as our species does. Richard Wrangham and Dale Peterson, both experts on wild primates, concluded in their book *Demonic Males: Apes and the Origin of Human Violence* that, if we wish society to become more peaceful, we should find more ways to emulate bonobos' female-centeredness and work to increase women's power.

The Power of Art

Humans began making art and music early in their evolutionary journey. Cave paintings and bone flutes started appearing roughly 40,000 years ago—about the same time as language and stone tools likely took a leap forward. Paleolithic paintings of animals and humans are exquisite and expressive. Ancient flutes, made from mammoth and vulture bones, produce tones with the same tonal relationships (whole tones, half tones) as the tones produced by modern musical instruments like the piano or guitar. We also see evidence of body art and decoration around that time. Moreover, as soon as we had language, we probably started using it playfully and creatively to make songs, poems, sagas—and jokes. Little of this had to do with survival needs. Music, art, and wordplay implied a power not of domination, but of shared celebration and transcendence.

It's common to speak of the power of a work of art—a painting, a musical performance, a dance, a play, or a piece of literature. But what kind of power is this? It's a different kind from those we have addressed so far, and to ignore it would be to overlook an essential aspect not just of humanity, but of nature itself. The evolution of life via the maximum power principle seems straightforward: those organisms that are able to harness and exert more power leave behind more offspring and displace other organisms. Viewed this way, life appears grimly competitive. Yet, when we look at the result of billions of years of evolution, it's amazingly beautiful.

As we saw in Chapter 1, animals and plants devote enormous effort toward creating beauty as part of reproduction. And, as we have just seen, aesthetic preferences have, via sexual selection, also likely sculpted the human body and altered our instinctive behavior. With sexual selection, evolution was no longer tied just to fitness.

Once beauty assumes a priority, it often becomes an end in itself. Though it may begin with courtship displays, beauty can take on a significance of its own. Birds appear often to sing for pure pleasure, even when there's no need to attract a mate. With humans, as cultural evolution accelerated, our quest for beauty also became largely detached from sexual selection. Aesthetic appreciation and the production of beauty co-evolved ever more quickly to become overwhelming obsessions.

Human Aesthetic Decadence

Investment in display behavior to attract mates can proceed to such extremes that it leads to "aesthetic decadence," contributing to a species' decline and even extinction, as we saw in Chapter 1. The question must arise as to whether at least some of humanity's exuberant aesthetic production is "decadent" in the evolutionary sense, in that it reduces our species' survival prospects. Unfortunately, the notion of aesthetic decadence is weighted with prejudice and with some rather awful history: Hitler thought virtually all modern art was decadent and ordered many important paintings and sculptures destroyed; during the 1930s and early '40s, artists like Ernst, Mondrian, and Duchamp fled their homelands to avoid harassment or worse—as did composers such as Hindemith and Schoenberg. Though Stalin's politics were diametrically opposed to Hitler's, the Soviets likewise regarded modernist composers as decadent, periodically making life hellish for Shostakovich and Schnittke. One person's decadence is another's masterpiece.

However, in my opinion the contemporary world does provide persuasive evidence of aesthetic decadence. What I have in mind is music and art produced specifically for commercial purposes. Advertising art can be clever and entertaining: that helps it sell products. But, as we'll see in Chapter 4, advertising is one of the pillars (along with cheap energy and consumer credit) of consumerism, and consumerism is in turn the engine of economic growth. As the economy grows, it chews up and digests ever-greater swathes of the natural world, leaving depletion, pollution, and habitat destruction in its wake. Our survival is very much in peril as a result.

In a capitalist society, commerce influences nearly all art and music, though to greatly varying degrees. Professional hip-hop artists and singer-songwriters struggle to find a unique "voice" that will appeal to agents, concert bookers, and fans willing to buy tickets or downloads. This constant striving for uniqueness is specific to the modern commercial milieu: in pre-capitalist or pre-industrial societies, multi-generational tradition guided aesthetic preference to a much greater degree. One could say that capitalism produces greater artistic variation, thereby speeding up aesthetic evolution. But if the species is rapidly evolving toward decadence and possible extinction, then its proliferation of increasingly varied forms of art and music is destined to be short-lived, however ingenious and enjoyable those forms may be.

The result, today, is that we live in an aesthetic human world. Nearly every surface in a modern city is *designed*. Cars, houses, office buildings, and tools of all kinds—from motorcycles to fountain pens—have become canvases for the creative process. And we are immersed in entertainment of every imaginable variety—from background music to novels to television dramas. The typical modern human "consumes" art almost from the moment they wake up till the moment they fall asleep at night.

The obsessiveness with which we pursue artistic production and appreciation can be illustrated by one activity to which (for better or worse) I have devoted tens of thousands of hours: playing the violin. Try watching and hearing violinist Hilary Hahn perform Paganini's 24th Caprice on YouTube. A dozen or more notes may fly by each second, each perfectly in tune, and each perfect also in articulation and tone color. Hahn's two hands are engaged in entirely different tasks that must somehow be exactly synchronized. And the point of the exercise isn't just to make no mistakes while doing several nearly impossible things simultaneously, but to confidently create beautiful and moving music. With all due respect to brain surgery, I can say with some assurance that no activity by a human or any other animal requires as much digital precision as top-level fiddling does. And it's not just the violinist's fingers that are involved, but the wrists, arms, and back muscles—and, first and foremost, the brain. Finger exercises (which every serious violinist spends endless hours on, to the weary aggravation of all within hearing distance) are useless without a trained "ear"—which really means a highly trained *brain*—that can recognize tiny variations in pitch and rhythm, and make nearly instantaneous corrections on the fly.

This digression underscores a point I've already made: that the dexterity of the human hand cannot be explained solely by anatomical factors. The evolution of violin playing, and of tool use in general, took place on many levels—anatomical, mental, social, and aesthetic—more or less simultaneously.

However, my example primarily illustrates many humans' utter devotion to aesthetics, to a degree that is difficult to justify in terms of either natural selection or sexual selection. Yes, many young men buy a Stratocaster and take guitar lessons in order to impress the girls. But I find it difficult to imagine that such a motive would compel a five-year-old child to begin practicing a supremely difficult musical instrument several hours a day and to continue doing so all through adulthood. Nor is pursuit of fame or financial reward an adequate explanation. Are violinists more fit than other humans? Do they leave more offspring? Do they tend to earn higher salaries? Do they attract more desirable sexual partners? My past experience as a volunteer board member of a local musicians' union, of which most members were professional symphony orchestra players, leads me to doubt that any of these is reliably the case (though learning a musical instrument does seem to give children an advantage in math and reading). Why should people devote so much more effort to developing the skill of violin playing than any of a hundred other skills that are much less demanding and that might have a better chance of leading to wealth or social prominence?

The sports lover will recognize a similar obsession. Devotees of soccer, surfing, and sumo exhibit a similar level of compulsion, and the subjective experience of an outstanding performance in any sport can be described as aesthetic. Commitment to athletics can perhaps more readily be explained in terms of competition, selection, and fitness (and the financial rewards for professional-level performance are sometimes astronomical), but the pursuit of excellence in sports and the arts is, in both cases, quasi-spiritual.

Evolution gave us power over the natural world. But in doing so it gave us abilities that could be used for things that had little or nothing to do with power in the conventional evolutionary sense—i.e., the ability to get food or find a mate. The power to communicate aesthetic pleasure and thereby to feel profound affinity with other people, including individuals of other species (as we do when we enjoy the beauty of birdsong or thrill to the sight of a pod of dolphins "dancing" in perfect synchrony through the waves, to cite just two possible examples), propel human culture forward in ways that are hard to measure, but that are impossible to ignore. These are powers that, as we'll see later in this book, provide hope for our future.

✦   ✦   ✦

With tools and fire, we humans gained unprecedented power to control our environments. We reshaped landscapes. We drove other animals to extinction. And with language and social coordination, our species of humanity may have wiped out competing human species. Still, within the one surviving human species—*Homo sapiens*—power relations probably remained fairly simple, if the study of surviving hunter-gatherer peoples is a reliable guide. No single individual could dominate everyone else without provoking a coordinated backlash, and no group of *sapiens* could systematically exploit another group for long. However, starting about 11,000 years ago, at the dawn of the Holocene epoch, that changed. The result was history.

POWER IN THE HOLOCENE

The Rise of Social Inequality

*Competition within groups destroys cooperation;
competition between groups creates cooperation.*

— PETER TURCHIN, *Ultra Society*

*Being powerful is like being a lady. If you have to tell
people you are, you aren't.*

— MARGARET THATCHER

*All governments suffer a recurring problem: Power attracts
pathological personalities. It is not that power corrupts
but that it is magnetic to the corruptible.*

— FRANK HERBERT

Political power is built on physical power.

— RICHARD WRANGHAM and
DALE PETERSON, *Demonic Males*

WHILE CHAPTER 2 MOSTLY EXPLORED HUMANS' EARLY DE-velopment of physical powers, this chapter focuses more on social power. If physical power is the ability to do something, social power is, as I have already suggested, the ability to get *other people* to do something. The invention of language, discussed in the previous chapter, was the ultimate source of most of our uniquely human social power. But social power has evolved since then, in dramatic and often disturbing ways.

As mentioned at the end of the last chapter, a significant shift in human social evolution started roughly 11,000 years ago, probably set in motion by yet another change in the global climate. The Pleistocene epoch had begun 2.6 million years previously and had featured 11 major glacial periods punctuated by occasional warm intervals. Now glaciers over much of North America and Eurasia melted, ushering in the Holocene epoch, during which all of recorded human history would unfold.

SIDEBAR 9
Measures of Social Power

Social power is inherently more difficult to measure than physical power. Money can be thought of as quantifiable social power, and can be measured to the cent; but wealth can be hidden in various kinds of property and investments. And money isn't everything: Mahatma Gandhi had virtually no personal fortune, yet he wielded immense social power.

In early states, measuring power was simple. He who had the most gold and silver, as well as the largest number of slaves, subjects, wives, and children, had the most power. However, even then social power also rested on the ability to persuade and motivate others: religious functionaries, who were often excluded from the worldly accumulation of wealth, nevertheless enchanted the masses through their ability to perform ceremonies marking a variety of occasions, to re-enact myths, and to foretell the future.

Today the ability to sway public opinion is still a significant form of power. It is possible to change other people's minds not only by the offering or withholding of money, but also by reasoned argument, inspiring example, and subliminal forms of persuasion commonly used in the advertising and public relations professions. Social power of this kind can be measured by votes, sales of products, popularity polls, applause, retweets, Facebook followers, citations in academic journals, and mentions in news feeds and history books.

We unconsciously measure and compare social power constantly. We signal our status through the size and location of our dwelling, the quality and appearance of our clothes, the foods we eat, and the car we drive; status can also be communicated more subliminally through smells and speech accents.

Some people deliberately send confusing status signals as a way of communicating their disdain for the very practice of assigning status; their doing so can itself become a marker of status within the ranks of the similarly rebellious.

Gardening, Big Men, and Chiefs: Power from Food Production

One might assume that the end of an ice age must have made life easier for humanity. In some places, that was no doubt the case, especially as the warmer climate stabilized. But warming was not gradual and uniform; it occurred in fits and starts. The shift from a colder climate to a warmer one resulted in rapidly rising sea levels, as well as localized floods alternating with dry conditions. Many animals went extinct during this period, and it appears that not all of them were victims of humans' improving hunting skills. Times must have been tough not just for these animals, but for at least some humans as well.

All human species other than *sapiens* were now gone. For countless millennia, everyone had subsisted by hunting and gathering, a way of life characterized by *horizontal* power relations: all adults in a given group generally had a say in organizing whatever work was done. But as climatic and ecological conditions changed, human societies were rapidly adapting to a variety of new niches. With the dawn of the Holocene, people started finding new ways of obtaining their food; and these innovations in food production would gradually lead to the development of *vertical* social power, in which an individual or small group determined what would be produced and how, and to whose benefit.

If the shift from scavenging, hunting, and gathering to food production was invited partly by the opportunity to stay in one spot and enjoy a seasonal surplus, it was also mandated by necessity. At this point, all human groups had language and tools. Therefore, their ways of life were malleable and capable of adapting rapidly via cultural evolution. Adaptation was now needed, not only because of shifting environmental circumstances, but also because *Homo sapiens* were just about everywhere. Africa, Eurasia, and Australia had been peopled for many thousands of years, and North America and South America were rapidly becoming populated as well.[1] Only the Pacific islands would remain free of humans for another few thousand years, awaiting the development of better navigation skills. Competition for living space was increasing. And the evidence suggests that levels of between-group violence were increasing too, as groups with adjacent or overlapping hunting territories saw herds of prey animals thinning.

Intensifying warfare meant that people had to do something or risk losing everything. The evidence suggests they did two things: they started living in larger groups (so as to better resist raids, or to more effectively carry them out); and, in order to support their larger group size, they began to grow food.

The first permanent settlements were likely located in ecologically rich and diverse wetlands, where people could continue to hunt and gather wild foods while also domesticating grains and useful animals. The earliest evidence of this settled way of life—dating from roughly 8,500 years ago—has been found in the Tigris-Euphrates region, where many tiny Mesopotamian kingdoms would later flourish. It would take another 4,000 years before sedentism and field crops would lead to the beginning of what might be called civilization.

Part of the motive for settling down had to do with the opportunity provided by geography. These centers of early settlement were, after all, regions of great ecological abundance. But that made them desirable to many people. As population densities grew, so did conflict. People had to defend their territories as they increasingly tended to stay in one place. Gradually, more food calories came from planting, fewer from hunting. People had previously figured out the basic life cycle of food plants, but they'd had insufficient motive to settle down to the hard and unhealthy life of planting and tending crops. Increasingly, that motive was inescapable.

As communities grew in size, keeping them socially unified required occasional large-scale meetings and rituals. Suddenly in the early Holocene—at first in what is now Turkey, but later throughout Europe and the British Isles—we see examples of monumental architecture that seem to mark ceremonial centers where thousands gathered seasonally. People's food, work, political organization, and spirituality were all being transformed.

Anthropologist Marvin Harris was one of the first scientists to grasp the key relationships between how people get their food and how they organize themselves and make sense of the world. He observed that all societies operate in three realms. First is the *infrastructure*, which consists of the group's ways of obtaining necessary food, energy, and materials from nature. Second is the *structure*, consisting of decision-making and resource-allocating activities—the group's political and economic relations. Finally, there is the *superstructure*, which is made up of the ideas, rituals, ethics, and myths that enable the group to explain the universe and coordinate individuals' behavior.

Change in any one of these three realms can affect the other two: the emergence of a new religion or a political revolution, for example, can alter people's material lives in significant ways. However, observation and analysis of hundreds of societies shows that the way people get their food is a reliable predictor of most of the rest of their social forms—their decision-making and child-rearing customs, their spiritual practices, and so on. That's why anthropologists commonly speak of hunter-gatherer societies, simple and complex horticultural societies, agricultural societies, and herding societies. Harris called his insight "infrastructural determinism"; we might simply say, *food shapes culture*.[2]

The transition to agriculture had many stages and jumping-off points. While some hunter-gatherer societies with permanent settlements, stone houses, social complexity, and trade were appearing in alluvial plains—at first in the Middle East, but later in China, India, and the Americas—others remained small and simple. Archaeologists are still piecing together the evidence of when and how planting and harvesting emerged.[3]

It's clear, though, that humanity's first major innovation in food production consisted of planting small gardens, typically using the same patch of land for several years until the soil became depleted, then moving on. That patch might be left fallow for 20 or 30 years before being cultivated again. Hunting still supplied some nutrition, as did fishing in the case of groups living near sea coasts, lakes, or rivers. Increasingly, people lived in villages, but these were small and movable. Anthropologists call groups who live this way "simple horticultural societies."

Gardening enabled the production of a seasonal surplus, which could be stored in preparation for winter or as a hedge in case of a poor harvest next year. But production of a significant surplus required that people work harder than they otherwise needed to just to satisfy their immediate needs. How to unify everyone's resolve to work harder than was absolutely necessary? Societies in many parts of the world came up with the same solution: the Big Man. Within the group, one individual (nearly always male, according to the available evidence) would set an example, working hard and encouraging his relatives and friends to do so as well. The payoff for all this toil would come in the form of huge parties, thrown occasionally by the Big Man and his crew.

Being a Big Man led to influence and prestige: people listened to you and respected you, and you could represent your group at inter-group ceremonies, where you competed for status with other regional Big Men. But getting to be a Big Man required that you give away nearly all of your wealth every year; moreover, other males in your group were always aspiring to be Big Men, too, so there was never any assurance that you could maintain your status for many years or pass it along to your sons. And a Big Man couldn't force others to do anything against their will. Even though one individual had acquired status, at least temporarily, society was still highly egalitarian.

It sounds like a pretty good way to live. However, in places that were becoming more densely populated, horticultural societies engaged increasingly in warfare—as surviving horticultural societies have continued to do right into modern times. Take for example, the Mae Enga, a Big Man horticultural society in New Guinea that was researched extensively in the 1950s and '60s. New Guinea is, of course, an island, and by the time anthropologists arrived it already hosted as many people as could be supported by the prevailing means of food production.[4]

The Mae Enga then numbered about 30,000, living at a density of 85 to 250 people per square mile. The society as a whole was divided into tribes, and those were subdivided into clans of 300–400 people. Pairs of clans often fought with one another over gardening space, and the intensity of warfare was high, with up to 35 percent of men dying not of old age, but of battle-inflicted wounds. Losing a war might mean having to abandon your territory and disperse to live among other clans or tribes.

The Mae Enga, like many other societies that lived in groups closely packed together, all subsisting in nearly identical ways, but separated by seas, mountains, or forests from the rest of the world, continued to pursue the simple horticultural way of life for millennia.

However, in some less naturally bounded areas, another shift in food production and social organization occurred, starting roughly 8,000 years ago. With increasingly frequent raids and other conflicts, communities began to band together into even larger units in order to protect themselves or to increase their collective power vis-à-vis other groups. Living now in big, permanent villages, and with population still growing, people had to find ways to further intensify their food production. While still using simple digging sticks for cultivation, they more often kept domesticated animals; meanwhile, they left land fallow for shorter periods, five years or fewer, and sometimes fertilized it with human or animal manure. Many of these groups employed terracing and irrigation as well.

As these transformations were taking place, the role of Big Man gave way to that of chief. Since warfare was becoming more intense and frequent, leadership took on more authoritarian qualities. Many societies designated different chiefs for peace and wartime, with differing kinds and levels of authority. But gradually, in many societies, war chiefs assumed office on a permanent basis. Chiefs often had the authority to commandeer food and other resources from the populace, and in some cases were able to keep an unequal amount of wealth for themselves and their families. Over time, their position became hereditary. Also, an increasing portion of production was directed toward exchange rather than immediate consumption, as trade with other groups expanded.

The Cherokee of the Tennessee River Valley, prior to European invasion and conquest, were a fairly typical chiefdom society. They traced their kinship via the mother, and women owned their houses and fields. Each family maintained a crib for maize to be collected by the group's chief, who maintained a large granary in case of a poor yield or for use during wartime. The chiefs of the seven Cherokee clans convened in two councils, where meetings were open to everyone, including women. The chiefs of the first council were hereditary and priestly, and led religious activities for healing, purification, and prayer. The second council, made up of younger men, was responsible for warfare, which the Cherokee regarded as a polluting activity. After a war, warriors needed to be purified by the elders before returning to normal village life.

Anthropologists call chiefdoms like this "complex horticultural societies." Compared to simple horticulture organized by Big Men, complex horticulture required working even harder. But, by this time, many regions had become so densely populated that it was impossible for people to return entirely to the easier life of hunting and gathering or even simple horticulture, and, in many cases, there was little choice but to continue intensifying food production.

From the standpoint of social power, the advent of chiefs represented a significant development. For the first time in the human story, a few individuals were able to commandeer more wealth than anyone else around them, and able also to pass their privileges on to their hereditary successors. Social power was becoming vertical. But a far more fateful shift would occur around 6,000 years ago: it culminated in the formation of the first states, which produced food via field crops and plows (i.e., through agriculture rather than horticulture), and which were ruled by divine kings.

Plow and Plunder: Kings and the First States

Cultural evolution theorists describe the process we've been tracing, in which competition between societies drove intensifying food production and population growth, by way of a controversial idea: group (or multilevel) selection. Whereas evolution typically works at the level of individual organisms, group selection kicks in when a species becomes ultrasocial. Then, whenever variations arise between groups within the species, competition between the groups selects for the fittest group. Some evolutionary theorists, such as Steven Pinker, resist the notion of group selection, arguing that new traits in highly social species can still be accounted for with the tried-and-true concept of individual natural selection. But, in my view, when seen through the lens of group selection, the development of human cultures through the millennia becomes far easier to understand.

Three ingredients are necessary for the engine of group selection to rev up: there must already be high levels of intragroup cooperation; there must be sources of variation within groups (new behaviors, tools, institutions, and so on); and there must be sources of competition between groups. In the early Holocene, human societies in several regions had all three ingredients in place. These were places where large numbers of human groups already existed, jostling against one another's boundaries. All of these groups were themselves the products of earlier cultural evolution that had led to the development of language and high levels of intragroup cooperation. Cultural variations were present, perhaps having appeared in further-flung regions, then migrating to these edges of territorial overlap, in which resources were relatively abundant. Competition inevitably arose over access to these resources, often leading to war. The result was even more cooperation *within* societies, and the tendency for societies to become larger, more technologically formidable, and more internally complex so as to more successfully compete against their neighbors.

According to the latest thinking among scientists like Peter Turchin, who specialize in studying cultural evolution via group selection, shifts in food production and social organization were motivated primarily by population growth and warfare.[5] Continually vulnerable to raids from neighboring groups, communities had little choice but to find ways of increasing their scale of cooperation. That meant building alliances or absorbing other groups. But as societies became larger, their governance structures had to evolve. When it became no longer possible to know everyone else in your group and to make collective decisions face-to-face, a form of rigid hierarchy tended to spring up. This eventually took the form of the archaic state.

All of this took time. Between four and six millennia would elapse between the appearance of societies relying on gardening, using simple digging sticks, and the advent of early agricultural states depending on field crops and plows. The latter would constitute a milestone in cultural evolution. States arose independently in several places, and at somewhat different times: around 6,000 years ago in Mesopotamia and southwestern Iran; around 2,000 years ago in China, the Andes, Mesoamerica, and South Asia; and 1,000 years ago in West Africa. The state was a social innovation on five levels—political, social-demographic, technological, military, and economic—and it could concentrate far more power than simple and complex horticultural societies.

Politically, the state was defined by the advent of kings. With increasing levels of between-group conflict, war chiefs had begun not only to remain in office permanently, but also to assume the ceremonial duties and religious authority of the former peace chiefs. Gradually, they became hereditary monarchs with ultimate, divinely justified coercive power over all members of the group. The king no longer simply had the duty to maintain customs, but could make laws, which were enforced under threat of violence exercised by officers of the state on the king's behalf. Those who broke the law, however arbitrary that law might be, were, by definition, criminals. The king claimed ultimate ownership of his entire governed territory, and also claimed to be the embodiment of the high god; divine privileges extended also to his family.

Socially and demographically, the state came to be defined by its territorial boundary. Successful groups were able to defend larger and more stable geographic perimeters, and began to identify themselves as much by place as by language or other cultural traits. The state gradually became multi-ethnic and multilingual.

As societies grew in population size, increasingly incorporating members absorbed from other groups, kinship-based informal institutions for meeting people's material, social, and spiritual needs no longer worked well. This helped drive the evolution of royal law as a substitute for tradition and custom in regulating behavior.

In addition, the agricultural state was characterized by full-time division of labor. The great majority (often 90 percent or more) of the populace were peasants working the land, or captive slaves working in mines, quarrying stone, felling timber, dredging, rowing ships, and engaging in similar sorts of forced drudgery. The rest of the population at first consisted of the royal family, their attendants, priests, and full-time soldiers. Gradually, the professional classes expanded to include full-time scribes, accountants, lawyers, merchants, craftspeople of various kinds, and more.

Finally, agricultural states were defined demographically by the existence of cities—permanent communities with streets, grand public buildings usually made of stone, and a high density of occupants numbering in the hundreds or thousands. Only a minority of the population lived within cities, but cities were the political, commercial, ceremonial, and economic hubs of the state. Food was produced as close to the city as possible, so state territories initially often comprised only a few square kilometers.

Technologically, the agrarian state depended on one innovation above all others—the plow, which enabled the planting of field crops, typically grains such as wheat, barley, maize, or rice. Working hard, a farming family could produce a surplus of grain that the king claimed on behalf of the state as a tax. In return, the state provided protection from raids and also held grain in storage in case of poor harvests.

States also competed with each other to develop more effective weapons—from more powerful bows and sharper and stronger swords to armor. Competition for better weapons in turn drove innovations in metallurgy—leading to the adoption first of copper, then of bronze, and finally of iron blades and other tools.

Militarily, states benefited (vis-à-vis other human groups) by employing full-time specialists in violence—i.e., soldiers, as well as full-time specialists in motivating, managing, and strategically deploying soldiers, and full-time specialists in making and refining weaponry. The development of organized military power occurred by trial and error, and entailed many setbacks. Disciplined, highly motivated, and well-equipped soldiers were formidable on the battlefield, but smaller forces could still prevail if they relied on the element of surprise (guerilla-style warfare has continued to be effective up to the present). The net result was a continuous arms race and frequent, deadly warfare.

Economically, the state functioned as a wealth pump, with surplus production from the peasantry continually being funneled via taxes upward to the king, the king's family, the priesthood, and the aristocracy—which at first consisted primarily of elite soldiers and their families. Like taxes on the peasants, increasing trade (mostly of luxury goods) primarily benefited the upper classes. The operation of the wealth pump, over time, tended to generate social cycles: as elites captured an ever-larger share of the society's overall wealth, peasants became increasingly miserable. Meanwhile, the number of elites and elite aspirants tended to grow until they all could not enjoy the privileges they wanted, thus leading to competition among elites. The society would become unstable, a condition that tended to persist and worsen before culminating in civil war or bloody coup. Periods of internal strife thus alternated with times of consolidation and external conquest.

In addition, the idea and practice of land ownership—unknown previously—became an essential means of organizing relations between families, and between people and the state. Ownership demanded an entirely new way of thinking about the world, one in which sympathies with nature would recede and numerical calculation would play an increasing role.

It is no accident that all early autocratic states adopted grains (primarily wheat, barley, millet, rice, and maize) as their main food sources. Grains have a unique capacity to act as a concentrated store of food energy that is easily collected, measured, and taxed. Grains' taxability was especially consequential: without taxation, it's likely that the wealth pump could not have emerged, nor could centralized, vertical-power government.

Innovations from any one of these five realms (political, social-demographic, technological, military, and economic) could spur innovations in other realms. For example, full-time division of labor tended to stoke technological change, as professionals had the resources with which to experiment and invent, refining weapons and farm tools. New weapons enabled the growth of military power. And new technologies also contributed to economic change by generating new professions and new sources of wealth.

Meanwhile, innovations in one kingdom were quickly adopted by neighboring ones, as states continually competed for territory. Failure to keep up might doom one's group to conquest, captivity, or dispersal. Cultural evolution had many ingredients, but war was its main catalyst.

Environmental impacts were a source of instability. Previously, people had often been able to escape undesirable environmental consequences simply by moving (recall that horticultural societies left their fields fallow for many years before returning to cultivate the same spot). But farmers living in agricultural states had no option but to stay put. Plows loosened and pulverized the soil surface and enabled the farmer to easily remove weeds. Over the short term, this boosted crop yields; however, over the longer term, it led to erosion. Also, irrigation led to salt buildup in soils (as it still does today). Meanwhile, the cutting of trees to build cities depleted local forests. As a result of these practices, the cradle of civilization, known as the "fertile crescent" of the Middle East, is now mostly desert. Ecological ruin was common in other centers of civilization as well.

The earliest civilizations (a term derived from the Latin *civis*, meaning "city") sometimes enjoyed unusually favorable natural environmental conditions. Notably, Egypt benefitted from the annual overflow of the Nile, which replenished topsoil for farmers. As a result, even though Egypt had its political ups and downs, it never succumbed to the same degree of ecological devastation that Mesopotamia eventually did. Gradually, relatively sustainable farming practices emerged here and there, such as the Chinese systematic use of human and animal manure to maintain soil fertility, or the Native American practice of intercropping (planting nitrogen-fixing beans together with squash and maize).

For the first few centuries of their existence, states had art and architecture but no writing. Because they left no histories, we have limited insight into how people in these societies thought and how they treated one another; what we do know comes from excavating their settlements and cataloging their more durable weapons and ornaments. These signs point to the brutality of rigid hierarchies, in which kings ruled with absolute power. Slavery was universally practiced, and one of the main incentives to warfare was the capture of men, women, and children for purposes of forced labor and breeding. Signs of widespread human sacrifice are unmistakable. The world had never before, and has rarely since, seen societies that were so starkly unequal and tyrannical. Indeed, the great empires of Greece, India, Rome, Persia, and China that most of us remember from the early pages of high-school history books were places of moderation and freedom by comparison. The earliest surviving texts from archaic kingdoms convey the distinctive flavor of despotism. Here's an excerpt from a cuneiform text found by archaeologists in Ashur, the capital city of the Assyrian Empire:

> I am Tiglath Pileser the powerful king, the supreme King of Lashanan; King of the four regions, King of all Kings, Lord of Lords, the supreme Monarch of all Monarchs, the illustrious chief who under the auspices of the Sun god, being armed with the scepter and girt with the girdle of power over mankind, rules over all the people of Bel; the mighty Prince whose praise is blazoned forth among the Kings: the exalted sovereign, whose servants Ashur has appointed to the government of the country of the four regions... The conquering hero, the terror of whose name has overwhelmed all regions.[6]

Of course, not all societies pursued the same path.[7] On the fairly isolated continent of Australia, in southern Africa, in the Arctic, and in parts of the Americas, the hunter-gatherer way of life persisted until Europeans arrived with guns and Bibles; and horticultural Big Man and chiefdom societies thrived on Pacific Islands, in parts of North and South America, and Africa. However, agriculturalists constantly sought to expand their territory. Hunter-gatherer and horticulturalist groups resisted; after all, who would want to live in a society so unequal as an archaic state? But they were no match militarily for full-time soldiers with the latest weapons. And so, when they came into contact with states, hunter-gatherers and horticulturists were usually defeated and absorbed. States rose and fell; sometimes, after the collapse of a civilization, kingship would give way temporarily to chiefdom. However, over the millennia, there would come to be little or no alternative to dependency on food from field crops and on managed herds of oxen and cattle, also fed at least partly on field crops.

Eventually, specialization and innovation would lead to two remarkable inventions—writing and money—which played pivotal roles in the creation of the modern world, and which we will discuss below. Meanwhile, throughout the development of civilization, an influential minority of society, with ever more tools available for the unequal exercise of power, were able to live in cities and pursue occupations at least one step removed from nature. As a result, society as a whole began to see the natural world more and more as a pile of resources to be plundered, rather than as the source of life itself.

SIDEBAR 10

Pandemics and the Evolution of States and Empires

Though disease microbes are so small that some could not be seen until the advent of the electron microscope in the late 20th century, they have exerted power over human societies, and enabled some societies to wield power over others. In his classic work *Plagues and Peoples*, William McNeill explored humanity's ecological relationship with infectious bacteria and viruses throughout history. McNeill tells how, as people began living in permanent settlements in close contact with domesticated animals, various disease organisms endemic to those animals (measles, smallpox, plague, influenza, etc.) adapted themselves to human hosts. Disease thereafter served as a mostly unintended weapon of the civilized against the uncivilized, and also as an occasional limit on the growth of civilization.

For early states, death rates from disease were high, and cities continually lost population as a result. The ultimate success of civilization therefore rested on two factors: the gradual biological process of mutual adaptation on the part of both humans and the disease organisms infecting them, resulting in a gradual reduction in these diseases' lethality; and successful, continual efforts to promote population growth and the movement of people from the countryside into cities.

Disease organisms also played roles in several key historical watershed moments, most notably the European takeover of the Americas. Virtually from the moment Columbus and his men set foot on American shores, they began infecting Native Americans with diseases to which Europeans had become partially accommodated, but that were entirely foreign to their new hosts. The result was catastrophic.

Between 1492 and 1600, 90 percent of the indigenous populations in the Americas (or roughly 55 million people) had disappeared. Some were deliberately killed by the invaders, but the vast majority succumbed to disease. This die-off of natives made it seem to Europeans that the land was virtually uninhabited, and made the process of conquest relatively easy. For Native Americans, the trauma would have intergenerational reverberations.[8]

Other microbial power plays have included: the Justinian plague (ca. 541 AD, causing an estimated 50 million deaths worldwide), which squelched Emperor Justinian's plans to reunite the Roman Empire; the Black Death in medieval Europe (ca. 1350, causing the death of 50 million people, or 60 percent of Europe's population at the time), which dramatically altered economic relations on the continent, raising average incomes due to a shortage of workers; the 1918 influenza epidemic (55 million deaths globally), which killed far more people than died in combat during World War I; and of course, the COVID-19 pandemic that began in 2020, which appears to have contributed to the eclipse of US economic and geopolitical power.

Herding Cattle, Flogging Slaves:
Power from Domestication

A complex or stratified human society (that is, one with full-time division of labor and differing levels of wealth and political power) can be thought of as an ecosystem. Within it, humans, because of their differing classes, roles, and occupations, can act, in effect, as different species. To the extent that some exploit others, we could say that some act as "predators," others as "prey."[9]

Prior to the emergence of agrarian states, human "predation" occurred primarily *between* groups, rather than *within* groups. When food was scarce, some groups raided other groups' stores of food, often wounding or killing fellow humans in the process, and sometimes taking captives. Raids could also be organized to defend or expand hunting territories. Revenge raids or killings often followed. However, within groups, no one's labor could be systematically exploited (for the moment I am not counting war captives as group members), so there was little internal "predation." However, as states arose, ruling classes began both to extract wealth from peasants via taxes and also to treat them as inherently inferior beings. Rulers were, in this sense, "predators"; peasants and slaves were "prey." The whole of history can be retold in these terms, illuminating the motives and methods of the powerful as they continually fleeced the relatively powerless. I'll return to this line of thinking shortly.

However, at this point, a related metaphor—domestication—may be further illuminating. As humans settled down and began to domesticate animals, they likely applied some of the same techniques and attitudes they were using to tame sheep and goats to the project of enslaving other humans—another, more sinister form of "domestication."[10]

We've already seen how a long process of human self-domestication may have led to the development of an array of new traits throughout the Pleistocene. But domestication also spread beyond humans in both deliberate and inadvertent ways, which are worth describing briefly. Roughly 40,000 to 15,000 years ago, somewhere on the Eurasian landmass, a nonhuman species—the wolf—became domesticated. The way this happened is a matter of speculation. Some scientists propose that humans captured wolf pups and kept them as pets. Others suggest this was another instance of self-domestication: perhaps the friendliest wolves gained a survival advantage from voluntary association with humans. Either way, the dog lineage gradually emerged, displaying all the traits of the domestication syndrome (reduced aggression, floppy ears, white patches of fur, and so on).

Later, when humans settled down to become farmers, the process of domesticating plant species became a deliberate preoccupation. People bred wheat with stems that wouldn't shatter easily, so grain could be harvested on the stem rather than having to be picked up off the ground; they also selectively planted wheat that ripened all at once, again to make harvesting easier. Over centuries and millennia, people domesticated dozens of plants, including fig, flax, vetch, barley, pea, maize, and lentil. The process mostly just required selectively saving and replanting seeds from plants with desirable traits.

Domesticating animals was a somewhat different process, motivated by three distinct purposes. The first was to produce animals that would live with humans to provide companionship and various services—including the herding of other domesticates (dogs), rodent suppression (cats), and entertainment (chickens, bred at first primarily for cock fights).[11] The second purpose was to modify prey animals (sheep, goats, cattle, water buffalo, yak, pig, reindeer, llama, alpaca, chicken, and turkey) to provide a docile and controllable food source. The third purpose was to tame large animals (horse, donkey, ox, camel) to provide motive power and transportation services.

The process of domestication of all these animals was a template that could be at least partly transferred to intrahuman relations. The order in time is fairly clear: animal domestication began before, or concurrent with, the development of social stratification and complexity—not after it.[12]

The steps involved in domesticating animals include capturing the animals, restraining or confining them, feeding them, accustoming them to human presence and to confinement, and controlling their reproduction in order to promote the traits wanted and discourage traits not wanted. Human slaves were similarly confined, fed, and forcibly accustomed to their new role and status. While the process of animal domestication centered on the deliberate genetic modification of the animal, human "domesticators" didn't succeed in genetically altering their human "domesticates" (that's the main reason human-on-human "domestication" is only metaphoric, not literal). Experiments have shown that it typically takes 50 generations or more of selective breeding to domesticate a wild animal such as a fox. Humans take longer than most other animals to reach reproductive age; thus, it would have been difficult to impossible for a restricted breeding group of humans to be kept enslaved long enough for genetic changes from selective breeding to appear—if selective breeding efforts were even undertaken.[13] Nevertheless, the enslaving or colonizing of human groups with obvious differences in skin color or other physical, genetically-based characteristics would only feed attitudes of superiority among the slavers and colonizers—especially much later in history, during the European-organized African slave trade beginning in the 15th century.

Anthropologist Tim Ingold argues that hunter-gatherers regard animals as their equals, while animal-herding tribes and agriculturalists tend to treat animals as property to be mastered and controlled.[14] According to many historians and archaeologists, cattle constituted the first form of property and wealth. Archaeologist Guillermo Algaze at the University of California in San Diego finds that the first city-states in Mesopotamia were built on the principle of transferring methods of control from animals to fellow humans: scribes employed the same categories to describe captives and temple workers as they used to count state-owned cattle.[15]

SIDEBAR 11

Slavery and Power

For people in the modern world it is often hard to understand the degree to which slavery was a fixture of life from ancient times until the mid-19th century. All early state societies practiced slavery in some form, and, as societies grew in complexity and wealth, their reliance on slave power tended to increase as well.

In the Roman Empire, at any given time, a quarter to a third of the population consisted of enslaved persons. Slaves were sold at markets, where they were fattened prior to sale, then stripped naked for display. On the Island of Delos, at one of the largest of the Roman slave markets, as many as 10,000 persons were sold each day. The sale of enslaved persons usually entailed separating them permanently from their parents, offspring, siblings, and spouses. While an upper-middle-class Roman household of free persons might typically own up to ten enslaved persons, wealthy landowners might own hundreds or even thousands. Living and working conditions on these large estates were such that enslaved persons were not expected to live many years, and owners carefully calculated the costs of ownership against the economic benefits of labor.

The Atlantic slave trade, beginning in the 16th century and continuing to the early 1800s, produced profits for Portuguese, Dutch, and British shipping companies, but at an enormous and hideous human toll. Millions of Africans were killed in the process of capture in their homelands and in forced marches to ports, and between 1.2 and 2.4 million more died in the Atlantic passage. Altogether, about 17.5 million persons survived the voyage; of these, perhaps another seven million died in "seasoning camps" located throughout the Caribbean.[16]

At the time of the Declaration of Independence, slavery was legal throughout the American colonies, and, until the Civil War, slavery was a primary contributor to the US economy. Cotton and tobacco were the nation's primary exports, and its economy was export-driven. Northern cities derived a considerable portion of their wealth from investing profits from the cotton and tobacco trade.[17]

The plantation owners of the American South presented themselves to the world as genteel landowners luxuriating in Greek-revival mansions, enjoying sophisticated art, music, food, and drink. But their wealth was extracted from forced labor camps policed by thugs with whips, clubs, and guns.

While the physical and psychological damage to enslaved persons was clearly of far greater severity, the institution of slavery also exacted a moral toll on white southerners. Alexis de Tocqueville, a French aristocrat and social critic who visited America in 1831, wrote that "The [slave-owning] American of the South is fond of grandeur, luxury, and renown, of gaiety, of pleasure, and above all of idleness: nothing obliges him to exert himself to subsist."[18] Some religious white southerners acknowledged that slavery was morally corrosive. Still, as a slave master wrote in 1835, "If we do commit a *sin* owning slaves, it is certainly one which is attended with *great conveniences.*"[19]

Though the institution of slavery was ended in the United States by the Civil War and Lincoln's Emancipation Proclamation, it has had continuing intergenerational impacts via persistent, systemic racism. Slavery was not a consequence of racism; instead, racism was the result of slavery. As Guyanese historian Walter Rodney noted, "...No people can enslave another for four centuries without coming out with a notion of superiority, and when the color and other physical traits of those peoples were quite different it was inevitable that the prejudice should take a racist form."[20] This persistent racism resulted in decades of white-on-Black terrorism in the form of thousands of lynchings, as well as Jim Crow segregation laws, and, more recently, police shootings of African Americans, continuing economic inequality, and continuing unequal rates of incarceration.

Instances of slavery appeared sometimes among complex horticultural societies, including pre-Columbian cultures in North, Central, and South America. However, with the advent of states and kings, slavery became universal, notably in the civilizations of Mesopotamia, Egypt, Greece, Rome, Persia, India, and China. Slavery

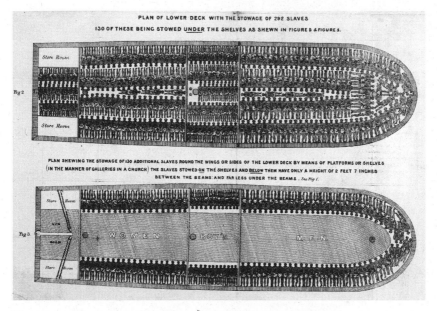

Figure 3.1. Placement of people on a slave ship. Credit: Public domain.

in many ways epitomizes power inequality in human relationships. We will have more to say about slavery in Chapter 6, where we will explore the abolition of slavery as an example and template of humanity's ability to rein in excessive power.

Women's and children's status saw a marked deterioration during the shift from horticulture to agriculture. As we've seen, women in hunter-gatherer societies generally enjoyed full autonomy (though societies did vary somewhat in this regard); they did in most horticultural societies as well, and many of these societies were matrilineal or matrifocal. Hunter-gatherers treated their children indulgently, regarding them as autonomous individuals whose desires should not be thwarted, even allowing them to play with sharp knives, hot pots, and fires. As a result, children tended to develop self-reliance and precocious social skills (as well as scars from cuts and burns). But, in farming communities, plowing was largely men's work, and men became owners of land and animals. People were the wealth of the state, and states were continually losing population to war, disease, and desertion, so women were encouraged to have as many children as was physically possible. Women's realm of work became clearly separated from that of men, and was restricted to household toil and child rearing. The authors of the Bible, like most other literate men in archaic kingdoms, regarded women and children as *chattel*, defined as property other than real estate (the word's origin is tied to the word *cattle*). Children were put to work as soon as they were able to perform useful tasks. Metaphorically, women and children were household human "domesticates."

Domesticated nonhuman animals can be said to have benefitted from their relationship with humans in some ways: by losing their freedom, they gained protection, a stable supply of food, and the opportunity to spread their population across a wider geographic range (as for plants, Michael Pollan discusses the advantages they reaped from domestication in his popular book *The Botany of Desire: A Plant's Eye View of the World*). "Domesticated" humans didn't get such a good deal, and there is plenty of evidence that they fought back or escaped when possible, as we'll see in a little more detail in Chapter 6.

As ideologies increasingly justified the power of upper classes over lower classes, and masters over slaves, psychological reinforcement of "domestication" spread, in many cases, even to those being "domesticated." Masters restricted slaves' knowledge of the world and tamped down their aspirations, encouraging them to accept their condition as proper and even divinely ordained.

Predation could be said to foster its own pattern of thought. Predators (nonhuman and human) do not appear to view their prey with much compassion; predation is a game to be played and enjoyed. If orcas and cougars could speak, they might echo the typical words of the professional bill collector: "It's nothing personal; it's just business." The psychology of domestication follows suit: sheep, cattle, pigs, and poultry are often viewed simply as commodities, rather than as conscious beings with intention, imagination, and feeling. For a family living on a farm, daily proximity with livestock entails getting to know individual animals, which may earn sympathy and respect (especially from children). But when any animal comes to be seen as food, compassion tends to be attenuated. Attitudes fostered by human-on-human "predation" and "domestication" likewise generated psychological distance.

As noted earlier, domestication wasn't always motivated by the desire for food. Some animal domesticates were pets, which provided affection, companionship, amusement, and beauty. In class-divided states, rulers developed similar relationships with "pet" artists, musicians, and other creative people, some of whom were war captives or slaves, as well as concubines and sex slaves, who similarly supplied companionship, beauty, and amusement. Today, many people lavish extraordinary care on their pets, as ancient Egyptians did with their cats. Likewise, society heaps attention and riches on popular musicians, actors, artists, athletes, models, and writers.

Predator-prey relationships are essential to healthy ecosystems. Are human "predator-prey" relationships similarly essential to complex human societies? Political idealists have replied with a resounding "no," though the evidence is debated. Non-hierarchical, cooperative institutions, such as cooperative businesses, are numerous and successful in the world today. Further, some modern societies feature much lower levels of wealth inequality than do others. Clearly, humans are able to turn "predation" into cooperation, shifting vertical power at least partly back to horizontal power—a point to which we will return in chapters 6 and 7. Nonetheless, in the world currently, inequality and "predation" remain facts of life in every society.

Stories of Our Ancestors: Religion and Power

In discussing the origins of language in Chapter 2, I suggested that our species' earliest use of words led to telling stories and asking existential questions, which in turn led to myths and belief in supernatural beings. Perhaps the most fateful new words in any nascent human vocabulary were *why*, *how*, and *what*. Why does night follow day? How did the rivers and mountains come to be? What happens to us when we die? Such questions were often answered with origin stories that were retold and re-enacted in rituals that entertained, united, and sustained communities from generation to generation.

While it was seeding minds with metaphysical questions, language was also driving a wedge between the left and right hemispheres of the human brain. Psychologist Julian Jaynes argued in his influential book *The Origin of Consciousness in the Breakdown of the Bicameral Mind* (1976) that, in the ancient world, the less-verbal right brain's attempts to be heard were interpreted by the more-linguistic left brain as the voices of gods and spirits. Today, neurologists understand that injuries or seizures to the brain's left temporal lobe can trigger hallucinations of the "divine."[21]

Among hunter-gatherer peoples, the specialist in matters of the sacred was the shaman (the word comes from Indigenous peoples of Siberia, but it denotes a nearly universal role in pre-agricultural societies). The shaman had a talent for entering altered states of consciousness, and giving voice to the spirits of the dead and of nature. Shamans treated individual and collective ills by intervening in the spirit world on behalf of the living, and by mending souls through cathartic ritual.

The subjective world of people in hunter-gatherer societies was magical—in a very specific sense of the word. Today, we usually think of magic as sleight-of-hand trickery, but in the minds of people in foraging and horticultural societies it was a technology of the imagination often involving the use of psychedelic plants and fungi, a set of practices to temporarily reconnect the brain hemispheres (though people didn't think in such terms), and a way of exerting power in the realm of spirits. The spirit realm (Aboriginal Australians called it the Dreamtime) was timeless and malleable, the place of all beginnings and endings.

Calling the spiritual practices of hunters and gatherers "religions" confuses more than it clarifies. Shamanic spirituality had little in common with the universal religions with which we are familiar today—such as Christianity and Islam—which we will discuss shortly. The latter constitute a moralizing force in mass societies and feature Big Gods that watch over us 24/7, whereas the nature spirits of shamanic cultures had personalities and foibles like humans. But they also had power, and the purpose of contacting them in organized rituals was to access that power.

The spiritual beliefs and practices of horticultural societies remained magical, but focused more on the *totem* (an Ojibwe term that many anthropologists have generalized to refer to any spirit being, animal, sacred object, or symbol that serves as an emblem for a clan or tribe). Seasonal ceremonies became grand affairs that were attended by many clans or tribes, and presided over by ceremonial chiefs. Sacrifice—including human sacrifice and ritual torture—appeared in complex horticultural societies, but became far more common in early states.

With the advent of the state, kings, and agriculture, we see the emergence of gods and goddesses—personalities separate from the human sphere who are to be worshiped with prayer and sacrifice. Gods were typically thought of as hierarchically organized, with a high god at the apex of a pyramid of divine power. The king claimed either to be an embodiment of the high god, or to have a special and direct avenue of communication; this was the justification for his wielding absolute social power among fellow humans. Rituals and ceremonies became grand affairs, taking place in monumental plazas and temples, and were presided over by a numerous, socially elite, and hierarchically organized priestly class.[22]

The trend is unmistakable: as societies became larger, sacred presences became more abstracted from nature and from people's immediate surroundings; people's attitude toward them turned toward worship and sacrifice; and spiritual intermediaries became organized into hierarchies of full-time professionals.

We've already seen that societies became bigger in order to become more powerful—to resist raids or invasions and to be able to seize people and wealth from surrounding societies. But now the biggest societies had reached a critical stage in terms of scale. They were already unstable: kingdoms were always vulnerable to economic cycles, and thus also to revolutions, civil wars, and epidemics. Yet, after the state, as a type of social entity, had existed for at least a couple of thousand years, with individual kingdoms rising and collapsing frequently, still another shift in scale became necessary.

Roughly 3,000 years ago, after the first written documents had appeared, new military challengers began to menace states along an arc running from the Mediterranean Sea all the way to China. Animal-herding tribes inhabiting the great Eurasian Steppe, stretching from modern Ukraine to Manchuria, began using the horse (domesticated around 5,500 years ago) in warfare. Two horses could be hitched to a small cart, carrying a driver and one or two archers. This simple vehicle, the chariot, was the most important military innovation in the ancient world. Adding to this advantage, the tribes of the Steppe later invented saddles, and supplemented chariots with cavalry—horse-riding warriors. Foot soldiers were simply no match for the deadly mobility of chariots and cavalry, and early states fell one by one to invaders from the Steppe.

There were only two effective defensive strategies that early states could muster. The first was to improve their own military technology by acquiring horses from the Steppe dwellers, adopting chariots and cavalry, and developing better armor to deflect the arrows of mobile archers. The second was to increase the scale of society still further, so as to recruit bigger armies and pay for them with a bigger tax base. With enough tax revenues, it would even be possible to build long and high walls to keep the invaders out. Kingdoms became empires, encompassing large geographic areas and many disparate cultures and languages. One of the first, the Achaemenid Persian Empire, lasted from 550 to 330 BCE and had 25 to 30 million subjects. But sheer size made these early empires even less internally stable than the archaic kingdoms that preceded them. They were nations of strangers—who, in many instances, couldn't even understand one another's speech. How to bind all these people together? [23]

SIDEBAR 12

DNA Evidence for Steppe Invasions

In his book *Who We Are and How We Got Here: Ancient DNA and the New Science of the Human Past*, Harvard geneticist David Reich summarizes recent breakthroughs in understanding the early history of humans; his most dramatic discoveries center on the last 5,000 years.

Reich's work addresses the tantalizing question at the heart of the science of linguistics. In 1786, a British civil servant named Sir William Jones discovered surprising similarities between Sanskrit and ancient Greek. His research led to the naming of the Indo-European language family—which includes the Germanic, Celtic, Italic, near eastern (Iranian), and north Indian languages (Hindi, Urdu, Bengali, Punjabi, Marathi, etc.). However, there has never been a consensus among the experts regarding the speakers of the original language that gave rise to these offshoots, or how that language came to be spread so far and wide. Using genetic markers, Reich has shown that the Indo-European languages coincide with genetic similarities between the peoples of Europe and north India, and both the languages and the genes appear to stem from migrations around 5,000 years ago from the vast Steppe, the grass-covered plains stretching from southeastern Europe to central Asia.

The inhabitants of these savannahs and shrublands were pastoral nomads who rode domesticated horses, drove wheeled vehicles, and used dairy products. Roughly five millennia ago, they swooped down upon the early states of the Middle East, the Mediterranean, and north India, forcing them to adapt their technologies and diets, and ultimately to grow into empires. We can glimpse the way of life of these fierce nomads in the writings of Homer, who told of warlords seeking prestige and wealth through plunder and rape. As Reich dispassionately puts it: "Males from populations with more power tend to pair with females from populations with less."

Cultural evolution found a solution, and it hinged on religion. Recall that cultural evolution requires cultural variation and competition. As the old palace-and-temple organization of social life was challenged by the political disruptions of empire-building, new religions were appearing everywhere throughout the ancient world, thus providing a significant source of cultural variation. For example, monotheism cropped up as a religious "mutation" in Egypt 3,300 years ago when Pharaoh Amenhotep IV changed his name to Akhenaten and instituted the worship of Aten, the Sun God, and banned the worship of the traditional pantheon of Egyptian deities. It was an unsuccessful innovation: after Akhenaten's death, Egypt reverted to its old polytheism and efforts were made to erase all traces of Aten and Akhenaten from stone inscriptions and monuments. However, in succeeding centuries other similar religious variations appeared and some succeeded spectacularly well, attracting millions of followers and persisting for many centuries.

How did the new religions differ from what went before, and how did they help promote the competitive success and durability of empires? The most successful new religions featured Big Gods who were often universal—that is, they transcended cultural differences.[24] Because they weren't tied to a specific cultural context, universal religions could seek converts anywhere and everywhere (Christianity, Islam, and Buddhism would become missionary religions, in that they actively sought converts; Judaism didn't take that route, and, while it still constitutes a Big God religion, it doesn't have nearly as many adherents).

Even more importantly, the Big Gods were supernatural monitors who watched over everyone continually, spotting any immoral behavior.[25] This internal divine surveillance proved to be exactly what was needed to guarantee trust among believers everywhere. And trust had many practical implications. For example, trade demanded trust between merchants. If all were believers in the same Big God, levels of trust increased significantly.

The new religions also often featured the notion of hell, or divine punishment, for those whose behavior was bad (antisocial), and heaven, or future rewards, for those whose actions were good (prosocial). Hell was effective at discouraging people from leaving the religion, while heaven was more effective in winning converts. Heaven was also a useful foil for the problem of economic inequality: it promised that, after a few brief years of poverty and suffering here on Earth, the good would enjoy an eternity of bliss. In the postmortem worlds, rich and poor would be re-sorted according to their inner virtues, with the righteous servant reaping a far better reward than the sinful master. As songwriter and union organizer Joe Hill would later put it:

Long-haired preachers come out every night
Try to tell you what's wrong and what's right
But when asked how 'bout something to eat
They will answer in voices so sweet

You will eat, bye and bye
In that glorious land above the sky
Work and pray, live on hay
You'll get pie in the sky when you die.[26]

Finally, Big God religions demanded personally costly displays of commitment. Such displays served to prove that individuals were not merely giving lip service to the religion, but could actually be trusted to maintain prosocial behavior. Such displays could include pilgrimages, public prayer, donations, fasting, occasional abstinence from sex or certain foods, and attendance at frequent scheduled worship events—or, in more extreme cases, self-flagellation, life-long celibacy, or voluntary poverty.

In his book *The Evolution of God*, journalist Robert Wright traces the story of how one quirky tribal war deity was gradually elevated to become a universal, moralizing, and all-surveilling Big God, the object of worship not just of all Jews, but Christians and Muslims as well. Moralizing Big Gods also appeared in China and India, but in the context of polytheistic religions. In all cases, these new religions, or new versions of old religions, possessed characteristics needed to ensure large-scale cooperation (supernatural monitoring, divine punishments and rewards, and costly displays of commitment).[27]

Several cultural evolution theorists, including Peter Turchin and psychologist Ara Norenzayan, have converged on the Big Gods explanation for the emergence of ultrasocial cooperation in what philosopher Karl Jaspers called the "Axial Age"—i.e., the turning point of human history, when empires and universal religions first appeared (roughly the eighth to the third century BCE).[28] Their argument is based not just on historical research, but on hundreds of experiments in psychology, behavioral economics, and related fields. In his book *Big Gods: How Religion Transformed Cooperation and Conflict*, Norenzayan describes ingenious psychological experiments that have confirmed that the expectation of divine surveillance does indeed lead people to cheat less, to be more generous toward others, and to punish others for antisocial behavior even if doing so is personally costly. Experiments have also shown that the idea of hell is better than that of heaven at motivating prosocial behavior.

It's important to remember that moral behavior was not a *spiritual* preoccupation previously. Among hunter-gatherers, spiritual practice had little or nothing to do with promoting prosocial actions. In those societies, everyone was constantly encouraged to share, while obnoxious individuals could simply be ostracized or killed. In archaic states, kings controlled their people's behavior through state-sanctioned coercive power justified by proclamations issuing from the chief god by way of the king and the priesthood. In contrast, the Big God religions put a watcher inside each individual's head. This was an economical and effective new way to control the behavior of millions of people who might have little else in common.

Of course, that's not all the new religions did. Contemplation of the divine, and of the mythic biographies of the religions' founders, gave individuals a sense of purpose and meaning. Religious art, architecture, and music expanded humanity's access to, and appreciation of, beauty in many forms. Further, the Axial Age resulted in at least a partial reduction of inequality throughout most subsequent civilizations. Emperors, kings, and members of the aristocracy had to convert to the new religions in order to maintain their credibility; but once they did so, the ways they exerted power shifted. Emperor Ashoka of the Maurya Dynasty, in what is now India, ruled from about 268 to 232 BCE. Early in his reign, Ashoka pursued successful conquests, resulting in perhaps 100,000 deaths. But then he converted to Buddhism and spent the rest of his reign spreading Buddhist doctrine. Here is a passage from one of his edicts:

> Beloved-of-the-Gods speaks thus. This Dhamma [i.e., universal truth] edict was written twenty-six years after my coronation. My magistrates are working among the people, among many hundreds of thousands of people. The hearing of petitions and the administration of justice has been left to them so that they can do their duties confidently and fearlessly and so that they can work for the welfare, happiness and benefit of the people in the country. But they should remember what causes happiness and sorrow, and being themselves devoted to the Dhamma, they should encourage the people in the country to do the same, that they may attain happiness in this world and the next.[29]

Compare Ashoka's words with those of Assyrian king Tiglath Pileser, quoted earlier. The archaic king Tiglath, unlike Ashoka, evidently wasn't much concerned with happiness of his people, but much more so with their obsequious contributions to his own glory and vertical power. Nevertheless, it would be the empires that were devoted to happiness and cooperation that would win in the end. The maximum power principle was still at work in social evolution, but in a new and unexpected way: by moderating their power in the short run, rulers had found a way to stabilize and extend social power in the longer run. This is a pattern we'll examine in more detail in Chapter 6.

Universal religions also had their dark side. The body count for Christianity alone runs into the tens of millions. Believers committed untold atrocities against nonbelievers as yet another way of demonstrating their commitment to the faith. The victims—primarily Indigenous peoples of the Americas, Africa, Australia, and the Pacific Islands—often faced the choice either to convert to a religion they could not understand, or risk torture, death, or enslavement. Whatever advantages they provided for believers, Big God religions were also instruments of vertical social power and domination vis-à-vis nonbelievers.

SIDEBAR 13

Justifying Colonialism

Beginning at the close of the 15th century, Europeans embarked on history's biggest land grab. Over the next four hundred years, they would take possession of North and South America, Asia, Africa, and the Pacific Islands including Australia and New Zealand—essentially, the whole rest of the world. They did this because they could: they had improved weapons and seaborne transport, and they carried diseases that wiped out millions who would otherwise have resisted conquest.[30] But how does one justify taking other people's land, often enslaving those people, and killing them by the millions?

Power tends to create its own justifications, and the human mind is resourceful. In this case, justification began with an official document, a Papal Bull issued by Pope Alexander VI on May 4, 1493. Its innocuous title, "*Inter Caetera*" ("Among Other Things"), hardly hinted at its vast historical ramifications.

The Bull stated that any land not inhabited by Christians was available to be "discovered," claimed, and exploited by Christian rulers. It also ordered that "the Catholic faith and the Christian religion be exalted and be everywhere increased and spread, that the health of souls be cared for and that barbarous nations be overthrown and brought to the faith itself." This "Doctrine of Discovery" became the basis of all European claims in the Americas and elsewhere, and also justified the United States' western expansion. The US Supreme Court, in the 1823 case *Johnson v. McIntosh*, held in a unanimous decision "that the principle of discovery gave European nations an absolute right to New World lands." In essence, Native Americans could claim only a right of occupancy, which could be legally abolished.[31]

Often, native peoples were given the choice between death and conversion to Christianity. Even if they converted, their land would still be taken and their children would live in poverty and servitude. The conquistadors were absolved of any moral qualms: the Big God of Christianity had declared (through his intermediaries) that this was all good.

The universal religions were so successful at motivating cooperative behavior, gaining new adherents, and encouraging believers to have more children (as we'll discuss in Chapter 5) that they have persisted and grown through the centuries right up to the present. Many empires based on Christianity or Islam (the Byzantine, Holy Roman, Ottoman, British, Spanish, French, Portuguese, and Russian Empires, to name just a few) rose and fell during the past two millennia, largely shaping world history. Today, 2.3 billion people call themselves Christians, 1.8 billion identify as Muslims, and half a billion practice Buddhism; altogether, these three missionary religions influence the behavior of nearly two-thirds of humanity.

Are Big Gods necessary to a moral society? Not necessarily. Atheists can be ultrasocial too, as we see in atheist-majority Scandinavian countries today. That observation opens the door to a discussion of the future of social power, which we'll take up in chapters 6 and 7.

Tools for Wording: Communication Technologies[32]

In Chapter 2, I described tools as prosthetic extensions of the body, and language as the first soft technology. Bringing tools and language together, via communication technologies, enabled the emergence of a series of new pathways to social power. As we'll see, the powers of communication technologies transformed societies, even playing a key role in the evolution of Big God religions.

Writing is the primary technology of language. The first true writing is believed to have appeared in Sumeria around 5,300 years ago, in the form of baked clay tablets inscribed with thousands of tiny triangular stick impressions (i.e., cuneiform writing). Earlier protowriting took the form of jotted reminders that made sense primarily to the writer or to a small community; these scribbles weren't capable of recording complete thoughts and in most cases cannot be deciphered by modern linguists. The cuneiform writing of Mesopotamia, in contrast, turned into a full symbolic representation of a human language, and through the texts that Sumerian scribes left behind we can glimpse the thoughts of people who lived thousands of years before us. Writing was invented independently about 4,000 years ago in China, and again in Mesoamerica by about 2,300 years ago; these writing systems achieved the same purpose as Sumerian cuneiform, but have almost nothing else in common with it.

The first written texts were inventories and government records produced in the archaic state of Sumeria by full-time scribes working for the king and the priesthood. Since only a tiny fraction of the populace was literate, there was no expectation that these records would be read by anyone outside the palace or the temple. But as the habit of writing took root, and as literacy spread, so did the usefulness and power of recording thoughts on clay or other media. This process took some time: several centuries separate the appearance of the first writing system and its use as anything other than a way of keeping track of grain yields, tax payments, numbers of cattle, and numbers of war captives.

The Epic of Gilgamesh, a Sumerian epic poem dating to almost 5,000 years ago, is widely regarded as the first piece of literature. It has a few themes and characters in common with the later Hebrew Bible, including Eden and Deluge narratives, with protagonists who resemble Adam, Eve, and Noah. Thus, the very first written story still resonates throughout much of humanity to this day.

Religious myths were probably transmitted orally for untold generations before being put into writing. But once recorded on clay, parchment, or papyrus, these stories took on a permanent and objective existence. Early religious texts were regarded as sacred objects, and achieved a durability and portability that assisted in the spread of faith and belief. This would be a key advantage in the case of the Big God missionary religions, as they sought converts over ever-wider geographic areas.

Writing conferred many practical advantages; "the power of the written word" is a phrase familiar to nearly everyone, and for good reason. Among other uses, writing enabled laws to be publicly codified. The Code of Hammurabi, written in 1754 BCE, was the world's first law document, specifying wages for various occupations and punishments for various crimes, all graded on a social scale, depending on whether one was a slave or free person, male or female. In pre-state societies all such matters were decided by custom and consensus; with written laws, justice became less a matter for *ad hoc* community deliberation, and more one of adherence to objective standards. Writing also assisted enormously with commercial record keeping, thus facilitating trade. It even altered military strategy, as spies could send coded notes about enemy troop movements.

Like all technologies, writing changed the people who used it. Oral culture was rooted in memory, multisensory immersion, and community; in contrast, written culture is visual, abstract, and objectifying. Poet Gary Snyder tells of his journey, many years ago, in Australian Aboriginal country with a Pintupi elder named Jimmy Tjungurrayi. They were traveling at about 25 miles per hour in the back of a pickup truck, the elder reciting stories at breakneck speed, beginning a new one before finishing the last. Later Snyder realized that these stories were tied to specific places they had been passing. The stories were meant to be told while walking at a much slower speed, so the elder had to struggle to keep up. Aboriginal orally transmitted place-stories were a multidimensional map that included ancestral relations and events; intimate knowledge of specific land formations, plants, and animals; and myths of the Dreamtime. Writing can give us a more rational representation of a territory, one we can refer to at our own pace. But it cannot capture the immediate, full-sensory engagement with nature and community that oral culture maintained. And so, as we became literate, entire realms of shared human experience tended to atrophy.[33]

Early writing consisted of what were essentially little pictures, and it took years to learn to read and write with them (as it still does with Chinese ideograms). But an innovation in communication technology—the phonetic alphabet—made it possible to learn to read and write at a basic level in roughly 30 hours. With this ease of use came a further level of abstraction in human thinking.

Alphabetic writing originated about 4,000 years ago among Semitic peoples living in or near Egypt. It was used only sporadically for about 500 years until being officially adopted by Phoenician city-states that carried on large-scale maritime trade. It's probably no historical accident that the alphabet first caught on with people who traded and moved about a great deal (other early alphabet adopters included the Hebrews, who were wandering desert herders; and the Greeks, who were traders like the Phoenicians). Alphabetic writing was more efficient than picture writing, and, because it was phonetic, could more readily record multiple languages, thereby making translation and trading easier. Significantly, for adherents of missionary Big God religions, the alphabet could aid in spreading the good word.

With its short roster of intrinsically meaningless letters, the alphabet could viralize literacy while speeding up trade and other interactions between cultures. It also led to a further disconnection of thought from the immediate senses (media theorist Marshall McLuhan called alphabet users "abced-minded"). Picture writing was already abstract, in that the animals, plants, and humans used as symbols were stylized and divorced from movement, sound, smell, and taste; but the alphabet was doubly abstract: even though its letters originated as pictures of oxen, fish, hands, and feet, this vestigial picture content was irrelevant to the meaning conveyed through words being written and read. To know that the letter "B" originated as a picture of a house tells us nothing about the meaning of words containing that letter.

Further innovations in communication technology happened much later, but brought equally profound social, economic, cultural, and psychological consequences.[34] The printing of alphabetic languages with moveable type, originating with Johannes Gutenberg's Bible in 1439, sowed the idea of interchangeable parts, and therefore of mass production.[35] Printing was also a democratizing force in society: soon, nearly any literate person could afford a cheaply printed book, and so at least in principle anyone could gain access to the world's preserved stores of higher knowledge. Literacy rates rose as increasing numbers of people took advantage of this opportunity. With printing, the expanding middle class could educate itself on a wide variety of subjects—from law to science to the daily churn of gossip and current events described in newspapers (the first of which appeared in Germany in 1605). A century after Gutenberg, printed Bibles underscored Martin Luther's argument that the common people should have direct access to scripture translated into their own language. The printing press thus played a key role in the origination and spreading of Protestantism.

The telegraph, invented in the 19th century, enabled nearly instantaneous point-to-point transmission of words and numbers. News could travel in a flash, but it had to be reduced to a series of dots and dashes—an early form of digital (though not strictly binary) communication. This could lead to mistakes: when a telegrapher in Britain announced the discovery of a new element named "aluminium" to an American counterpart, the second "i" in the word wasn't received and recorded; as a result, the element is still known in the United States as "aluminum."

The patenting of the telephone by Alexander Graham Bell in 1876 led to instant two-way communication by voice. While printing was a static and visual medium, telephonic communication engaged the ear and required the full, immediate attention of users. This made the phone a source of rude interruption to the print oriented, but an enthralling lifeline to teenagers eager to chat with boyfriends and girlfriends.

The writing of music (i.e., musical notation, which originated in ancient Babylonia, was revived in the Middle Ages, and was perfected in the 17th century) exposed music to analysis and enabled the authorship and ownership of compositions. Analysis and authorship then created the incentive for the development of greater complexity and originality in music, leading to Bach, Beethoven and beyond. Sound recording (starting in the 1890s) made music almost effortlessly accessible at any time and exposed listeners to exotic musical styles. Now performances, as well as compositions, could be owned by musicians, record companies, and listeners. Recording led quickly to the jazz age, then to rock & roll, folk, country, rhythm & blues, disco, punk, grunge, hip hop, world music, electronic music, hypnagogic pop, and on and on, each generating its own cultural world of in-group fashion and lingo.

Meanwhile movies (which became popular in the 1910s) borrowed their content from printed novels and theatrical plays, and their style from still photography (invented in the 1820s). Films were, metaphorically, the projections of dreams. As such, they riveted the imaginations of moviegoers, creating a giant industry owned by powerful trend-setting moguls and populated by "stars" whom members of the public longed to know as intimately as their own family members.

Radio, like printing, could deliver a message to many people scattered widely, each in the privacy of their home; and like the telephone, it was auditory and instantaneous. But unlike the telephone, radio (which came into widespread use in the 1920s) would become primarily a one-way means of broadcasting voice and music. Because it was auditory, radio involved listeners in a more psychologically intensive way than print (McLuhan called radio the "tribal drum"). In the mid-20th century, millions gathered around their radio cabinets to listen raptly to political leaders like Franklin Roosevelt and Adolph Hitler—as well as preachers like Pastor Charles Fuller, comedians like Bob Hope and Gracie Allen, and musicians ranging from Ella Fitzgerald to Jascha Heifetz. Most radio programs featured commercials to generate revenue to pay for content. The power of radio thus produced both mass propaganda (it's questionable whether fascism could have taken hold in Europe without radio) and mass advertising—signal developments in the political and economic history of power.

Television (widely adopted in the 1950s) brought a low-res version of the movie theater into nearly every home. Because the TV image was made up of thousands of blinking dots requiring unconscious neurological processes to make sense of them, it entrained the brain in a way that movies didn't. Further, because it engaged viewers with both sound and image, and in their own homes, television tended to turn viewers into hyperpassive receivers of broadcast content. In America, for the first three decades of the television era, there were only four options for what to watch at any given time— programming on three commercial networks and one commercial-free educational network (other nations had even fewer choices). As a result of all these factors, television enormously enhanced the power of corporate advertisers—and, in authoritarian states, that of the government. Indeed, television may have been the most powerful technological means of shaping mass thought and behavior ever developed, until quite recently.

During the past three decades television has changed. High-definition LCD screens, the proliferation of cable networks, and the emergence of on-demand streaming services have altered TV's fundamental nature. Today it is even more of a hybrid medium than film or early television. Not only does it include sight and sound at high levels of fidelity; not only does it present content derived from movies, print novels, theater, newspapers, radio sports and weather reports, and even comic books; but television now also incorporates elements of the computer, the internet, and social media. TV is a more democratized medium than it was formerly, in that it offers consumers more options. But it is also more socially and politically polarizing, as is the case with related new developments discussed below.

The internet-connected personal computer, which came into general use in the 1990s, was, like printing, a source of horizontal power for many society members. Anyone could become a publisher by creating a web page and writing a blog, or by posting videos. Anyone could open an online store. Meanwhile, the user could access enormous amounts of information about every imaginable topic via a search engine or Wikipedia. It's too bad McLuhan wasn't around to see it; he might have called the internet-connected computer a prosthetic extension of the brain itself.

Computers and digitization have led to shifts in human cognition and social organization that are still in progress; the same with smartphones and social media, which fashion an information world tailored to each individual's set of expectations, interests, and prejudices. These devices and their software offer a portable, even wearable, and infinitely customizable portal into an immense digital world of news, entertainment, and personal communication with friends, relatives, and business contacts scattered across the world.

However, the newest communication technologies are means not just of mass expression but also of mass persuasion, in that they give their corporate managers opportunities for digital surveillance and marketing, and offer political actors (not only governments, but fringe groups of every stripe) means for the spread of precisely targeted propaganda often cleverly disguised as news. Digital face recognition, among other automated intelligence technologies, combined with the internet-connected computer and smartphone, supply governments with the means for nearly total information control. The real-life equivalent of Big Brother (the totalitarian leader in George Orwell's prophetic dystopian novel 1984, published in 1948) now has the ability to know who you are, where you are, and what you are thinking and doing at almost any moment. He can manipulate your beliefs, desires, and fears.

In this briefly recounted history of communication technologies, one can't help but note a steep acceleration of cultural change, and a correspondingly swift intensification and extension of social power. Each new communication technology has magnified the power of language (and images), and some of these technologies have given us vastly more access to information and knowledge. But each has also imposed costs and risks. At the end of the story we find ourselves enmeshed with communication technologies to such a degree that many people now use the word *technology* to refer solely to the very latest computerized communication devices and their software, ignoring millions of other tools and the tens of thousands of years during which we developed and used them. We are increasingly captured in the ever-shrinking *now* of unblinking digital awareness.

Numbers on Money

Economists typically describe money as a neutral medium of exchange. But, in reality, money is better thought of as quantifiable, storable, transferrable, and portable social power. I have defined social power as the ability to get other people to do something. With enough money, you can indeed get other people to do all sorts of things.

To understand money, it's essential to know the context in which it emerged. Money facilitates trade, and in early societies trade was an adversarial activity undertaken primarily between strangers. Within small pre-state societies—that is, among people who knew one another—a gift economy prevailed. People simply shared what they had, in a web of mutual indebtedness. Trade was a tense exchange of goods in which men from different tribes sought to get the better of one another; if there was suspicion of cheating, someone could get hurt or killed. In these societies, cooperation was essential to everyone's survival, so any intra-group activity that smacked of trade was unthinkable.

However, with the advent of the multi-ethnic state and full-time division of labor, trade necessarily began to occur within the broadened borders of society. In order to keep the process of exchange from eroding community solidarity and occasionally erupting into violence, the state created and regulated markets. The palace and the temple made laws specifying where and when markets could operate, and set the terms on which sellers should extend credit to buyers. The state also created and enforced standard units of measure.

Credit (i.e., debt) and money arose simultaneously within this milieu, and were, and are, two sides of the same metaphorical coin. As anthropologist David Graeber recounts in his illuminating book *Debt: The First 5,000 Years*, you can't have one without the other.[36] Debt and money have co-evolved through the millennia, with each unit of any currency implying an obligation on someone's part to pay or repay a similar or greater amount, in the forms of money, goods, or labor.

Money also co-evolved with number systems, arithmetic, and geometry. Keeping track of money and credit (and the stuff they bought—often cattle or land) required numbers and the ability to manipulate them. Counting, the earliest evidence of which is preserved in tally marks on animal bones, began long before verbal writing, perhaps up to 40,000 years ago. By 6,000 years ago, people in the Aagros region of Iran were keeping track of sheep using clay tokens. Cities within the archaic state of Sumeria each had their own local numeral systems, used mostly in accounting for land, animals, and slaves. Geometry evolved for the purpose of surveying land. As trade expanded, so did numeracy, and so did the sophistication of mathematics. Numbers and their manipulation would, of course, eventually give humanity impressive powers in other contexts—primarily in science and engineering, where they would enable the discovery of exoplanets, the sequencing of the human genome, and a plethora of other marvels.

The earliest forms of money consisted of anything from sheep to shells, but eventually silver and gold emerged as the most practical, universally accepted mediums of exchange—while also serving as measures and stores of value. However, in early states, ingots or coins themselves were not typically traded. Instead, traders kept credit accounts using ledgers (at first on clay tablets; traders in later societies would use paper books or, most recently, computer hard drives). Precious metals were stored—often more-or-less permanently—in the palace treasury by the state as an ultimate guarantee of the state's ability to regulate trade, especially with other states. Later on, gold and silver coins came into more common use by merchants and nobles, but most peasants still had little contact with them, conducting whatever business they engaged in using grain, domesticated animals, and credit, with money serving mainly as a measure of their indebtedness.

People invented money because they wanted to make trading easier, simpler, safer, and more flexible. Once money came into use, the exchange process was freed to grow and to insert itself into aspects of life where it had previously never been permitted. Over time, the gift economy shrank and the exchange economy grew. Today, the gift economy survives within families, religious communities, and the nonprofit sector of the economy, as well as in informal interactions among friends and in the wake of wars and natural disasters. Elsewhere, virtually everything has a price, and nearly all human activities are measured by and for their monetary value.

During eras when slavery was common, human beings were routinely bought and sold; fortunately, those days are mostly gone. But most people are still forced by circumstance to sell themselves on the installment plan via wage labor. Is it correct to call today's child worker in India, making pennies per day disassembling smartphones for recycling, a slave? In the technical sense, no—unless the child is being bought or sold. But the child's life circumstances may arguably be just as miserable as they would under conditions of true slavery.

Debt slavery, common in the ancient world, also has reverberations in the present. Tracing the evolution of that practice requires us to resume our exploration of the history of money and debt.

It's difficult to identify when the practice of charging interest on loans originated; while it probably existed in Mesopotamia up to 7,000 years ago, clear written documentation only goes back to about 2,500 years ago.[37] Interest gave the agricultural wealth pump new gearing with which to operate at ever finer levels. Interest on loans continually, silently, and subtly siphoned wealth to those in a position to loan money. People whose main occupation was investment or trade gradually became more powerful; while those whose main occupation was growing food, making things, or taking care of others gradually became relatively poorer.

The problem with the wealth pump, as we've already seen, is that it leads to repeating cycles of societal expansion, consolidation, and collapse (as Peter Turchin and his colleagues have abundantly documented, using a database encompassing hundreds of societies).[38] Interest on loans plays a role in this cycle: once interest has siphoned most of a society's money to the lender-investor class, then a large proportion of people tend to be highly indebted and will eventually be unable to make further payments on their loans. At that point, the financial economy shudders to a halt. That's essentially what has happened in every financial depression over the past century or so, but it was already occurring in ancient Sumeria, Israel, and Egypt.

The only workable solution then was to declare the wholesale cancellation or forgiveness of debt (in modern debt crises, widespread
bankruptcies and defaults likewise clear unrepayable debt, though
with considerable social distress). Hence the Law of Jubilee, set forth
in Leviticus 25:10–13:

> You shall make the fiftieth year holy, and proclaim liberty
> throughout the land to all its inhabitants. It shall be a jubilee
> to you; and each of you shall return to his own property, and
> each of you shall return to his family. That fiftieth year shall be
> a jubilee to you. In it you shall not sow, neither reap that which
> grows of itself, nor gather from the undressed vines. For it is a
> jubilee; it shall be holy to you. You shall eat of its increase out
> of the field. In this Year of Jubilee each of you shall return to
> his property.

What does this have to do with cancellation of debt? Everything.
In ancient times, debt was usually repaid not in money, but in animals and grain. If the harvest was bad and payments could not be
made, farmers' children, wives, and land could be taken as payment.
Debt was a serious, painful, psychologically and socially wrenching
business. And so, the periodic cancellation of debt, the return of
family members to their families, and the return of families to their
houses and fields, was cause for celebration throughout the land.

Later, in the European Middle Ages, debt bondage took the form
of serfdom. Within the feudal system, the lords of manors held
ownership rights not just of land, but also the serfs who worked that
land. The latter were regarded as being permanently in debt to their
lords. In return for working the lord's fields (and mines and forests),
serfs were entitled to protection, justice, and the right to cultivate
certain fields for their own subsistence. Like slaves, they could be
bought, sold, or traded (though they generally could only be transferred together with the land). Serfs could be abused by the lord at
will, and could marry only with their lord's permission.

Another system of debt bondage, called peonage, was epitomized
in the southern United States in the 19th century. Poor white farmers

and formerly enslaved African Americans, who could not afford to purchase land, would work the land of someone else in exchange for a share of the crops. Sometimes this arrangement, also called sharecropping, was mutually beneficial. However, many landowners abused the system, forcing the sharecropper to buy seeds and tools from the owner's store at inflated prices. Landowners and their employees also often took advantage of sharecroppers' illiteracy by keeping fraudulent accounts. The net result was to keep the peon or sharecropper in perpetual debt to the landowner.

Any discussion of the history of debt and money must include a mention of banking—a practice that probably started in ancient Babylonia, with various innovations appearing in Greece (where we see early bank "vaults," in the form of temple treasuries), Rome (whose language provides us with the root of the word *bank*, from *bancu*, or "bench"), Europe in the Middle Ages (where we see the first double-entry bookkeeping), Britain in the 17th century (which created the first central bank, the Bank of England), 19th-century America (where the first credit cards appeared), and late 20th-century America (where electronic banking emerged). Lending money at interest enabled some bankers to become as rich and powerful as kings. Famously or infamously, the Rothschild banking family of Europe made one of the world's great fortunes by financing both sides in the Napoleonic wars.[39]

Debt shifted wealth back and forth between borrowers and lenders (with lenders almost always benefitting the most over the long run), but it also shifted consumption from future generations to the present. With credit, we consume now but pay later. In ancient times, this temporal shift in consumption had limited practical impact aside from encouraging indebtedness, as the scale of the entire human enterprise was relatively small. Today, the time-shifting, consumption-enabling power of debt has enormous implications for a globalized human society that is overusing Earth's resources and overfilling planetary waste sinks (as we will explore further in Chapter 5).

Capital, a word that derives from the Latin word for "head" (as in head of cattle), refers to money and other goods set aside for the production of more money and goods. If you use your whole paycheck to pay for rent and food, you may have money, but no capital. But if you set some money aside to invest, or save it in order to start your own business, you have capital. Capitalism is a system consisting of the private ownership of capital, and of laws to protect private property and to encourage private investment. Since markets are set up and regulated by governments, the latter always play a role in supporting capitalist economies—even though capitalists often say they dislike government economic intervention. Starting in the Renaissance, European societies began to develop the basis for capitalist economies by encouraging capital investment and trade through laws protecting the rights of owners and investors over those of employees and informal workers.

A signal event in the history of capitalism was the development of the corporation, which enabled investors to pool resources for a specific commercial purpose, while shielding themselves from liability in the cases of bankruptcy or harm to others in the course of doing business. State-authorized business entities existed in ancient Rome and the Maurya empire, but the East India Company (England, 1600) and the Dutch East India Company (Holland, 1602), both founded for the purpose of extracting wealth from maritime trade with European colonies, are generally acknowledged as the first modern corporations. By the late 19th century, American law would affirm corporate "personhood," granting corporations the same right to free speech enjoyed by individual people, while shielding them from liabilities encountered by actual human persons who cause serious injury to others (liabilities that include punishments such as imprisonment). Corporations have become one of humanity's primary means of aggregating economic and social power both to accomplish a dizzying array of purposes, and to funnel wealth to corporate managers and shareholders.

The formal study of how money and capital function began in the last couple of hundred years—a highly anomalous period in world history, in which economies were generally expanding at an unprecedentedly rapid pace. Economics, with its incomprehensible incantations and equations, became an unassailable citadel of modern ideological power, though economists continually issue predictions that deserve about as much credibility as those produced by ancient divination methods. As rebel economist Steve Keen has pointed out, basic concepts in economics such as theories of supply, demand, and rational markets are logically inconsistent and flatly contradicted by experience. Further, as economic heretic Herman Daly has spent his career arguing, conventional economics assumes continual, unlimited growth of production and consumption, whereas in reality we live on a finite planet, where overconsumption of resources must ultimately result not just in an economic crash, but ecological collapse. These are not small inconsistencies; they are core flaws in a set of beliefs that calls itself a science but is actually more of a religion, with an origin myth (the discredited notion that, prior to the invention of money, everyone lived by barter), a priesthood (credentialed economists), and a deity (the "invisible hand" of the market).

SIDEBAR 14
The Original Sins of Mainstream Economists

The absurdities that have been incorporated into orthodox neoliberal economic theory are numerous; two deserve to be mentioned in the context of this book.

First: Today, economics claims to be a science, yet its earliest theorists thought of themselves as moral philosophers. In the 17th century, John Locke proclaimed that property rights stem from "natural law," asserting that a person has a natural right to own what he "produces," and that by mixing his labor with natural resources (such as land), that land becomes his property. This became known as the "labor theory of property," which still lies at the heart of neoclassical economic theory. However, Jonathan Nitzan and Shimshon Bichler have argued convincingly that property isn't a "natural right"; it's an act of *power*.[40]

Property, according to Nitzan and Bichler, is an act of exclusion. If a person claims ownership of something, this entails excluding others from using that thing. This exclusionary power lies at the heart of the notion of privatization. Property is therefore a social convention maintained through institutionalized power (laws, courts, and police). And so economics, at its core, is a set of moral assertions about power relations, rather than a science.

Second: Early political economists identified three "factors of production": land, labor, and capital. But, over time, land was slowly dropped, leaving only labor and capital, under the justification that (in the words of economists William Nordhaus and James Tobin), "...Reproducible capital is a near perfect substitute for land and other exhaustible resources."[41] This seemed sensible because capital was becoming so productive that there was less apparent economic need for land. Why was capital becoming so productive? Essentially, because society had stumbled upon extraordinary new energy sources in the forms of fossil fuels, which made every productive process (including agriculture) much cheaper. The proper response of economists should have been to start including energy as a factor of production, but economists didn't do that; instead, they omitted land from their equations.

This line of thinking led to the absurd assertion, by celebrated economist Robert Solow, that "The world can, in effect, get along without natural resources."[42] More recently, prize-winning economist William Nordhaus has said that climate change isn't a terribly serious problem because it just affects land, and only capital and labor really count (more specifically, he argues that even if a destabilized climate makes agriculture impossible, that's not so bad because agriculture accounts for only a small fraction of global GDP).[43] Nordhaus fails to make the obvious connection: without agriculture there would be far fewer humans, no civilization (at least in its current form), and therefore no capital or labor in any meaningful sense.[44]

The past century has seen the emergence of ever-more abstract forms of money, along with ever-more sophisticated investment instruments. Stocks, bonds, derivatives, and cryptocurrencies now enable investors to make (or lose) money on the movement of prices of real or imaginary properties and commodities, and to insure their bets, and even their bets on other investors' bets. More than half of all stock trades are now made by computers using complex algorithms, with the actual ownership of a particular stock occurring for the duration of only minutes or even fractions of a second. As money and debt have become more abstract, they rule our lives ever more completely. The student loan debt that today weighs down millions of American young people's lives is essentially just a set of numbers on computer hard drives—but the real-life consequences of those numbers are profound.

Today, money and debt are still joined at the hip. Most money is created through interest-bearing bank loans. While the formal link between money and precious metals was severed in 1971 by President Nixon, the United States federal government still keeps gold and silver in its treasury vaults as a backstop to ensure international trade. And governments still regulate markets by setting interest rates, and through legal tender laws, contract laws, property laws, and banking rules. Governments also specify standard units of measure. We are worlds away from the cultural milieu of ancient Sumeria, whose money/credit system was probably the world's first; but from a basic structural perspective our modern money system still functions in much the same way.

In effect, the increasing adoption of a symbolic medium of exchange and basis for the creation and payment of debt (i.e., money) has led to a near-universal, self-reproducing system of wealth inequality. The power of money is entirely a construct of the human mind, yet it enables and disables us all on a daily basis.

Pathologies of Power

As we've seen, social power is not all the same. Horizontal social power, which predominated prior to the emergence of the state, could be epitomized by the phrase, "Let's do this together." Vertical social power always implies a bribe or threat: "If you do this, I'll give you that," or "You must do this or else." Whether one or the other form of social power predominates is partly a matter of social scale: horizontal decision-making methods that work well within small groups of people are harder to administer and maintain within larger groups. Also, when concentrated forms of power—like money or military force—become available, some people tend to get power-greedy.

We've lived with vertical social power for several thousand years now, long enough for cultural evolution to make it a feature of individual and group psychology. The literature on the psychology of power has burgeoned in recent decades, and currently rests on a plethora of observations, studies, and experiments reported in thousands of peer-reviewed papers and dozens of books.[45] The findings of the research are, in many cases, unsurprising, but certainly worthy of a brief summary here.

Vertical social power (for the purposes of this discussion, let's just say "power") can turn perfectly normal people into monsters. People with a little power want more. People with low self-esteem often abuse what little power they have. And people whose power is threatened often lash out. These tendencies have enlivened the plots of endless novels, plays, and movies—from Shakespeare's "Macbeth" to HBO's "Succession."

Research shows that people with more power than others feel freer to act; they also talk and interrupt more. They engage in more goal-directed behavior and process information more selectively.[46] Power, in the words of one researcher, "appears to foster a strong orientation to rewards and opportunities."[47] Experiments organized by Solomon Asch in the 1950s showed that ordinary people will agree to ridiculously incorrect assertions by authority figures in order to conform.[48] Stanley Milgram's famous and disturbing studies of obedience to perceived authority, carried out in the 1960s, showed that ordinary people look to those with power for direction, even when asked to do things that are morally questionable, and assume that those with power will take responsibility for whatever followers do in response to orders or cues.[49] Milgram's findings were later bolstered by Philip Zimbardo's even more controversial Stanford Prison Experiments (in which volunteers took on the roles either of prison guards or prisoners), which showed that a system of vertical power can quickly and dramatically distort ordinary people's moral compasses and change their behavior toward others.[50]

Metaphorically speaking, power is a psychosocial drug, analogous in some ways to cocaine or methamphetamine. Some people appear to be more innately inclined toward power addiction; for them, power can be irresistible and the loss of it intolerable. Some addicts get their fix merely by being in the presence of power, or by serving those who have it with obsequious enthusiasm. Others scrabble for any vestige of power they can achieve, whether wealth or position, and wear it as a badge of self-worth.

Our bodies and brains are shaped by rank and status. We use a lot of brain power in continually judging the place of every person in our immediate environment in terms of power hierarchies. Having a low self-perceived place on the totem pole can result in lower long-term levels of health and cognitive ability.[51]

Having higher social status can also be debilitating, though in different ways. Dacher Keltner, a psychology professor at UC Berkeley, found through lab and field experiments that subjects who enjoyed vertical social power acted as though they had suffered a traumatic brain injury: they were more impulsive, less aware of risks, and less adept at seeing things from other people's point of view. Sukhvinder Obhi, a neuroscientist at McMaster University who studies brains rather than behavior, found something similar. When Obhi viewed the brains of more-powerful and less-powerful individuals using a transcranial magnetic-stimulation machine, he found that power seemed to impair a neural process called "mirroring," which is foundational to the experience of empathy.[52] Keltner points to what he terms the "power paradox": a person with power eventually tends to lose some of the capacities that were needed to obtain that power in the first place.[53]

Biological evolution sensitized us to differences of status; cultural evolution resulted in an enormous widening of variation in status.

My tiny flock of four hens spend a great deal of their time establishing and reinforcing a pecking order. Chickens are social animals, and, like chimps, wolves, and other social creatures, they insist on having a clear hierarchy. Mimi, who is a big and beautiful hen, happens to occupy the lowest rung on our flock's ladder of social power. Any of the other three hens may peck her on the back of the head, sometimes to discourage her from eating the tastiest morsel available, other times just because. Mimi is not as happy or confident a chicken as Buffy, who's at the top of the ladder. But what Mimi has to put up with is minor compared to the indignity and suffering that humans heap upon one another in the name of status and power.

Humans in modern societies exist in multiple overlapping hierarchies, and we have to navigate our way through all of them more or less simultaneously. We tend to minimize (if we can) our participation in those hierarchies in which we feel disempowered, while we overvalue those hierarchies in which we excel.

SIDEBAR 15

Institutional Power

Institutions and organizations focus and magnify social power in order to solve problems and achieve collective goals. Typically, institutions (whether governmental, commercial, or religious) are partially or entirely organized vertically, so that top officers wield outsized authority backed by the resources of the organization as a whole. Kings, presidents, priests, CEOs, and even heads of some criminal organizations all wield institutional power, and help shape the institutions they control. Institutions can persist for generations and, because they unite the efforts of anywhere from a few people to millions, they possess a degree of power that no individual on her own could ever wield. Individuals who are invested with institutional power are therefore, in a sense, superhumans.

Institutions are systems. Acting essentially like big organisms, they have porous boundaries and maintain a kind of metabolism—with inputs, outputs, and feedbacks. They also create and maintain methods for making decisions, and implement those decisions with rules, incentives, and punishments.

Without institutions, we couldn't build complex modern societies. Institutions can do undeniable good: for example, nonprofit organizations help to end abuses of environment and human rights. Paul Hawken has estimated that there may be as many as two million such organizations active in the world today.[54]

However, institutions can focus collective power in ways that are destructive and hurtful. The possible examples are nearly endless—from the role of the Catholic Church in the genocide and enslavement of native peoples in the Americas, to that of the North Korean communist government in the famine that gripped that country in the 1990s.

The people who comprise institutions face a continual requirement to maintain mutual loyalty to their establishment. Motivations for loyalty can range from fear to expectation of reward to conviction that the institution is doing beneficial work. Some organizations, such as Green America (formerly Co-op America), pursue horizontal power within the institution as an explicit goal, and promote greater equality broadly. Others, such as the Ku Klux Klan, not only have impressive-sounding internal positions of vertical power (Imperial Wizard, Grand Dragon), but exist for the primary purpose of maintaining unequal social power in society at large.

A few social theorists have asked whether there should be a maximum size for organizations, or a maximum lifetime. Thomas Jefferson opined that governments should be periodically abolished and reconstituted (he wrote that "a little rebellion now and then is a good thing, and as necessary in the political world as storms in the physical"). Kirkpatrick Sale, in his influential 1980 book *Human Scale*, argued that all social units are best kept at a size that enables face-to-face interaction and individual control—the village, town, or bioregion.

Organizations, like people, tend to fight for survival and growth, and the pathologies of power are pitfalls for collective organisms, just as they are for human individuals. Therefore, according to Sale, not just governments, but corporations and other formal institutions as well should have limits with regard to size and longevity.

Every institution that's characterized by vertical power is populated with many members who merely wish to make their way in the world, but it will also likely contain a few who are pathological power addicts and power enablers. Unfortunately, such power structures often reward the power addicts, who quickly learn to make, use, or abuse rules to their advantage. Power addicts may wield their advantages to gain sexual favors or to impose upon those they find sexually desirable; or to profit from and take credit for the efforts of others. Organizations often eventually learn the corroding effects of power addiction and create mechanisms of accountability; however, as long as vertical power structures remain in place within organizations, the tension between power-seeking and its pathologies, on one hand, and means of power restraint, on the other, tends to persist or return.

Those who crave power often prop themselves up by putting others down. Racism and sexism are common reflexes of the pathological power-seeker. Those with more power rationalize unfair and even reprehensible behavior toward those with less by assuming or asserting that the powerless are lazy, corrupt, incapable, unintelligent, or otherwise undeserving.

SIDEBAR 16

Genocide as Ultimate Exclusionary Social Power

While the term *genocide* is of historically recent coinage, it refers to an ancient practice: the intentional effort to destroy a people—usually defined as an ethnic, national, racial, or religious group—in whole or in part. It is the most brutal form of exclusionary power: excluding a people not just from resources, but from the very opportunity to exist.

Historical instances of genocide (depending on the definition used) date back at least to the time of early state societies. For example, in the Hebrew Bible, God commands the Israelites "not to let a soul remain alive" among the indigenous inhabitants of Canaan (Deut. 20:16).

Many instances of genocide are more recent. Few modern residents of California are aware of the genocide that resulted in the Indigenous population of that state decreasing from perhaps as many as 150,000 in 1848 to 16,000 in 1900. This decline resulted not just from disease and starvation, but also from a series of horrific massacres.

The following is only a very partial list of other historically recent genocides:[55]

- The Congo under King Leopold of Belgium (1885 to 1908): 2 to 15 million deaths
- French conquest of Algeria (1830 to 1875): 500,000 to 1 million deaths
- British conquest of Australia (1789–1901): 500,000 to 700,000 deaths
- Turkish genocide of Armenians (1915): 1 million to 1.5 million deaths
- German genocide mainly of Jews, but also of Romani, disabled persons, and other groups (1933–1945): 11 million deaths
- Cambodia under Pol Pot (1975 to 1979): 1.5 to 3 million deaths

Anthropologists and psychologists who have studied genocide attribute the practice to extreme ingroup-outgroup bias. Experiments have shown that people readily form groups and are easily induced to be aggressive toward outsiders.[56] If the process snowballs, outsiders can be dehumanized, and aggressive behavior can, in the worst instances, become almost absurdly brutal. When mob psychology becomes rampant, even normally mild-mannered people can be swept up in a collective violent fervor that participants sometimes later describe as having been like a drug experience.[57] In some instances (as in Nazi Germany), this mentality can be continually retriggered and re-intensified not just for hours or days, but over a period of years.

Due to evolutionary and historical factors discussed in chapters 1 and 2, men (rather than women) have generally tended to be the designers and beneficiaries of vertical power systems, and men also tend disproportionately to exhibit pathologies of power. In addition, for centuries, Europeans—as colonizers, settlers, soldiers, businessmen, and missionaries—used technological and economic advantages to overpower others in much of the rest of the world; today their descendants frequently exhibit characteristics and behaviors typical of the over-empowered.[58] One of those characteristics is a sense of entitlement—an unrealistic and unmerited expectation of desirable living conditions and favorable treatment at the hands of others. Moreover, most of humanity has recently adopted an attitude of entitlement vis-à-vis the rest of the natural world as a result of its access to the power of fossil fuels (as we will see in Chapter 4).

There is a clear correlation between physical power and vertical social power. Physical power can derive simply from being larger, more muscular, or more aggressive; it can also arise as a result of the possession of weapons, or the ability to command others who are skilled in the use of violence. Access to energy sources—from animal or human muscle-power to the ability to harness electricity—is still another source of physical power, which is often mediated by money—the currency of power. Such physical power confers social power, which, as it grows, tends to become more vertical than horizontal.

Threats to power provoke predictable responses in individuals and whole societies, including anger and aggression, defenses and defensiveness, and projection—the attribution to others of one's own unacknowledged and undesirable characteristics, feelings, or behaviors.

Of course, the psychology of power has a flipside: the psychology of powerlessness. The powerless and the disempowered have an increased tendency to feel humiliation and shame, submission and subservience, and exhibit a hesitancy to act. However, feelings of powerlessness can also have paradoxical manifestations in narcissistic personality disorder, hostility, and aggression. All of these responses and tendencies can have intergenerational impacts.[59]

Whether in family, school, work, or politics, we're all immersed in the pathologies of power. If we're lucky, we learn to navigate these waters without being harmed irreparably, and without harming others. Many are not so fortunate.

To our credit, we have also evolved ways of managing status for the common good. In democracies, candidates compete for power and status by way of their various plans for the purported betterment of society as a whole (as we'll discuss in more detail in Chapter 6). But even in highly developed democracies, abuses of power are common.[60]

SIDEBAR 17
Key Writers on Power

Nicolo Machiavelli (1469–1527) is often regarded as the father of modern political science; today he would be called a geopolitical realist. In his book *The Prince*, he advised his elite readers on how to obtain, extend, and maintain political power. Machiavelli proposed that immoral behavior, such as the use of deceit and the murder of innocents, was normal and effective in politics. He recommended killing off entire rival aristocratic families if necessary, and argued that, in motivating the masses, it is more important to maintain their fear than their love.

Machiavelli's defenders describe him as an empiricist who was simply interested in what works. Writing in the mid-20th century, German philosopher Ernst Cassirer called Machiavelli the Galileo of politics, in that he distinguished between the "facts" of political life and the "values" of moral judgment.

Philosopher **Friedrich Nietzsche** (1844–1900) regarded the *will to power* as the main driving force in humans, though he never systematically defined the concept. Many of his interpreters regarded the will to power as simply a psychological principle, but there is some evidence to suggest that Nietzsche himself saw it more broadly as a force underlying all reality, analogous to Schopenhauer's *will to live*.

Some Nazi apologists saw Nietzsche's *will to power* as the motivation behind social Darwinism and used it to justify their quest for geographic expansion. Existentialist philosopher Martin Heidegger, himself a member of the Nazi party, criticized this reading, suggesting that physical and political power weren't what Nietzsche had in mind, pointing to this passage from Nietzsche's notebooks:

> I have found strength where one does not look for it: in simple, mild, and pleasant people, without the least desire to rule—and, conversely, the desire to rule has often appeared to me a sign of inward weakness: they fear their own slave soul and shroud it in a royal cloak (in the end, they still become the slaves of their followers, their fame, etc.).

Alfred Adler (1870–1937), founder of the second Viennese school of psychology and the school of individual psychology, was the first clinician to identify the *inferiority complex* as a component in personality development. Adler, influenced by Nietzsche, adopted the *will to power* as the underlying organizing principle of the human personality, in contrast to Sigmund Freud's *pleasure principle*. Adler hoped to help the individual patient overcome the superiority-inferiority dynamic and thereby become a fully self-actualized individual.

In this chapter I've not attempted to provide a detailed chronology of social power throughout history. That task was accomplished in sociologist Michael Mann's magisterial four-volume series, *The Sources of Social Power*, whose 2,300 pages trace the development of ideological, economic, military, and political power from the origin of states through the 20th century. My aims have been differently focused. First, my goal is to explore physical power as well as social power, and the relationship between the two. Second, our examination of social power in this chapter did not require a blow-by-blow account of the rise and fall of specific states and empires; instead, we have focused on the basic processes by which humans have obtained power over one another, including warfare, "predation," the wealth pump, state formation, the "domestication" process, the development of communication tools, and the invention of money and credit.

My only criticism of Mann's treatise, which I recommend to readers interested in the details of human power relations through history, is that, like most social science literature, it fails to elucidate the pivotal role of energy in human society, and thus fails to adequately explain or account for the extraordinary and perilous increase in power over nature that humans have acquired in the last 200 years. In the next chapter, we will zero in on precisely that subject.

POWER IN THE ANTHROPOCENE

The Wonderful World of Fossil Fuels

Civilization is the economy of power, and our power is coal.

— JUSTUS VON LIEBIG

Every basket is power and civilization. For coal is a portable climate. It carries the heat of the tropics to Labrador and the polar circle; and it is the means of transporting itself whithersoever it is wanted. Watt and Stephenson whispered in the ear of mankind their secret, that a half-ounce of coal will draw two tons a mile, and coal carries coal, by rail and by boat, to make Canada as warm as Calcutta, and with its comfort brings its industrial power.

— RALPH WALDO EMERSON

Power, speed, motion, standardization, mass production, quantification, experimentation, precision, uniformity, astronomical regularity, control, above all control—these became the passwords of modern society in the Western style.

— LEWIS MUMFORD

*Then the coal company came with the world's largest shovel
And they tortured the timber and stripped all the land
Well, they dug for their coal till the land was forsaken
Then they wrote it all down as the progress of man.*

— JOHN PRINE, "Paradise"

EVEN WITH THE ADVANTAGES OF TOOLS AND LANGUAGE, IT took *Homo sapiens* 300,000 years to grow its population to one billion (which it achieved around 1820). It took merely another 200 years (from 1820 to 2020) to reach almost eight billion. This last unprecedented growth spurt—let's call it the Great Acceleration—also saw the overwhelming majority of all the environmental damage that humans have ever caused.[1] Whereas other developments in social evolution took thousands of years to come to fruition, the Great Acceleration has occurred in a comparative eyeblink of time.

The consequences for the planet of the Great Acceleration have been so stark that, in the opinion of many scientists, we humans have initiated a new geological epoch—the *Anthropocene*. Geologists in the distant future (assuming there are any) who look back on rock strata that began forming since the start of the Great Acceleration will note a series of sudden, unmistakable shifts: signs of a substantial alteration of atmospheric and ocean chemistry, evidence of widespread extinctions of animals and plants, higher radioactivity (from atomic weapons tests in the 1950s, if not from nuclear war), and the presence of widely dispersed human artifact traces (notably plastics). The fact can no longer be disputed or ignored: humans are driving unmistakable changes to global natural systems. And those changes are speeding up.

In short, the past two centuries have seen the most rapid, wrenching cultural, ecological, and even geological reordering in the history of our species. And, if the current rate of acceleration continues, it will also constitute a turning point for all life on Earth.

There are two popular explanations for this whirlwind of transformation. One centers on science and technology—which together kicked off a self-reinforcing feedback of innovation, in which each new tool or discovery created conditions in which even more new tools and discoveries could emerge. The other explanation centers on capitalism, recounting the story of how a flood of investment capital and re-invested profits drove innovation upon innovation. Both explanations are partly right: without science, technology, and capital investment, the Great Acceleration would not have occurred. But both explanations are also largely wrong.

There is a third ingredient to the Great Acceleration that technologists and economists often miss or take for granted, and that is the energy derived from fossil fuels. If coal, oil, and natural gas had somehow not existed, we would be living very differently today; far fewer of us would be alive; and our impact on the biosphere, while likely significant, would be dramatically smaller than is currently the case. While I am often described as an advocate for reducing society's reliance on fossil fuels, I believe we must give credit where it's due, and I also believe that fossil fuels deserve most of the credit for the Great Acceleration.

Without energy we can do nothing; we are powerless. In early phases of human history, energy came from food, animal muscle, firewood, wind, and water. But there were limits to how much energy we could derive from these sources, and by the mid-19th century we were pressing against those limits.

Fossil fuels changed everything. They presented us with sources of power that were concentrated, storable, and portable. A single barrel of oil contains chemically stored energy that can do the same work as roughly 25,000 hours of human muscle-powered labor. The ability to obtain the equivalent of several years of labor-energy for $55 (the averaged 2019 per-barrel oil price) largely explains the widespread effort, during the past century or two, to mechanize nearly every productive process. Further, for many decades, finding and extracting fossil fuels required comparatively little energy and effort; the energy returned on the energy invested in oil exploration and production was often a spectacular 50:1 or higher (we'll return to the subject of energy profitability shortly). With so much energy available so cheaply, the limits to what we could do seemed to flee before us.

The end of slavery, the triumph of modern democracy, motorized transportation, the industrial food system, the growth economy, urbanization, consumerism, and a dramatic expansion of the middle class all resulted from the application of ever more fossil energy to resource extraction, manufacturing, and distribution. Population grew far faster than in any previous phase of human history, due to the greater availability of cheaper food, and to better public health enabled by pharmaceuticals and sanitation chemicals often made with fossil fuels. At the same time, warfare became more mechanized and deadlier. A proverbial genie was released from its bottle, and seemingly nothing could put it back.

In this chapter we will explore the extraordinary powers humanity has derived from fossil fuels, and how those powers have changed us and our circumstances. But before we do, we must set the stage with a short discussion of how humans obtained and used energy before the Great Acceleration, so that we can appreciate just how much of a difference fossil fuels have made.

It's All Energy

As we saw in Chapter 1, every organism lives by harnessing a natural flow of energy. Also, recall that physicists define power as the rate of doing work or transferring energy. Without energy, there can be no power.

Ironically, obtaining energy requires energy. We have to expend energy to hunt prey, to grow and harvest crops, to cut firewood, or to drill an oil well. The biological success of any species, and of any individual within a species, depends on its ability to obtain more energy than it expends in its energy-gathering efforts. The relationship between energy obtained and the energy expended in an energy-harvesting activity can be expressed as a ratio—Energy Returned on Energy Invested, or EROEI, also known as the energy profit ratio. I've already noted that the energy profit ratio for oil exploration and extraction historically was 50:1 or higher. Anthropologists and physicists have calculated the EROEI figures for a wide range of human and animal energy-obtaining activities. Unfortunately, while energy profitability is literally a life-or-death issue, and should play a role in public discussions about our energy and food choices, the subject is complex and it is easy to skew numbers one way or the other either by ignoring an important energy input or by overestimating it (output energy, whether barrels of oil or the caloric content of a certain number of bushels of wheat, is usually easier to measure directly). Nevertheless, despite the inherent uncertainties in EROEI figures, I'll reference a few of these below so that the reader can make rough mental comparisons.

For nonhuman animals, energy inputs other than ambient heat are nearly all in the form of food. Of course, humans likewise derive energy from food, but we have also found ways of obtaining and using non-food energy (recall the discussion in Chapter 2 of "The Fire Ape"). The expansion of humanity's powers has thus entailed two fundamental strategies: intensifying food production, and increasing our harvesting and harnessing of non-food energy. Let's consider pre-fossil-fuel food intensification first.

Hunting and gathering was often an easy way of life with good energy profitability. While EROEI estimates for foraging range widely due to the varied nature of the environments people inhabited and food sources they accessed, Vaclav Smil estimates that for gathering roots, "as many as 30–40 units of food energy were acquired for every unit expended." For other plant gathering, he figures that typical returns were 10:1 to 20:1. Hunting large animals again yielded a ten-to-twenty-fold energy profit, while hunting small animals was often hardly worth the trouble from an energy standpoint.[2] As a rough confirmation of these estimates, a study in the 1960s by anthropologist Richard Lee showed a 10:1 energy profitability for the overall food procurement efforts of the !Kung bushmen of the Kalahari.[3]

The growing of crops in early agricultural societies also yielded an energy profit, which was again highly variable. In some famine years, not enough grain could be produced even to provide seed for next year's planting, so no net food energy yield resulted. However, in good years surplus food could be set aside. Once all energy inputs were accounted for, on average the EROEI of traditional agriculture was likely significantly less than 10:1, and possibly lower than 3:1.[4] As we saw in Chapter 3, agriculture wasn't adopted in order to make life easier, as it would have been if it had a high energy profit ratio; rather, it was taken up largely out of necessity in order to support a bigger population.

The route to further food production intensification branched into five interweaving pathways: working harder, using animal labor to supplement human labor, irrigation, fertilization, and growing a greater variety of crops. Let's explore each of these briefly.

Hard work was inherent in agriculture itself: plowing, harrowing, planting, hoeing (i.e., weeding), harvesting, threshing, winnowing, hauling, and managing stored grain were seasonal activities that required long days of tiring effort. Intensification of food production typically meant cultivating more land, or adding new tasks, such as managing draft animals, to the already long list of farming chores.

Animal labor could supplement human labor for some of the most physically demanding farming activities, such as plowing and hauling. Species and breeds of animals varied in terms of how much power they could apply, how long they could work, and what kinds and quantities of food they ate. For example, a horse could provide up to twice as much useful power as an ox, but the ox could graze and eat straw, while the horse required grain that had to be grown and harvested. Human labor costs and the requirement to grow additional food to feed draft animals (in late-19th-century America, a quarter of all agricultural land grew food for horses) reduced their net advantage, but, if managed well, draft animals more than paid for themselves in energy terms.

Irrigation increased yields, sometimes dramatically, in arid areas. However, lifting water from even the shallowest wells entailed work. Human-powered irrigation demanded especially tedious and tiring effort. Animal muscles were useful for this purpose, and were employed for moving water beginning in the earliest civilizations (though not in the Americas). For example, the ancient Greeks moved water by means of a system of clay pots on loops of ropes, powered by a blindfolded animal walking in a circle.[5] Some places (e.g., China and Central America) had the problem of too much water, as the result of seasonal flooding. Growing crops under these circumstances required drainage—which again entailed energy inputs. But water-moving efforts could pay off: irrigation and drainage could return up to 30 times their food-energy cost through increased agricultural yields.[6]

Fertilization was needed due to the depletion of nitrogen, potassium, phosphorus, and trace minerals in soils that were continually cultivated. Nitrogen was the element most quickly lost, and since it determines grain size and protein content, most fertilization systems focused on it. Every imaginable sort of organic matter was used (food waste, straw, stalks, husks, chaff, and leaves were collected and composted), but manures were the primary traditional nitrogen fertilizers. In societies ranging from ancient China to 19th-century France, efforts were made to divert and compost human urine and excrement for this purpose—as well as animal wastes, which tended to be more plentiful. Because nitrogen in manures and composts was easily lost to the atmosphere, large quantities of traditional fertilizers had to be applied in order to achieve significant benefits. Thus, significant amounts of farm and urban labor had to be devoted to the unpleasant activities of collecting, hauling, and applying human and animal wastes. It was only the pivotal importance of nitrogen, and the looming potential loss of productivity without it, that could have motivated such efforts.

The employment of greater crop variety provided another, and often more effective, solution to the nitrogen dilemma. The ancient Romans learned that rotating grain crops with legumes like lentils or vetch replenishes the soil. Native Americans similarly planted the "three sisters" (maize, beans, and squash) together, having long ago discovered the synergistic benefits of what modern organic gardeners call "companion planting." However, it was only after 1750 that European farmers began systematic legume-grain crop rotation. The payoffs in terms of increased yields were dramatic, doubling and even tripling productivity and thus adding to Europe's general affluence.

Pre-industrial sources of non-food energy included human muscle, draft animals, firewood (and other biomass), wind, and water power. Each of these was slowly adapted through better organization (finding ways to link many small power sources together, such as by harnessing teams of horses); and through technical innovations that focused power or made energy transfer processes more efficient (such as the invention and improvement of harnesses, gears, pulleys, wheels and axles, levers, and inclined planes). As time passed, these early power sources and machines were applied to an increasing variety of tasks.

Draft animals were useful not just for agriculture, but for a range of other activities as well, including hauling, human transport, moving water, and providing motive power for mills. As already mentioned, opportunities for the employment of animal power had to be balanced against the cost of managing the animals and providing them with food and living space. By 1900, Britain's horse population stood at 4.5 million, and the nation was forced to import grain to feed so many animals. Also, effort was needed in collecting animal feces, not only to benefit agriculture, but also to manage pollution: at the turn of the 20th century, the clogging of the streets of cities like New York, London, and Paris with organic wastes from horses was considered one of the top urban problems.

From ancient times, wood was valued both as fuel and building material. It could be fashioned into tools—from scythe handles to musical instruments—as well as houses and ships. As a fuel, it had the advantage of relatively high energy density (a kilogram of seasoned firewood contains about 15 megajoules of energy; if translated directly to electrical energy without conversion losses, that would be enough to run a large television most of the day). Wood was also widely available—at least in some countries. And its heat, when it was burned, could be used for warming living spaces, heating water, cooking, and even for making glass.

Wood could be partially combusted in an oxygen-poor chamber to produce charcoal, which burned much hotter. This was key for developments in metallurgy, leading to the use of more iron and steel, along with copper, lead, tin, silver, and gold. Superior iron and steel in turn drove improvements in armor, weapons, cooking implements, horseshoes, nails, plowshares, gears, pulleys, and other tools. The downside of charcoal was that a great deal of wood had to be used to produce a modest amount: under average production conditions, a ton of wood yields a quarter-ton of charcoal. In late medieval Europe, charcoal production was a major industry employing hundreds of thousands, and was a significant cause of deforestation, especially in and around the ancient forests near the center of the continent.

The very advantages of wood led to its overuse. Between the years 400 and 1600, Europe's forest cover shrank from 95 percent of land to only 20 percent, and by the end of that period wood was so hard to come by that blast furnaces could only operate every third year. As wood became scarce, not only did its price rise, but its effective energy value declined: traveling by horse cart for many miles to obtain a load of wood might expend as much energy as the wood contained. Today, this tradeoff still limits wood-fired electricity power-generation plants: if they're not located close to the forest from which they derive their fuel, transport energy costs reduce the operation's net energy yield—and over-harvesting can cause the forest boundary to gradually retreat beyond the range of overall energy profitability.[7]

Other kinds of biomass were also burned for energy, including straw, dung, and peat for heat; and oil, wax, and tallow for illumination. By the early 19th century, parts of Europe faced fuel scarcity; however, eastern North America, endowed with enormous forests, avoided that problem: by the time deforestation was becoming an issue there in the late 19th century, fossil fuels (which the continent also possessed in abundance) were already coming into use.

The use of wind power to move ships via sails started at least 8,000 years ago in Mesopotamia. Sails and rigging went through a long process of refinement for greater efficiency and control, leading eventually to the tall and elegant five-mast clipper ships of the late 19th century. Due to the unpredictability of winds, sailing has always entailed skill and luck, and even the most sophisticated sailing ships can be stalled at sea for days or even weeks at a time due to lack of breeze.

On land, windmills began to be used for grinding grain and pumping water 1500 years ago, with widespread adoption of this technology occurring in the Middle East, China, Europe, India, and Central Asia. In North America, several million windmills dotted 19th-century farms and towns, where they were used mostly for pumping water. In England, Germany, and the Netherlands, during the same period, tens of thousands of windmills provided up to ten horsepower each when working at full capacity, powering mills and pumping water.

Water power was used for grinding grain from the time of ancient Greece, and was also adopted for this purpose in China and Japan. The Romans contributed gears, which enabled millstones to be placed vertically and to turn up to five times faster than the propelling water wheel. Water was used as motive power in early textile factories both in Britain and in the American Northeast, where rivers provided both a means of running looms and a way of moving raw materials and finished goods.

While watermills and windmills were at first primarily employed for grinding grain, they were gradually adapted, via gears and belts, to the operation of a variety of tools such as saws and looms, and to the accomplishment of an array of tasks that included making paper and crushing ores. The economic transformation brought by these powered machines, starting in the late 17th century, is sometimes called the First Industrial Revolution. While its impacts were minor in comparison with those of the fossil-fueled industrialization process that would follow, the latter depended in large measure on the technical innovations that had previously harnessed the power of water and wind.

By the early 19th century, the limits of pre-fossil-fuel industrialism had not been reached, but were within view. Forests in Europe and North America were shrinking or disappearing. There were only so many rivers and streams that could provide power, and available materials and technology limited the height and efficiency of wind-mills, thus constraining the amount of power they could harness. Food production had expanded, but limits to soil fertility were confounding agricultural scientists (a temporary reprieve appeared with the discovery of huge deposits of guano—nitrogen-rich excreta of sea birds—on islands off the coasts of Chile and Peru, but these were soon mined and exhausted).

Through the centuries, the shape, size, composition, and operations of societies had adapted themselves to slowly diversifying energy sources. As world population arrived at one billion around the year 1820, there seemed little possibility that this number could be doubled. In most nations, 75 to 90 percent of people worked at farming, thus limiting the size of the manufacturing and managerial workforce. Peking, China, the world's most populous city at the time, held 1.1 million people, while London's population was roughly a million, Paris held 600,000, and New York encompassed just 100,000. Within these relatively advanced urban centers, most people lived in poverty. A tiny fraction of society enjoyed great wealth, and was able to fund the production of art, music, and literature of the highest quality, some of which could be enjoyed by the rest of society in public galleries and concerts. But for even the richest person, the pace of travel was limited to the speed of sailing ships and horses. Energy use per capita then stood at about 20 gigajoules per year—perhaps twelve times the energy captured and used (via food) by a theoretical pre-human primate of similar body weight, and roughly five times what could be captured and used by a hunter-gatherer using fire but having no agriculture or domesticated animals.[8]

In short, the energy regime of the pre-fossil-fuel world was renewable, but it was not necessarily sustainable. Europe had already become crowded with humans, whose demands for food- and non-food energy were imperiling ecosystems. For North America, a similar quandary lay within sight, as Native peoples were herded onto reservations so their land could be divided, deforested, farmed, mined, and built upon.

The Coal Train

When the ancient Romans invaded Britain and began to survey their new colony, they found an unfamiliar velvety black stone that could easily be carved into attractive jewelry. It also had the odd characteristic of being flammable. There is no record from that time of coal being burned as a heat source by the Britons or Romans; however, it was used for that purpose in China at roughly the same period. Around the year 1300, Marco Polo conveyed to his Venetian readers some of the oddities of the exotic Chinese civilization, among which was "...A sort of black stone, which they dig out of the mountain, where it runs in veins. When lighted, it burns like charcoal, and retains fire much better than wood." The Chinese were using coal to heat water for their numerous stoves and baths; and while the country also had an abundance of wood, that fuel "...Must soon prove inadequate to such consumption; whereas these stones may be had in abundance, and at a cheap rate."[9] (See sidebar 18, "An Arrested Industrial Revolution in China.")

However, let's return our attention to Britain, because that's where much of the subsequent story of coal would unfold in all its dark glory. By the 13th century, forests were retreating quickly from British towns as trees were cut for firewood, for construction materials, and for shipbuilding. "Sea coal" (so called because it was often found on beaches) began to be used as fuel instead of wood. But it was considered inferior to wood because it gave off foul, sulfurous smoke when burned, and there were only so many places where it could be found.

Coal was curious stuff. Early users even thought of it as living, assuming it would grow like a plant if given water and time. As 20th-century science would show, coal started forming toward the end of the Carboniferous period, roughly 320 million years ago. Phases of formation were intermittent, lasting until as recently as 50 million years ago. Climate change probably played a key role in the process: when glaciers melted, waters rose and buried fallen trees and giant horsetail plants, which, instead of decomposing, decayed only partly, leaving behind black carbon. Miners and geologists often found fossils in early coal mines, ranging from tree-like ferns to footprints of cow-sized amphibians.[10] These added to the lore and lure of the fuel, conjuring visions of an ancient world of giants, along with speculation about how this lost world might fit into the Bible's undisputed historical narrative.

Britain grew to be the epicenter of the coal revolution for two main reasons. First, coal was, as we have seen, adopted early on out of necessity, as domestic forests were being decimated, and as people needed an alternative to wood. Second, it came into wide use in Britain of all places by lucky accident, since the nation was endowed with significant coal resources, including coal that was close to the surface and that could easily be dug out with simple tools. When Britons started using coal for heat, in the early medieval period, their island was a cultural backwater in comparison with China, India, the Middle East, and some other areas of Europe. As the British gradually adapted themselves to using coal, their nation became not only a global economic power, but also a center for the development of science and industry.

Increasing demand for coal drove technological developments that in turn drove still more coal demand, in a self-reinforcing feedback loop. When surface coal resources were exhausted, coal mines were opened, and these gradually deepened. But as miners neared the water table, mines tended to fill with water, requiring bucket brigades to empty them temporarily so that mining could proceed. A pump was needed that could move large amounts of water quickly. In the 17th century, Denis Papin had experimented with a toy model of a steam engine, and Thomas Savery later built a small and extremely inefficient steam-driven pump. By 1712, Thomas Newcomen had introduced a 3.5 kilowatt engine to pump water from mines; however, less than one percent of the heat trapped in the coal that fired the engine was converted to work. In 1769, James Watt gained a patent for a new steam engine design that raised efficiency dramatically. Now the steam engine was a practical device that could be used not only to pump water from coal mines, but to provide motive power for a wide range of other projects.

Transporting coal presented another challenge. Mine operators experimented with systems of rails to hold horse-drawn wagons on tracks as they hauled coal from mines to canals or ports, where it could be loaded on barges or ships for the journey to London and other industrial centers. George Stephenson had the idea of combining a steam engine with rails, thus creating the first railroad. Now an engine running on coal could also haul coal across the country. Soon coal-burning steamships were plying the seas, and factories were employing coal to make steel, glass, and a burgeoning array of other materials and products. By the mid-19th century, Britain was burning over half the world's annual coal budget to power an industrial economy whose sourcing of raw materials and distribution of products stretched around the globe.

With the discovery of coal-tar dyes in 1854, coal seeded the beginnings of the chemical industry. Not only were nearly all early synthetic dyes derived from coal tar, but chemical compounds like carbolic acid, TNT, and saccharin as well. And new chemical corporations like IG Farben grew rich providing them.

Coal could be gasified by heating it in enclosed ovens with an oxygen-poor atmosphere. The result was a mixture of hydrogen, methane, carbon monoxide, and ethylene, which could be burned for heat and light. Called manufactured gas or town gas, the product was fed into a network of pipes to provide street lighting. The first gas utility, the London-based Gas Light and Coke Company, incorporated by royal charter in April 1812, would set the pattern for private utility monopolies that still dominate the distribution of gas and electricity in many nations.

Ironically, coal was now lighting cities at night while also cloaking them in smoky gloom for weeks or months at a time. The following passage from *The Smoke of Great Cities* by David Stradling and Peter Thorsheim conveys the atmosphere of coal towns:

One visitor to Pittsburgh during a temperature inversion in 1868 described the city as "hell with the lid taken off," as he peered through a heavy, shifting blanket of smoke that hid everything but the bare flames of the coke furnaces that surrounded the town. During autumn and winter this smoke often mixed with fog to form an oily vapor, first called smog in the frequently afflicted London. In addition to darkening city skies, smoky chimneys deposited a fine layer of soot and sulfuric acid on every surface. "After a few days of dense fogs," one Londoner observed in 1894, "the leaves and blossoms of some plants fall off, the blossoms of others are crimped, [and] others turn black." In addition to harming flowers, trees, and food crops, air pollution disfigured and eroded stone and iron monuments, buildings, and bridges. Of greatest concern to many contemporaries, however, was the effect that smoke had on human health. Respiratory diseases, especially tuberculosis, bronchitis, pneumonia, and asthma, were serious public health problems in late-nineteenth-century Britain and the United States.[11]

Coal mining, especially in its early days, was grimly unhealthy. Many miners succumbed to accidents resulting from asphyxiation by accumulated gas, as well as from explosions, fires, and roof collapses, and most eventually suffered from respiratory ailments, including pneumoconiosis, or black lung disease. Meanwhile, mining often polluted water and air, and degraded forests, streams, and farmland.

SIDEBAR 18
An Arrested Industrial Revolution in China

The Great Acceleration almost ignited 1,000 years earlier than it finally did—and not in Europe, but in China. The European Industrial Revolution rested on three pillars: the increasing use of coal as a fuel, technological innovation, and capital investment (which in turn depended on government support for private property rights).[12] These three factors were present in China as well.

At the start of the first millennium of the current era, the Chinese were using coal not just as a source of heat in buildings, but in metallurgy as well. Coal output had begun to surge, from an estimated 13,500 tons in the year 806 to 125,000 tons by 1078. Iron and steel output grew in tandem, yielding more and better farm implements, weapons, coins, and other tools.

During this period, inventors in China were heaping innovation upon innovation, including movable type printing (1000); the blast furnace (1050); mechanical water clocks (1090); paddlewheel ships (1130); the magnetic compass (1150); water-powered textile machinery (1200); and, most dramatically, huge oceangoing junks with watertight bulkheads, a carrying capacity of 200 to 600 tons, and a crew of about 1,000 (1200).[13]

During the Song Dynasty (960–1297), the government gave peasant farmers the right to buy and sell land. Land privatization resulted in the consolidation of smaller farms and the appearance of commercial agriculture. At the same time, the government began to systematically encourage trade. China's economy became highly monetized as a result. Paper money proliferated, population levels expanded, the economy grew, and so did cities.

By the 1260s, China had achieved a level of technological sophistication and economic development that would not be seen in Europe until the early 19th century. But, rather than sweeping ahead, Chinese industrialization stalled, though the country remained ahead of Europe in technological and economic terms until about 1800.

Why did China's industrial juggernaut sputter? Chinese rulers cut off support for their growing iron and steel industries because they feared the emergence of a class of industrialists that might ally itself with treasonous military leaders. In short, the government saw industrialization as more of a threat than a source of increased power.[14]

America at first lagged behind Britain in coal usage. But the United States was cutting and burning its eastern hardwood forests at a furious pace, and it happened to have abundant underground coal supplies—even larger ones than Britain, as it turned out. By 1885, coal was America's dominant energy source, and, by the early 20th century, the US was the world's top coal producer and consumer.

The widespread use of coal in industry quickly swept aside or transformed not just existing technologies, but political and social relations that had been cemented in place centuries earlier. Surely the most salutary societal consequence of the adoption of coal in industry was the ending of slavery. As Charles Babbage, the inventor of the first computer, noted in his 1832 book *On Economy and the Machine and Manufacturers*, mechanical slaves were already surpassing human slaves in power and speed. For the first time in history, fossil "energy slaves" could supplant the forced labor of millions of human beings. Slavery was a brutal program of enforced social power relations, but its purpose was to organize and efficiently apply energy in order to produce wealth. And the simple fact was that coal-fed machines could produce and apply energy more effectively in a growing number of instances than human or animal muscles could, and thereby yield more wealth. As energy historian Earl Cook wrote in his 1976 book *Man, Energy, Society,* "The North defeated the South by a campaign of attrition supported by coal mines, steel mills, and railroads."[15]

According to historian Lewis Mumford, wage labor first appeared on a mass scale in coal mines.[16] Moreover, the first labor unions were created by coal miners, who were also responsible for the earliest and largest industrial strikes, beginning in the early 19th century. Increasingly, energy was emerging from small areas within nations (coal mines) and flowing out through narrow, humanly constructed channels (canals and railways) to operate powerful productive machinery. Unlike the previous agricultural economy, this new coal-powered industrial system employed specialized workers at key nodes along energy's paths of power, and these workers were frequently abused, underpaid, and subjected to harsh, dangerous, and unhealthy conditions. The coal economy thus proved to be the perfect breeding ground for a new kind of political power. As Timothy Mitchell writes in *Carbon Democracy*, "The power derived not just from the organisations [that miners, steel workers, and railway workers] formed, the ideas they began to share or the political alliances they built, but from the extraordinary quantities of carbon energy that could be used to assemble political agency, by employing the ability to slow, disrupt, or cut off its supply."[17] The word *sabotage* was coined to refer to mass work stoppages meant to call attention to intolerable working conditions in coal-mining and related industries. Mitchell goes on:

> Modern mass politics was made possible by the development of ways of living that used energy on a new scale. The exploitation of coal provided a thermodynamic force whose supply in the nineteenth century began to increase exponentially. Democracy is sometimes described as a consequence of this change, emerging as the rapid growth of industrial life destroyed older forms of authority and power. The ability to make democratic political claims, however, was not just a by-product of the rise of coal. People forged successful political demands by acquiring a power of action from within the new energy system. They assembled themselves into a new political machine using its processes of operation.[18]

The period from the 1870s to the First World War has been called both the age of democratization and the age of empire, and both trade union-propelled democracy and steamboat colonialism were shaped by coal. Britain and the European powers had been maintaining overseas colonies since the time of Columbus, but heightened industrial throughput and faster, more reliable seaborne transport opened the opportunity for new ways of exploiting hinterlands. Many raw materials, such as cotton, still depended on land and labor for their production; increasingly, whole nations were repurposed—via conquest, land seizure and privatization, and debt—to supply those materials for the mills of industrial nations.

But if coal triggered a social and economic revolution, it also marked a turning point in another sense: whereas previously the world's energy regime was renewable though not necessarily sustainable, now it was neither. Every increment of coal extraction was an increment of depletion of a finite store of fuel. Even though coal resources initially appeared vast, dramatically increasing consumption levels would turn what were at first estimated to be thousands of years' worth of coal into what will probably finally amount to about 300 years of supply, once production levels from the last mines dwindle to insignificance sometime in the next few decades. Britain's coal industry is already extinct due to the depletion of its resources. The once-abundant coal of America's northeast is mostly gone. And while China is now by far the world's top coal producer and consumer, that nation's extraction of coal is already declining as miners are having to dig deeper and expend more energy to access the fuel. In country after country, coal production is decreasing not just due to government policies to protect the climate, but also because of rising mining costs and declining resource quality.[19] There will always be enormous amounts of residual coal left in the ground, but beyond a certain point the amount of energy that will be required to recover the curious black stone will exceed the energy it can provide when burned. With its reliance on coal, humanity began building an ever-expanding castle on an eroding sand bar.

Still, as enormous and unenduring as was the coal-fired societal revolution of the 19th century, an even greater transformation lay in store in the 20th century, due to yet another energy transition.

Oil, Cars, Airplanes,
and the New Middle Class

Oil saw limited usage prior to the Industrial Revolution. The Babylonians caulked their ships with bitumen (the thickest of oils), and also used it to waterproof baths and pottery, and as an adhesive to secure weapon handles. Egyptians used it for embalming, and the Bible refers to bitumen being used as a coating for Moses's basket and Noah's Ark. The ancient Chinese used bamboo pipes to carry oil and natural gas into homes of the wealthy for heat and light. And ancient Persians and Native Americans used crude oil for purported medicinal benefits.

Petroleum ("rock oil"), bitumen, and natural gas were formed differently from coal during two long, intense periods of global warming, roughly 150 and 90 million years ago. Contrary to the popular notion that oil consists of dinosaur remains, in fact it started mostly as algae, which were buried so quickly as to halt decomposition. Slowly, heat and the pressure of sedimentation turned carbohydrates into hydrocarbons. Whereas coal is mostly carbon, petroleum molecules consist of chains of hydrogen and carbon atoms. These chains are of varying lengths—from methane, the simplest hydrocarbon molecule (which is the main constituent of natural gas), to ethane, propane, butane, pentane, hexane, heptane, octane (the main constituent of gasoline), and even larger molecules. Oil refineries and petrochemical plants use heat to separate these constituents and recombine them to form kerosene, gasoline, diesel fuel, lubricating oils, and the precursors of various plastics, among other chemicals.

The first modern, commercial use of oil was in the form of kerosene burned in lamps for illumination. This was a welcome replacement for whale oil, which was being used for the same purpose, and which was becoming scarce and expensive as a result of the decimation of whale populations worldwide by the whaling industry (as memorialized in Herman Melville's masterpiece, *Moby Dick*). The first commercial oil wells were opened in Pennsylvania around 1860; soon, refining and distribution systems also appeared. Customers typically bought small amounts of kerosene and machine lubricating oil in refillable metal cans at the local general store.

As electric lighting became common in cities around the world, the petroleum industry faced a crisis of declining demand. The industry's salvation came with the invention of the internal combustion engine and the automobile, for which gasoline proved to be an affordable and effective fuel. An explosive byproduct in making kerosene, gasoline was previously discarded as waste. But now, as thousands and soon millions of automobiles began to putter through city streets across the world, oil demand soared.

The automobile offered the power of a large team of horses, but without the requirement for land to grow food for them. As cars proliferated, horses disappeared; and as they did, less agricultural output had to be devoted to feeding draft animals, and more land could be used to grow food for people.

The romance and lure of the automobile were undeniable. It offered easy mobility, which effectively meant increasing power over space and time. Though early cars broke down frequently, their reliability gradually improved. Meanwhile, their cost declined significantly after Henry Ford developed the modern assembly line (the 1903 Ford Model A cost $800, but by 1925 a superior 20-horsepower Model T could be had for $260). While initially a toy for the wealthy, the automobile quickly became—in wealthy countries, at least—an affordable and common transportation appliance.

Cities were redesigned around the needs of the automobile. Whereas in the early 20th century most American cities featured streetcar systems and interurban railways, after World War II most of these were dismantled (partly as a result of purchases by shell companies set up by General Motors, Firestone tire company, and Standard Oil of California). While Europe and Japan kept and improved their train systems, the US concentrated on building highways. As new housing and shopping areas were designed for the convenience of drivers, cities became ever less walkable, thereby making car ownership a near-necessity. Even house designs were affected, as increasing amounts of indoor space were devoted to covering and protecting automobiles.

To gain an impression of the enormous social and economic changes wrought by the automobile, watch (on YouTube) the 1906 pre-earthquake movie taken from the front of a San Francisco streetcar as it made its way along Market Street toward the Embarcadero. Modes of transport (walkers, trolleys, automobiles, and horse carts) mix and jostle at a leisurely pace. In contrast, today's Market Street is managed to maximize order, efficiency, and safety, with pedestrians allowed on the street only for brief intervals signaled by prominent, colored lights. We assume that automobiles have priority, and stay out of their way if we value our lives.

Diesel engines (which, of course, burn diesel fuel, a heavier blend than gasoline because it contains larger molecules) produce more torque and generally offer better fuel economy than gasoline engines, and were adopted for large trucks, tractors, and most locomotives— except where rail systems were electrified. Ships generally burned an even heavier oil, one that is yielded by refineries after lighter hydrocarbons, including gasoline and diesel fuel, have been removed. The use of oil in long-distance shipping led to the movement of ever-increasing quantities of raw materials and finished goods. Rapid freight movement by rail led to the proliferation of catalog shopping in the early 20th century, while the internet, container shipping, air freight, and diesel trucking support online retail shopping today.

Kerosene, once used primarily for lighting, found a new purpose after the Second World War as fuel for jet engines. In 2019, US airlines alone used over 17 billion gallons of jet fuel to carry nearly a billion passengers and 43 billion pounds of air freight.[20]

The speed, reliability, and affordability of modern transport modes have been chief enablers of economic expansion. But other oil-fueled activities played essential roles, including fueled resource extraction (via powered shovels, drills, and other mining equipment), powered industrial processes (blast furnaces, cement kilns, foundries, mills, etc.), and powered, and increasingly automated, assembly lines. Altogether, the material throughput of civilization has grown from about seven billion metric tons per year in 1900 to approximately 90 billion metric tons in 2018.[21]

By about 1930, as oil became the world's primary energy source, the rapid increase in industrial throughput was leading to a crisis of overproduction: more goods were available than could be absorbed by buyers (the Great Depression was, in part, a result). Leaders of industry hit upon three general solutions: a dramatic expansion of advertising, planned obsolescence (deliberately making products that would quickly wear out or become aesthetically outmoded), and the deployment of consumer credit at an unprecedented scale so that people could consume now and pay later. "The economy" was now, for the first time, regarded as an independent and measurable entity whose maintenance was the business of government and industry. Following World War II, the US federal government, working in tandem with industry, introduced the term "consumer" as an alternative to the more customary terms "citizen" or "person." Together, manufacturers and government regulators were engineering a new system of societal management, called "consumerism," whose purpose was to maintain ever-rising levels of commercial activity and employment.[22]

Consumerism was both necessitated and enabled by the growing middle class. The industrialization of agriculture—with tractors, powered mechanical seeders, harvesters, combines, and trucks—meant that fewer people were needed to work at producing food in order to feed the overall populace. As economic opportunity migrated from farms to cities, people followed. The result was a steady, continuing trend toward urbanization. By 2009, for the first time in history, the world's urban population exceeded its rural numbers.[23] Urbanization and mechanization in turn led to the emergence of a bewildering plethora of jobs and occupations, from retail salesperson to office clerk, lawyer, janitor, computer programmer, or registered nurse. Today, the website CareerPlanner.com lists 12,000 different occupations, but does not claim to be comprehensive or exhaustive.[24]

Whereas agricultural life favored a division of labor between women and men, the overwhelming majority of urban factory and office work could be done equally well by people of any gender. Thus fossil-fueled industrialism also contributed to the liberation of women—though decades of ongoing political activism would be required to gain legal guarantees of equal opportunity and equal pay.

Meanwhile, in the face of rapid continual growth of production and consumption (as well as population), political and commercial leaders came to assume that the growth of the economy (measured in terms of the amount of money changing hands, but actually representing increasing material and energy throughput, as well as population growth) was perfectly normal, indeed something to be maintained by all possible means. By the mid-20th century, government revenues, returns to investors, and stable employment were all understood to depend on economic growth, and all politicians promised more of it.

All of this happened first in the world's core industrial nations (Western Europe including Britain, the US and Canada, Australia, and Japan; the Soviet bloc industrialized rapidly, but never caught up, pursuing a path separate from the capitalist system and relying to a far greater degree on centralized planning). Beyond these core industrial countries lay nations whose raw materials and cheap labor would be the foundation upon which the *global* economy would be built, starting in the mid-20th century. The project of managing the aspirations of people in these nations would acquire the name *development*. It was a project that provided a new rationale for colonialism, based less on the improvement of local economic conditions (its ostensible justification) than on the "development" (i.e., exploitation) of resources, a process managed by bankers eager to saddle impoverished regimes with mountains of debt.[25]

Also in the mid-20th century, in addition to coal and oil, a third fuel and major driver of industrialization entered the scene. Natural gas—initially a byproduct of oil production, but increasingly a target of specialized drilling and production efforts—had not only replaced town gas, but had become essential for home heating and as a cooking fuel (though some homes continued to heat or cook with coal until the 1960s, or with fuel oil up to the present). It also served as a heat source for industrial processes such as making glass, cement, bricks, and paper. Natural gas would also be used as a basis for the production of an enormous array of chemicals and plastics, and in many nations would largely replace coal for electrical power generation (which we will discuss shortly).

However, perhaps the most crucial use of natural gas, from the perspective of human power over nature, would be for the making of ammonia-based nitrogen fertilizers. Nitrogen had been the limiting factor for agricultural production prior to the advent of fossil fuels; now that limit could be pushed back ever further. As enormous quantities of artificial fertilizer began to be applied, crop yields soared. New hydrocarbon-based pesticides, and the development of high-yield crop varieties, along with nitrogen fertilizers, together comprised what came to be known as the "Green Revolution," which, starting in the 1950s, more than tripled world agricultural output. As a result, hundreds of millions, if not billions, of human beings were saved from starvation. Together with better sanitation and medicines, the Green Revolution led to the most rapid sustained population growth in history.

The coal age had produced enormous fortunes for owners of mines, steel companies, and railroads, as well as the bankers who funded them. However, the economics of oil flowed differently. Because oil was much easier to transport via pipeline and tanker, petroleum revenue streams were often global in nature. Further, while the United States was the world's superpower of oil production during the first half of the 20th century, pumping over half the world's petroleum in most years, even larger amounts of oil and gas happened to be located in poor nations in the Middle East. Thus, the unfolding of the story of oil would hinge on global geopolitics.

Oil was discovered in the Middle East early in the century, but production was delayed, mostly in order to keep global oil prices from collapsing. After World War II, as global petroleum demand soared, the Middle Eastern oilfields quickly came online, but their output continued to be managed by American and British companies and their governments' foreign policies. Increasingly, while the United States could no longer dominate the world in oil exports, it could still control the global oil market, and hence the global economy. Its primary means of doing so was through its currency, the dollar, which (after 1972) was effectively backed not by gold, but by oil—since nearly all oil trades were carried out in dollars, thus maintaining high demand for the currency and giving a subtle financial advantage to its issuer.

If coal inadvertently stoked democratic aspirations to sociopolitical power sharing in the late 19th and early 20th centuries, oil did the opposite. Oil production required a much smaller workforce than did coal mining; similarly, moving petroleum via pipelines and tankers required fewer human hands than it took to load and unload coal trains and barges. Further, oil production sites were often far from industrial and population centers. All of these factors made the oil industry less vulnerable to the kinds of industrial labor actions that had wracked the coal industry during the decades leading up to progressive reforms of the early 20th century (the eight-hour workday, the banning of child labor, the right to unionize, etc.). While industrial sabotage was a route to democratization and a principal impediment to elite power within wealthy nations during the coal age, consumerism and economic growth largely kept a lid on radical activism during the oil era. At the same time, oil companies were able to use their immense wealth and their pivotal role in the world's energy economy to bend governments to their will, funding influential think tanks and industry-friendly candidates to shape policies on agriculture, defense, trade, energy, and the environment.

Instead, it was efforts by Middle Eastern oil-producing nations to take control of their own resources, and the wealth those resources generated, that provided the principal obstacle to elite Anglo-American control of the global economy in the post-WWII era. It would be the tension between the powers of the oil companies and global financial capital on one hand, and the people and their leaders in resource-rich but monetarily poor nations on the other, that would fuel tension and intervention during the oil age.

Meanwhile, the overall contours of the petroleum interval were shaped from the start by depletion. As with coal and other non-renewable materials, the best-quality oil resources were generally identified, extracted, and burned first. This "low-hanging-fruit" pattern of depletion meant that world oil production would inevitably reach a maximum rate and then decline; indeed, all individual oilfields showed a waxing and waning curve of production over time. But virtually no policy makers planned for a future when the world's most precious resource—which now supported not only the global economy but the survival of billions of humans through the food and medicines it produced—would inevitably become scarce. The immense power of oil would eventually prove self-limiting, but not before it utterly transformed society and the planet on which we all depend.

Oil-Age Wars and Weapons

In Chapter 3 we saw why warfare and the adoption of new food and energy sources played key roles in cultural evolution during the past 10,000 years. War and energy have continued to shape society right up to the present. During the 20th century, wars were fought with, and over, the world's newly dominant source of energy, i.e., petroleum; at the same time, the use of oil dramatically increased the lethality of conflict. And wars—especially the two World Wars—shaped technologies and institutions that would in turn determine the contours of social power during longer intervals of peace and economic expansion.

Many historians agree that World War I resulted at least in part from Germany's determination to expand its colonial holdings to rival those of Britain. Germany had developed science and industry to unparalleled heights, but lacked many resources, including energy resources such as high-quality coal and petroleum (it did have low-quality coal that wasn't adequate for metallurgy). Britain had created a colonial system based on coal, but by 1910 it was clear that oil would be the energy source of the future. Yet Britain had very little oil, even in its colonial possessions (its largest reserves were in Burma). Germany understood the importance of oil, and was building a Berlin-to-Baghdad railway partly as a means of obtaining Middle Eastern crude.

For the first years of the Great War, both sides were hampered by lack of fuel. Britain converted its fleet of battleships to run on oil, but German submarines were sinking oil tankers from America on their way to England. Meanwhile, after Romania—Germany's principal petroleum supplier, since the Iraq rail route never came to fruition—sided with the Allies, German industry struggled with shortages of lubricants and gasoline. When the US entered the war on the side of the Allies, its main contribution was not soldiers (welcome though they were), but fuel.

World War I was the first oil-powered war. The tank, essentially an armored and armed gasoline- or diesel-burning tractor, proved itself able to overwhelm infantry, while oil-fueled battleships could out-run and out-maneuver ships running on coal. The Great War saw the first use of airplanes—as well as dirigibles—serving as both weapons and reconnaissance vehicles. Meanwhile the Haber-Bosch chemical process, which would in later decades produce millions of tons of am-monia for nitrogen fertilizer, was initially used to produce ammonia-based explosives.

Each of these new oil- or gas-based technologies would prove piv-otal from a strategic standpoint, and deadly from a human perspec-tive. Germany, Britain, and France lost a generation of young men. Altogether, 18 million human beings died in the Great War, more than in any previous conflict.

For all the principal nations involved (Germany, Austria-Hungary, France, Britain, Russia, and to a lesser extent the US) the war required a near-total commitment of resources and personnel. Germany spent 59 percent of its GDP for military purposes, France 54 percent, and Britain 37 percent. Food and fuel were rationed for noncombatants, and all participating nations engaged in widespread pro-war govern-ment propaganda and suppression of dissent.[26]

Michael Mann, in *The Sources of Social Power*, describes the conse-quences of the war:

> When it was finally over, the three dynastic empires that had started the war were all destroyed, and so was the Ottoman Em-pire. Nation-states, many of them embodying greater citizen rights, were established almost everywhere around Europe, but the overseas empires remained. British and French power were formally restored, although irreparably damaged, and only Americans and Japanese profited much. The United States

had passed from being a major debtor to being the world's banker, owed massive sums by all the European powers. Japan had acquired German colonies in the Far East, jumping-off posts for later expansion in China and across the Pacific.[27]

But key issues had not been settled. Germany still saw itself as deserving the status of global power, and still lacked sources of raw materials. And that country was now burdened with paying war reparations that even many observers in victorious nations regarded as ruinously excessive. Especially when seen in light of the scale of bloodshed it entailed, the Great War had been essentially pointless.

The aftermath of World War I saw communist revolution in Russia and the counter-revolutionary emergence of fascism—i.e., ultra-nationalist authoritarianism based in a corporate-state nexus—in Italy (with fascists gaining power in 1921), Germany (1932), and Spain (1936). Extreme ideologies, promising a utopian future, had been unleashed by extreme economic inequality in the case of Russia, and by defeat and humiliation in that of Germany.

In some ways World War II was simply a continuation of WWI, following a generational gap. Hitler scapegoated the Jews for Germany's defeat and financial situation and called for *"lebensraum"*—space within which to obtain resources necessary for further growth and industrialization. During WWII, the Nazis would pursue supplies of oil through invasions of Romania, North Africa, and the Soviet Union—which had significant petroleum resources. At the same time, they would seek to overwhelm Britain and France, which had kept Germany from attaining its destiny.

In the Far East, Japan had adopted a quasi-fascist style of government and was likewise pursuing empire and resources—again oil, though also food and cheap labor. The Japanese invaded China in 1937, committing brutal atrocities, with the Soviet Union and the US providing aid to China. The US was Japan's principal source of imported oil, but Franklin Roosevelt, despite wariness of Japanese expansionism, delayed an embargo in the knowledge that this would be interpreted as an act of war. When an oil and gasoline embargo was eventually declared on August 1, 1941, war became inevitable.

Prosecution of the war, again, largely hinged on oil. Germany initially sought swift, decisive victories through the use of surprise motorized attacks (*"Blitzkrieg"*). Similarly, Japan's attack on Pearl Harbor relied on oil-fueled ships, gasoline-burning airplanes, and the element of surprise. Further, the outcome of the war was determined largely by access to oil supplies. The US was again in a favored position, with its large domestic reserves, and was able to cut off Japan's access to Royal Dutch Shell's oilfields and refineries in the Dutch East Indies (now Indonesia). And, at an enormous cost in lives, the Soviet Union was able to frustrate Germany's acquisition of Russian oil. Both Japan and Germany literally ran out of gas, having insufficient fuel for tanks, ships, and airplanes.

Altogether, World War II was the deadliest conflict in world history. More than 60 million died in combat or in the Holocaust—Germany's industrial-scale effort to exterminate European Jews. While the phrase "total war" was first used in the Great War, its meaning was fully realized in the Second World War, with the indiscriminate bombing of cities (including London, Dresden, Hiroshima, and many others) killing millions of civilians. Virtually all the available resources of the nations involved were directed toward military purposes, propaganda efforts were scientifically designed and coordinated, and the lives of survivors were indelibly stamped with the experiences of wartime—from privation and horror to extreme self-sacrifice and cooperation.

SIDEBAR 19

Geopolitics: Global Power

The study of how geography influences politics and international relations goes back at least to the ancient Romans, and flourished in the colonial empires of Spain and Britain in the 16th to the 19th centuries. Geopolitical treatises are effectively advice books on how to conquer and rule the world. More recently, with the advent of fossil fuels, oil-powered flight and shipping, satellites, and nuclear-tipped missiles, geopolitical thinking has been explored and developed as never before, now by government agencies and nongovernmental think tanks. Some of the big ideas of geopolitics include:

- Travel by sea has historically been faster and cheaper than travel by land.
- Therefore, coastlines with natural ports and harbors are typically hubs for geopolitical power.
- A larger territory is an advantage, but only if it can be defended.
- A larger population is an advantage, but only if it can be fed.
- Rivers, mountain ranges, and sea coasts are natural borders for nations and cultures.
- New weapons (such as the advent of submarines or guided missiles) and tactics (like massed naval power) can utterly transform the world-game.
- New communication technologies (from printing to radio to social media platforms) can be used to amplify and extend geopolitical power.
- Economics and trade are forms of geopolitical power.
- Alliances—knowing how to form them and when to break them—can enable states to expand their power.

Some key geopolitical thinkers:

- Alfred Thayer Mahon (1840–1914) became famous for two books on the historical influence of sea power. He emphasized that naval operations were won chiefly by decisive battles and blockades.
- Friedrich Ratzel (1844–1904) coined the idea of *Lebensraum*, or "living space," which was to have enormous influence over Nazi geopolitical thinking beginning in the 1920s.
- Sir Halford Mackinder (1861–1947) is chiefly famous for a 1904 article, "The Geographical Pivot of History," in which

he argued that Central Asia (the "Eurasian Heartland") is the axle on which turns the fates of empires.

- Karl Haushofer (1869–1946) built on Mackinder's theories and proposed that the Industrial Age meant larger states and empires, with Mackinder's Heartland as the ultimate prize and seat of power.
- Henry Kissinger (1923–) and Zbigniew Brzezinski (1928–2017) argued that the United States should maintain its geopolitical focus on Eurasia and particularly on Russia, despite the breakup of the USSR and the end of the Cold War.
- Aleksandr Dugin (1962–), a Russian fascist and nationalist, wrote *The Foundations of Geopolitics: The Geopolitical Future of Russia* (1997), which has influenced the recent thinking of Russian military, police, and foreign policy elites.
- Robert Kaplan (1952–), whose *The Revenge of Geography* (2012) brings the subject of geopolitics into the 21st century, discussing the Arab Spring and the rise of social media, argues that the political upheavals of the 21st century will be determined by factors such as land forms, population density, and temperature.

During and after both World Wars, mass mobilization and the destruction of capital resulted in a dramatic reduction of economic inequality in all industrial nations. Indeed, according to Walter Scheidel, in his book *The Great Leveler*, the two World Wars did more to further economic equality than all peacetime government policies such as unemployment insurance, a minimum wage, or progressive taxation.[28] In 1938, Japan's "1%" made off with nearly 20 percent of all income; by seven years later, their share had fallen to 6.4 percent. Similar leveling occurred in victorious nations: students of economic inequality in the United States call the 30 years from 1914 to 1945 the

"Great Compression" due to the anomalous drop in income share for the highest earners—with the "1%" losing 40 percent of their national income share. There were many contributing factors, including rationing (more about that in Chapter 6) and the GI Bill in the US. Nations needing to pay for the war instituted high marginal tax rates on income and inherited wealth. The wealthy lost foreign assets, and some industries were nationalized. Altogether the economic history of the 20th century was reshaped profoundly by these two intense, prolonged eruptions of violence.

New technologies continued to play key roles in the Second World War, which saw much greater and more effective use of bombers and fighter planes, including jet fighters (developed in Germany), which now had longer range and carried far deadlier guns and bombs. Germany also introduced guided missiles, which rained terror on London and could not be countered with fighter aircraft. Computers were used for the first time in war, for code breaking (by the British) and for designing nuclear weapons—which the US developed and used against civilian populations in two Japanese cities. Each of these technologies would have enormous ramifications in the postwar economy, via civilian jet aviation, the space program (leading to satellites for telecommunications and global positioning systems, or GPS), digital computing, and nuclear electrical power generation.

The end of World War II saw many countries—Britain, Germany, France, Italy, most of the rest of Europe, Japan, China, and the Soviet Union—in ruins. The colonial systems of Britain and France were essentially finished. One nation, the United States, emerged stronger than ever, now in position to design and lead new global governance institutions, most notably the United Nations. The US was also in position to dominate the world's new financial system. Henceforth the world's currency of account would be the dollar, and new US-led institutions, principally the World Bank and International Monetary Fund, would maintain financial order among and between the "developed" (i.e., industrialized) nations and the "developing" nations tasked with supplying resources and cheap labor.

Immediately after the war, the US and USSR, allied during the conflict, divided Europe into spheres of influence. This led to a Cold War, lasting roughly 40 years until the collapse of the Soviet Union in 1991. Though it entailed remarkably few direct fatalities (that is, if proxy conflicts like the Korean War are not counted), the Cold War nevertheless spurred dramatic investments in new military technologies, including nuclear power and new generations of nuclear weapons, rapidly improving computers, and the internet. The microchip was invented for intercontinental ballistic missiles (ICBMs) and fighter planes, while GPS was initially developed for tracking troop movements.

Nuclear weapons raised the risk of conflict to the extinction level, and the continued existence of such weapons makes it likely that at some point they will again be used. During the Cold War, the threat of nuclear annihilation prevented direct confrontation between superpowers. Instead, most conflicts took the forms of guerilla warfare and terrorism, and efforts by powerful nations to suppress these asymmetrical uses of force. A principal example was the Vietnam War, which the US entered unwisely and lost bitterly to a much smaller nation that used guerilla tactics.

During the fossil-fuel age, the lethality of conflict increased steeply, while mortality from violence during peacetime declined in most nations (as Steven Pinker documents in his book *The Better Angels of Our Nature*). Having access to more energy created the opportunity for shared affluence (even if economic inequality persisted and sometimes worsened), and thus for greater social order. We are knit together increasingly by trade, transport, and communications, but when we fall into irreconcilable disagreement our new powers enable us to lay waste to entire cities, even continents, with guns, bombs, and poisons.

With the fall of the Soviet bloc, the United States emerged as the world's sole superpower. However, America proceeded quickly to undermine that status with pointless and costly invasions of Afghanistan and Iraq. Meanwhile, China's growing economic might led to that nation forming ad hoc alliances (such as with Iran and Russia) in order to undermine American hegemony. Corruption and authoritarianism in the Middle East sparked a series of revolutions (of which the most persistent was the Syrian uprising, which drew in Turkey, the US, and Russia), but these were eventually put down. In the US, weariness of interventions, along with domestic political polarization fed by a new form of soft warfare (pioneered by Russia) based on social media technologies, led to the surprising ascendancy of a right-wing populist federal administration of extraordinary incompetence and corruption.

Meanwhile, since 2005, world extraction of conventional petroleum has flatlined. Virtually all new production has come from unconventional sources—such as Canadian oil sands and US "tight oil" produced by hydrofracturing ("fracking") and horizontal drilling. During the decade starting in 2010, the latter boosted American petroleum production by 6.5 million barrels per day, enough to forestall global shortages and to enable the US to approach energy self-sufficiency. However, tight oil wells deplete especially quickly, raising the prospect that the fracking boom will be not only unprofitable for oil companies and investors (as it has largely been so far), but also extremely short-lived. The all-time peak of the rate of world oil production appears to have been reached in late 2018, and is unlikely ever to be exceeded.

And here we find ourselves. The era of US global dominance, and indeed the oil age itself, is coming to an end, heightening the risks of future economic turmoil, widespread anger provoked by increasing inequality, domestic political violence, and even war.

But we have not finished our exploration of the ways new energy sources have shaped physical and social power in recent history.

Electrifying!

Electricity is not a source of energy; it is a carrier of energy, a means of making it available for use in homes, offices, and factories. But as such, it has revolutionized society just as thoroughly as have coal and oil. Today, we are all plugged in on a near-constant basis, and an interruption in the supply of electrical power can bring society to a standstill.

After the first electricity generator was invented in 1834, it was decades before electricity began to find widespread commercial applications. That's because, to be useful, electricity requires a system of machines—including, at the very minimum, a generator, transformer, and motor—and all of these had to be invented and perfected before electricity could be put to practical use. Starting in the 1870s, Thomas Edison, a former railroad telegrapher, created what amounted to an invention factory in order to develop new technologies or improve existing ones. It quickly became clear that electricity would ultimately power most of these technologies (including light bulbs, sound recording, and motion pictures), so Edison and his team of engineers also worked to perfect electrical generators, transformers, motors, and distribution systems. While rival Serbian physicist Nikola Tesla (the inventor of the polyphase alternating current motor and generator) ultimately produced more successful designs, in the popular mind the American home-grown Edison was generally regarded as a technological hero, literally bringing light to the world (a pre-echo of the worshipful attention Bill Gates, Steve Jobs, and Silicon Valley would garner in the late 20th century).

Cities were electrified first, as the close proximity of structures made wiring connections relatively cheap. But gradually, over the course of the 20th century, all of rural and urban America (and dozens of other countries) were connected to grids—which distributed electricity from generating plants to end users via transmission lines and transformers. The end result was a giant network of wires and machines that were constantly coordinated and adjusted to make power uniformly available at any moment, day or night, in billions of locations simultaneously.

Marshall McLuhan called electricity an extension of the nervous system. It has revolutionized communications and commerce, speeding up nearly every human activity, both knitting us together with mass communications and ultimately polarizing us through social media. One telling example of electricity's impact concerns music. Prior to electrification, all music was acoustic, and the guitar was a quiet instrument suitable for the parlor. Today, a plugged-in guitar can shatter eardrums, guitar playing techniques have evolved to take advantage of this greater sonic power, and whole genres of music depend on it. Similar examples could be cited in fields ranging from medicine to gambling. As we saw in Chapter 3, most of the key developments in communication technology (the telegraph, telephone, radio, television, the internet-connected computer, and the smartphone) have occurred just in the last century and a half as a result of electrification.

Electricity must be generated using a primary energy source—whether coal, flowing water, oil, natural gas, nuclear fission, biomass (wood), geothermal heat, wind, or sunlight. When the generation process relies on heat from the burning of a primary energy source such as coal, natural gas, oil, or biomass, or from the fission of uranium atoms, the energy conversion process tends to be highly inefficient. For example, in a coal power plant only about 40 percent of the heat energy in the coal is converted to electrical power. In addition, the transmission of electricity entails energy losses of up to ten percent. We accept such inefficiencies because the product, electricity, is able to deliver energy in a highly convenient and versatile form. Generators of electricity that don't require conversion—such as wind turbines and solar panels—don't entail the same losses and inefficiencies. Moreover, electric motors are highly efficient in converting energy to motive power, as compared to fuel-burning engines. That's why electric cars tend to be less polluting, when all factors are considered, than equivalent-sized gasoline-burning cars, even if the electricity with which the e-car was charged was generated using coal; they are, of course, far less polluting if charged with power generated by solar, wind, or hydro.

Nuclear electricity generation began in the 1950s as a result of the Atoms for Peace program (which was, in part, an effort by the nuclear weapons industry to create a more favorable public image for itself). Initial promises that nuclear electricity would be "too cheap to meter" proved unfounded. Nuclear power plants have been expensive to build, requiring government assistance in financing and insurance. The power generation industry typically measures costs as averaged over the full life cycle of a power plant; in these terms, the "levelized" cost of nuclear power is higher than that of coal, natural gas, or hydro. Most nuclear power plants are, in effect, high-tech steam kettles, with heat from nuclear reactions used to create steam, which then turns turbines to generate power. The nuclear reactions do not create greenhouse gases; unfortunately, however, they do produce nuclear waste, which remains dangerously radioactive for centuries, creating storage issues that have proven difficult to solve. Thus, nuclear power has always been controversial, especially following widely publicized accidents such as the Three Mile Island disaster (Pennsylvania, 1979), the Chernobyl disaster (Ukraine, 1986), and the Fukushima reactor meltdown (Japan, 2011). The "World Nuclear Industry Status Report" of 2019 concluded: "Stabilizing the climate is urgent, nuclear power is slow. It meets no technical or operational need that these low-carbon competitors [i.e., solar and wind power] cannot meet better, cheaper, and faster."[29]

Hydro power generation costs tend to be extremely low—much lower than nuclear, natural gas, or coal; however, large hydro dams entail construction projects that require massive initial investment and often degrade or even destroy natural river ecosystems. And there are only so many rivers that can be dammed. A global survey suggests the available resources could provide for a doubling of world hydro power, but only by ignoring concerns about sensitive environments that would likely be impacted.

As a result of all these constraints and trade-offs, many energy analysts consider solar and wind power to be our best bets as future energy sources. But currently these sources supply only a few percent of world energy. (We'll explore the prospects for renewable energy transition in more depth in the next chapter.)

Maintaining our reliance on fossil fuels over the long run is simply not an option, given the twin problems of climate change and resource depletion. Maintaining our current patterns of producing, distributing, and using electricity is also problematic. The grid has been called history's biggest machine. As such, it is highly complex and vulnerable to an array of threats, several of which are foreseeable. The following are scenarios in which a general, persistent loss of electrical power could occur; some are near-certainties over the long run:

- Acts of war—including not just bombs and bullets, but electronic hacking of the software that keeps grids functioning.
- A Carrington event (i.e., a solar storm capable of knocking out electronic devices and transmission systems globally); the last one happened in 1859 and another could occur just about any time.
- Simple depletion of resources that keep grids up and running now: coal, natural gas, and uranium are all finite substances; supply problems are not anticipated immediately, but could occur within the next few decades. (Over the longer term, the depletion of copper, high-grade silicon, lithium, and other mineral resources will limit the build-out and repair of grids and the electronic technologies that depend on them.)
- Neglect of infrastructure and a deteriorating transmission grid (grid operators have been sounding the alarm on this score for the last couple of decades).
- Impacts of climate change, including but not limited to: disruption of seasonal rainfall that enables hydro power plants to operate; high summer heat that could warm lakes and rivers to the point where nuclear power plants couldn't be cooled properly; and the need to preemptively shut down grids to avoid starting seasonal wildfires in fire-prone regions (as I write, nearby regions of my home state of California are experiencing blackouts planned for this reason).

Most readers have probably had the experience of living through a power blackout lasting for hours; some may recall one lasting for days. Everyone who has had such an experience understands that, while electrical power may have started out as a convenience or a luxury, it has become essential to modern life. Without electrical power, the gasoline pumps at service stations do not operate. Credit card readers do not work. If backup generators run out of fuel, municipal water plants cease to function. Yet we have created an electrical power system that is not just vulnerable, but unsustainable in its current form.

The Human Superorganism

As noted in Chapter 1, humans are ultrasocial, similar in this respect to ants or bees. In chapters 2 and 3, we saw how human beings have developed ever more ways of cooperating. We explored the ways in which language and complex social organization (driven by warfare and new energy sources) have enabled more coordination among individuals. As cultures evolved, writing, money, and Big God religions knitted us even closer together. With fossil fuels and electrification, and thus greater mobility and instant communications, we have indeed become a human "hive."

Any one of us can walk into a store, a bank, or a hotel, and, assuming we have the appropriate symbolic tokens (cash or a credit card), we can immediately do business with complete strangers. We have faith that strangers will protect us in the case of physical threat, treat our injuries, and make sure we have water and electricity. When we enter a restaurant and sit at a table, we and the server don't have to spend many tense minutes determining whether one of us has hostile intentions, or whether we are of the same clan. The general shape of the interaction is already agreed upon, as the result of centuries of server-patron encounters at millions of restaurants. We look at the menu, order food, and within minutes we're eating. A tip is expected, and we leave one. That's cooperation. And it makes us, as a species, collectively powerful.

Acting together, we have the ability to commandeer and direct flows of energy originating in millions of years of stored ancient sunlight. We have the power to extract thousands of different resources scattered across the planet, transport them, transform them with thermal and chemical processes, and assemble them into a numbing array of consumer products. Because we act together in so many ways and contexts, we as individuals can travel around the globe in hours—even minutes in the case of astronauts. We can see events happening far away as they occur, communicate instantly with co-workers, or press a key and have a product delivered to our doorstep in hours. And we can devastate entire ecosystems without even realizing what we are doing.

We take these levels and kinds of cooperative power (whether creative or destructive) for granted. Indeed, we often think that our biggest problem, as humans, is that we aren't cooperative enough. After all, we squabble endlessly over politics—so much so that it's often hard for democracies to get anything done. We bicker and fight over inequality and access to resources. We're so fiercely competitive that we even create objectively meaningless opportunities for competition via sports teams and their ritualized clashes.

But even when we compete, squabble, bicker, and clash, it is within the context of extraordinary degrees of cooperation. Within our teams, we work in synchrony, and cheer on our teammates. Indeed, war—the most lethally competitive activity humans engage in—makes us even more cooperative than we are normally, willing to sacrifice even our lives for the sake of the "team."

We evolved from primates that lived in small groups and that cooperated little more than hyenas, wolves, crows, or dozens of other social animals do. But gradually, bit by bit, over millions of years, we developed teams that were bigger and more coordinated, until they comprised hundreds, then thousands, then millions, and now billions of increasingly specialized humans. The end result is a global team that acts, at least to some degree, as a collective unity, together making up the human Superorganism—which wields vastly more power than any other single organism in Earth history.[30]

The idea that humanity currently comprises a Superorganism is propounded by sociobiologists like Edward O. Wilson and cultural evolution theorists such as Peter Turchin. Kevin Kelly, founder of *Wired* magazine, popularized the notion among techno-geeks with his 1994 book *Out of Control*.[31] While there are skeptics, particularly among some traditional evolutionary theorists, the idea that ultra-sociality can lead to the emergence of superorganisms explains how and why groups of individual organisms can exhibit coordinated collective action, collective homeostasis (in which the group as a whole maintains self-regulation of various parameters, such as defense of group boundaries), and emergent behaviors (that is, behaviors of the group that emerge from the relationships of individuals to one another, and that cannot be predicted on the basis of a thorough knowledge of the individuals themselves). The superorganism is more than the sum of individual decisions. Indeed, once a superorganism emerges, individual choices tend to be constrained by the demands of the collective entity.

The global human Superorganism (with a capital "S," signifying the entity that now encompasses our entire species) is still quite young. We became ultrasocial during the last 10,000 years, especially as we began living among strangers in cities. For the next few millennia, nations acted as superorganisms, cooperating internally to accomplish their overarching goals. But the birth of the universal human Superorganism occurred much more recently, in the late 20th century, with the advent of the global economy and instant global communications. That certainly wasn't long ago in terms of biological evolution, or even cultural evolution.

The sheer power of the Superorganism is staggering. Each year, it cuts and uses up to seven billion trees; it excavates 24 million metric tons of copper and nine billion short tons of coal; and it produces (and brews and drinks) one billion tons of coffee. In order to mine resources and construct buildings and highways, it moves up to 80 billion tons of soil and rock—over twice the amount moved by all natural forces such as rivers, glaciers, ocean, and wind.[32] The Superorganism is taking control of Earth, and the Anthropocene is its era of dominance.

Some still argue that the power of the Superorganism is just the sum of the powers of eight billion humans. But if not for the coordination and cooperation that enable the existence of the Superorganism, there wouldn't be eight billion humans, and each of the non-networked humans who did exist would be wielding far less power per capita than is the case today. On one level, the Super-organism may be merely a concept or metaphor, but it is a useful one because it points to a reality that it is crucially important for us to understand and grapple with if our species is to survive the 21st century. We'll return to a discussion of the Superorganism in chapters 6 and 7.

◆   ◆   ◆

In sum: coal, oil, and gas have increased our individual and collective powers enormously. They have given us previously unimaginable wealth and comfort, enabled us to explore other planets, and much more. Through the development and exercise of these greatly expanded powers, we have changed not only the nature of human existence, but also the way our planet functions. And we have inadvertently created a collective global entity, the most powerful in Earth history, which we are only beginning to understand. One thing is clear: our world-shaping power is bringing us to a crossroads.

OVERPOWERED

The Fine Mess We've Gotten Ourselves Into

...[T]he energetic metabolism of our species has grown in size to be comparable in magnitude to the natural metabolic cycles of the terrestrial biosphere. This feature underlies almost all environmental challenges we face in the 21st century, ranging through resource depletion, overharvesting of other species, excessive waste products entering into land, oceans and atmosphere, climate change, and habitat and biodiversity loss.

— YADVINDER MALHI, *Is the Planet Full?*

Earth will be monetized until all trees grow in straight lines, three people own all seven continents, and every large organism is bred to be slaughtered.

— RICHARD POWERS, *The Overstory*

WE OFTEN THINK OF THE DOWNSIDE OF POWER JUST IN terms of its abuses—people using power with evil intent. There are many ways of abusing power, and the fields of clinical psychology, political science, and law are largely concerned with understanding and remedying abuses of social power. However, in this chapter I will focus primarily on power dilemmas that are less frequently discussed, but that may be even more important for us to understand, especially in the 21st century: problems that occur when sheer amounts of physical power overwhelm natural systems, and when concentrated vertical social power threatens individual and collective human well-being.

The problems of the abuse of power and of over-empowerment are not mutually exclusive. Indeed, especially in complex societies, over-empowerment often encourages the abuse of power. But the dilemma of too much power is not confined to the social arena, and may be easier to understand, at least in principle, by way of a few examples that have little to do with human relations.

Electrical engineers are well acquainted with the problem of too much power. An electrical power overload—in which electricity flows through wires that are too small to handle the current—can cause a fire. Similarly, physicists know that, while fuels and batteries store energy for useful purposes, when that energy is released too quickly an explosion can result.

Earth is constantly bathed in 174 petawatts of power from the Sun (a petawatt is one quadrillion watts). That solar power drives weather and life on our planet; everything beautiful and pleasant in our lives traces back to it. But sometimes this power moves through Earth's systems in a destructive surge. Storms cause the quick release of enormous amounts of energy. The wildfires in Sonoma County, California, where I live, can burn with many gigawatts of power. A gigawatt of electrical power that's controlled via power lines, transformers, and circuits can supply light, heat, and internet connections to a small-to-medium-sized city. A gigawatt of radiative power unleashed in a firestorm can torch that same community in just a few hours.

The same essential principle holds with social power. Cooperation, the basis of social power, gives us the means to accomplish wonders. But social power, channeled through elaborate economic systems featuring various forms of debt and investment, can also enable the pooling of immense amounts of wealth in just a few hands, resulting in needless widespread poverty and eventual civil unrest. That same social power, released in a sudden burst, can lead to the deaths of millions of people through war or genocide.

In this chapter I will summon evidence to make the case that we humans have, in many instances, amassed too much physical power and vertical social power for our own good, and that we are overpowering natural systems and social systems with that power, creating conditions potentially leading to unnecessary human mortality and suffering, as well as ecological disaster, on a massive scale.

Power is essential; without it, we would be literally powerless. But one can have too much of a good thing. How much power is enough? How much is too much?

The best way to judge how much power is too much is by assessing the consequences of its usage, taking care to untangle (as much as possible) issues caused by abuse of power from those resulting simply from the concentration or proliferation of too much of it. Here's a telling example: our ability to control fire has done us immense good over the millennia. But today nearly everything we do is associated with a little fire somewhere, perhaps in a factory or a power plant or the engine of our car. Engineers have taken great care, over the past few decades, to increase the efficiency of engines and furnaces, especially in the industrialized countries that do most of the burning. But each of those fires still inevitably emits carbon dioxide, and all these billions of fires added together are giving off enough CO_2 to imperil life on the planet. Fire is good; but too much fire, even intelligently controlled fire, can be very, very bad. (I'll discuss climate change in more depth below.)

We are about to survey evidence that humanity *as a whole* is overempowered, evidence that consists of harmful consequences already unfolding. To make a logically consistent case, I will show that these consequences are not merely random bad things happening. Difficulties, even tragedies, are inevitable parts of life; and war, famine, and inequality have long histories. I will show what's different today, focusing on impacts that are of a scale far larger than those occurring previously, that could result in severe injury for massive numbers of humans and in some cases the entire biosphere, and that are clearly traceable to humanity's recently enhanced powers. In addition, in each section below, we will see why solutions to our deepening problems cannot be achieved just by exposing and ending abuses of power, but are instead necessarily contingent on a reduction, or reining in, of specific human powers.

Climate Chaos and Its Remedies

Climate change is the biggest environmental crisis humans have ever faced, and it is probably the most crucial nexus of current issues concerning physical and social power. Let's briefly explore the problem in general terms; then we'll see why it is a result of too much power and why it can be solved only by limiting and reducing our power.

Climate change is primarily (but not solely) caused by humanproduced greenhouse gases. The main greenhouse gases and their main sources are:

- *carbon dioxide*, released from fossil fuel combustion;
- *methane*, released by farm animals and by the natural gas industry (natural gas is essentially methane); and
- *nitrous oxide*, released by agriculture and fossil fuel combustion.

We contribute to climate change when we burn coal or natural gas for electricity, or oil for transportation, thereby releasing carbon dioxide and nitrous oxide. The cattle we raise for food emit methane.

Industrial agriculture and timber harvesting release carbon stored in soils and vegetation into the atmosphere as carbon dioxide.

At the dawn of the industrial age, the carbon dioxide content of Earth's atmosphere was 280 parts per million. Today it's over 420 ppm and rising fast. Greenhouse gases trap heat in the atmosphere, causing the overall temperature of Earth's surface to increase. It has risen by over one degree Celsius so far; it is projected to rise as much as five degrees more by the end of this century.

Now, a few degrees may not sound like much, and in the middle of winter it might even sound comfortable. But the planet's climate is a highly complex system. Even slight changes in global temperatures can create ripple effects in terms of weather patterns and the viability of species that have evolved to survive in particular conditions.

Moreover, climate change does not imply a geographically consistent, gradual increase in temperatures. Different places are being affected in different ways. The American southwest will likely be afflicted with longer and more severe droughts. At the same time, a hotter atmosphere holds more water, leading to far more severe storms and floods in other places. Melting glaciers are causing sea levels to rise, leading to storm surges that can inundate coastal cities, placing hundreds of millions of people at risk.

As carbon dioxide emissions are raising the planet's temperature, they are also being absorbed by the Earth's oceans, increasing the water's acidity and imperiling tiny ocean creatures at the bottom of the aquatic food chain. Because the oceans are also being impacted by other kinds of pollution (nitrogen runoff and plastics) and are being overfished, some scientists now warn that entire marine ecosystems are at risk this century. To put it as plainly as possible, the oceans are dying.

Since the 1990s, the United Nations has sought to rally the world's nations to reduce greenhouse gas emissions. After 21 years of international meetings and failed negotiations, 196 nations agreed in December 2015 on a collective goal of limiting the temperature increase above pre-industrial levels to 2 degrees Celsius (3.6 degrees

Fahrenheit) with the ideal being 1.5 degrees or less. But actual emissions have so far not been capped.

Some of the difficulty in reaching consensus on climate action comes from the fact that most historic emissions came from the world's financially and militarily powerful nations, which continue burning fossil fuels at high rates; while some poorer and less powerful nations, which have less historic responsibility, are rapidly increasing their fossil fuel consumption in order to industrialize. Some of the poorest nations are also the ones most vulnerable to climate change. All of this makes it hard to agree who should reduce emissions and how fast.

SIDEBAR 20

Personal Carbon Output

Although a gallon of gasoline weighs around six pounds, combustion binds two molecules of oxygen to every carbon molecule, so that when burned, one gallon of gasoline yields 20 pounds of CO_2.

That's a pound for every mile driven in a car that gets 20 miles to the gallon. With roughly a billion people driving worldwide, those pounds of CO_2 accumulate dramatically.

It takes energy to make a car, pave roads, grow food, and produce everything else that supports life in the modern world. Each US citizen, on average, releases close to 20 tons of CO_2 into the atmosphere each year.

That's 40,000 pounds, more than a really big four-axle dump truck can haul. Fifty years of this energy-intensive way of life, or more than 50 of these dump trucks per person, make for quite a carbon mountain.

CO_2 is invisible and odorless, so we don't notice the enormous quantities spewing into the atmosphere. We notice the impacts from climate change, but many people are unable to trace these back to the source in terms of their own behavior.[1]

There are many areas of uncertainty in climate science, even though the basic mechanism of global warming is clear. So far, in many cases actual ecosystem impacts have come faster than scientists have forecast, but we don't know whether future impacts will adhere more closely to experts' models or continue to outstrip them. Also, there are uncertainties about how best to account for, and assign responsibility for, greenhouse gas emissions: for example, if China burns coal to produce goods for American shoppers, should the resulting emissions be credited to China or the United States? How should we count emissions from shipping and aviation (which are usually omitted from national greenhouse gas accounts)?

There is controversy as well about how we might capture carbon from the atmosphere and sequester it for a long time. As the threats of the climate crisis become increasingly clear and our remaining "carbon budget" (the amount we can emit before crossing dangerous thresholds) shrinks, ever more attention is being paid to what are called "negative emissions technologies." But most of these technologies—with the exception of natural ways to sequester carbon in soils and trees—are theoretical, unproven, or unscalable.

Altogether, climate change poses enormous challenges, including these:

- How can we rapidly reduce the use of fossil fuels without experiencing unacceptable consequences of economic contraction?
- How can we reduce the use of fossil fuels fairly?
- How can we protect nations and communities that are most vulnerable to climate change?
- Can we take carbon out of the atmosphere and store it so as to reduce the severity of climate change?

The difficulty of the project of reducing emissions is illustrated by the example of California. That state's climate pollution declined by just 1.15 percent in 2017, according to the California Green Innovation Index.[2] (For the US as a whole, greenhouse gas emissions fell less than one percent in 2017, but *rose* 3.4 percent in 2018.) At the 2017 rate, California won't reach its 2030 decarbonization goals (cutting emissions to 40 percent below 1990 levels) until 2061, or its 2050 targets (80 percent below 1990 levels) until 2157. The state has pioneered cap-and-trade legislation and ambitious fuel economy standards, among many other climate-friendly policies and initiatives. If California will be *up to a century late* in meeting its climate goals, what does this say for the rest of the nation, and the world as a whole? Further, it must be noted that California understandably undertook the easiest and lowest-cost actions first; as time goes on, further increments of carbon reduction will tend to be more expensive and politically fraught. Clearly, meeting global climate targets ("net zero" emissions by 2050 to limit warming to 1.5 degrees C, or by 2070 to stay below 2 degrees C) will require vastly more effort than is currently under way.

Climate change is often presented as a technical pollution issue. In this framing, humanity has simply made a mistake in its choice of energy sources; a solution to climate change entails switching energy sources and building enough carbon-sucking machines to clear the atmosphere of excess CO_2. But techno-fixes (that is, technological solutions that circumvent the need for personal, political, or cultural change) aren't working so far, and likely won't work in the future. That's because fossil fuels will be difficult to replace, and energy usage is central to our collective economic power.

Consider this brief thought experiment. Could the Great Acceleration in population and consumption have happened with wind or water power? As we've seen, these energy sources were already available at the start of the Acceleration, but they were by themselves inadequate to spark a massive industrialization of production and transport, or to enable a tripling of agricultural output; instead, it was the characteristics of fossil fuels that propelled the dramatic expansion of humanity's powers. Since we adopted fossil fuels, we have introduced new sources of energy (solar PV panels, giant wind turbines, and nuclear power plants), but we would never have been able to develop them if we hadn't had access to the power of fossil fuels to enable improvements in metallurgy and a host of other industrial fields that led to our ability to construct solar cells, wind towers, and nuclear reactors.

Moreover, the replacement of fossil fuels with other energy sources faces hurdles. As noted in the last chapter, many energy analysts regard solar and wind as the best candidates to substitute for fossil fuels in power generation (since, as we've seen, nuclear is too expensive and too risky, and would require too much time for build-out; and hydro is capacity constrained). But these "renewables" are not without challenges. While sunlight and wind are themselves renewable, the technologies we use to capture them aren't: they're constructed of nonrenewable materials like steel, silicon, concrete, and rare earth metals that require energy for mining, transport, and transformation.

Sunlight and wind are intermittent: we cannot control when the Sun will shine or the wind will blow. Therefore, to ensure constant availability of power, these sources require some combination of four strategies:

- Energy storage (by way of batteries; or by pumping water uphill when there is sufficient power, so that it can be released later to generate power as it flows back downhill) is useful to balance out day-to-day intermittency, but nearly useless when it comes to seasonal intermittency; also, storing energy costs energy and money.
- Source redundancy (building far more generation capacity than will actually be needed on "good" days), and then connecting far-flung solar and wind farms by way of massive super-grids, is a better solution for seasonal intermittency, but requires substantial infrastructure investment.
- Excess electricity generated at times of peak production can be used to make synthetic fuels (such as hydrogen, ammonia, or methanol), perhaps using carbon captured from the atmosphere, as a way of storing energy; however, making large amounts of such fuels will again require substantial infrastructure investment, and the process is inherently inefficient.

- Demand management (using electricity when it's available, and curtailing usage when it isn't) is the cheapest way of dealing with intermittency, but it often requires behavioral change or economic sacrifice.

Today the world uses only 20 percent of its final energy in the form of electricity. The other 80 percent of energy is used in the forms of solid, liquid, and gaseous fuels. A transition away from fossil fuels will entail electrification of much of that other 80 percent of energy usage, which includes transportation and industrial processes. Many uses of energy, such as aviation and high-heat industrial processes, will be difficult or costly to electrify. In principle, the electrification conundrum could be overcome using synfuels. However, doing this at scale would require a massive infrastructure of pipelines, storage tanks, carbon capture devices, and chemical synthesis plants that would essentially replicate much of our current natural gas and oil supply system.

Machine-based carbon removal and sequestration methods work well in the laboratory, but would need staggering levels of investment in order to be deployed at a meaningful scale, and it's unclear who would pay for them. The best carbon capture-and-sequestration responses appear instead to consist of various methods of ecosystem restoration and soil regeneration. These strategies would also reduce methane and nitrous oxide emissions. But they would require a near-complete rethinking of food systems and land management.

Recently I collaborated with a colleague, David Fridley of the Energy Analysis Program at Lawrence Berkeley National Laboratory, to look closely at what a full transition to a solar-wind economy would mean (our efforts resulted in the book *Our Renewable Future*).[3] We concluded that it will constitute an enormous job, requiring many trillions of dollars in investment. In fact, the task may be next to impossible—*if* we attempt to keep the overall level of societal energy use the same, or expand it to fuel further economic growth.[4] David and I concluded:

We citizens of industrialized nations will have to change our consumption patterns. We will have to use less overall and adapt our use of energy to times and processes that take advantage of intermittent abundance. Mobility will suffer, so we will have to localize aspects of production and consumption. And we may ultimately forgo some things altogether. If some new processes (e.g., solar or hydrogen-sourced chemical plants) are too expensive, they simply won't happen. Our growth-based, globalized, consumption-oriented economy will require significant overhaul.[5]

As we've seen throughout this book, energy transitions are never merely technical; they are transformative for society. The end of fossil fuels—even in the best case, with forethought and proactive investment—will result in cascading disruptions to every aspect of daily existence.

The essence of the problem is this: nearly everything we need to do to solve the climate problem (including building new low-emissions electrical generation capacity, and electrifying energy usage) requires energy and money. But society is already using all the energy and money it can muster in order to do the things that society wants and needs to do (extract resources, manufacture products, transport people and materials, provide health care and education, and so on). If we take energy and money away from those activities in order to fund a rapid energy transition on an unprecedented scale, then the economy will contract, people will be thrown out of work, and many people will be miserable. On the other hand, if we keep doing all those things at the current scale while also rapidly building a massive alternative infrastructure of solar panels, wind turbines, battery banks, super grids, electric cars and trucks, and synthetic fuel factories, the result will be a big pulse of energy usage that will significantly increase carbon emissions over the short term (10 to 20 years), since the great majority of the energy currently available for the project must be derived from fossil fuels.

It takes energy to make solar panels, wind turbines, electric cars, and new generations of industrial equipment of all kinds. For a car with an internal combustion engine, ten percent of lifetime energy usage occurs in the manufacturing stage. For an electric car, roughly 40 percent of energy usage occurs in manufacturing, and emissions during this stage are 15 percent greater than for an ICE car (over the entire lifetime of the e-car, emissions are about half those of the gasoline guzzler). With solar panels, energy inputs and carbon emissions are similarly front-weighted to the manufacturing phase; energy output and emissions reduction (from offsetting other electricity generation) come later. Replacing a very high percentage of our industrial infrastructure and equipment quickly would therefore entail a historically large burst of energy usage and carbon emissions. By undertaking a rapid energy transition, while also maintaining or growing current levels of energy usage for "normal" purposes, we would be defeating our goal of reducing emissions now (even though we would be working toward the goal of reducing emissions later).

One way to mostly avoid that nasty conundrum would be to slow the transition, replacing old power plants and emissions-spewing cars, planes, ships, trucks, and factories as they wear out with new versions that don't produce carbon emissions. But the process would take far too long, since much of our existing infrastructure has a multi-decade lifespan, and we need to get to zero net emissions by roughly 2050 to avert unacceptable climate impacts.

Some folks nurture the happy illusion that we can do it all—continue to grow the economy while also funding the energy transition. But that assumes the problem is only money (if we find a way to pay for it, then the transition can be undertaken with no sacrifice). This illusion can be maintained only by refusing to acknowledge the stubborn fact that all activity, including building alternative energy and carbon capture infrastructure, requires energy.

The only way out of the dilemma arising from the energy and emissions cost of the transition is to reduce substantially the amount of energy we are using for "normal" economic purposes—for resource extraction, manufacturing, transportation, heating, cooling, and industrial processes—both so that we can use that energy for the transition (building solar panels and electric vehicles), and so that we won't have to build as much new infrastructure (since, if our ongoing operational energy usage is smaller, we won't need as many solar panels and so on). Increased energy efficiency can help reduce energy usage without giving up energy services, but many machines (LED lights, electric motors) and industrial processes are already highly efficient, and further large efficiency gains in those areas are unlikely. We would achieve an efficiency boost by substituting direct electricity generators (solar and wind) for inherently inefficient heat-to-electricity generators (natural gas and coal power plants); but we would also be introducing new inefficiencies into the system via battery-based electricity storage and synfuels production. In the end, the conclusion is inescapable: actual reductions in energy services would be required in order to transition away from fossil fuels without creating a significant short-term burst of emissions. Some energy and climate analysts other than David Fridley and myself— such as Kevin Anderson, Professor of Energy and Climate Change at the University of Manchester—have reached this same conclusion independently.[6]

As we've seen, energy is inextricably related to power. Thus, if society voluntarily reduces its energy usage by a significant amount in order to minimize climate impacts, large numbers of people will likely experience this as giving up power in some form.

It can't be emphasized too much: energy is essential to all economic activity. An economy can grow continuously only by employing more energy (unless energy efficiency can be increased substantially, and further gains in efficiency can continue to be realized in each succeeding year). World leaders demand more economic growth in order to fend off unemployment and other social ills. Thus, in effect, everyone is counting on having more energy in the future, not less.

A few well-meaning scientists try to avoid the climate-energy-economy dilemma by creating scenarios in which renewable energy saves the day simply by becoming dramatically cheaper than energy from fossil fuels, or by ignoring the real costs of dealing with energy intermittency in solar and wind power generation. Some pundits argue that we have to fight climate change by becoming even more powerful than we already are—by geoengineering the atmosphere and oceans and thus taking full control of the planet, thereby acting like gods.[7] And some business and political leaders simply deny that climate change is a problem; therefore, no action is required. I would argue that all of these people are deluding themselves and others.

Problems ignored usually don't go away. And not all problems can be solved without sacrifice. If minimizing climate change really does require substantially reducing world energy usage, then policy makers should be discussing how to do this fairly and with as little negative impact as possible. The longer we delay that discussion, the fewer palatable options will be left.

The stakes could hardly be higher. If growth in emissions continues, the result will be the failure of ecosystems, massive impacts on economies, widespread human migration, and unpredictable disruptions of political systems. The return of famine as a familiar feature of human existence is a very real likelihood.[8]

The recent global economic recession resulting from governments' response to the coronavirus pandemic at least temporarily reduced energy usage and carbon emissions. It was not enough to avert the climate crisis, and no one was happy with the unequally distributed economic pain entailed—the health impacts (suffering and lost lives), the economic sacrifice (lost jobs), and the increasing levels of economic inequality (as the low-paid bore the brunt of both disease and layoffs). A serious, long-term response to the climate crisis will require us to reduce energy usage still further in industrialized nations, and to find ways to share the burden more equitably. It's difficult to overstate what is at stake: if we don't adequately address climate change, the fate not just of humans, but of many or even most of the planet's other species may hang in the balance.

Disappearance of Wild Nature

We humans are accustomed to thinking of the world only in terms of our own wishes and needs. The vast majority of people's attention is taken up with family, work, politics, the economy, sports, and entertainment—human interests. It has become all too easy to forget that nature even exists, as society has grown more urbanized, and as the processes of growing food, making goods, and producing energy are handled in large industrial operations far from view. There are millions of children in North America who have never seen the stars at night, visited a national park, or picked a wild blackberry, and who have only the vaguest idea where their food, fuel, and electricity come from.

Earth's wild flora and fauna provide an entertaining stage setting for nature documentaries and tourism, but appear to have little further significance. Yet we humans are part of a living system extending from viruses and bacteria all the way up to redwood trees and whales—one with myriad checks, balances, and feedbacks. We evolved within this living system, and cannot persist without it.

As our human populations and consumption habits grow, we displace other species. We turn wild lands supporting a great variety of animals, plants, and fungi into plowed fields dominated by a single crop. Our greenhouse gas emissions change the climate, reducing habitat still further. And we introduce toxic chemicals into the environment, which work their way up through the food chain all the way to mothers' breast milk.

SIDEBAR 21

Rising Risk of Disease and Pandemic

As of this writing, the novel coronavirus pandemic that began in early 2020 is still raging across the planet. For many years, public health experts had warned that a serious epidemic was increasingly likely to emerge, and to quickly become global, as a result of encroachment of civilization into wild habitat and rapid global transportation.

SARS-CoV-2 has proven to be highly infectious, and also to have a case fatality rate far higher than that of most yearly influenza viruses. Without dramatic measures to slow the coronavirus's spread, many millions would likely have died in the first year. Nations did undertake dramatic lockdown measures to suppress infection rates, so that their healthcare systems would not be overwhelmed, and to provide time for the discovery of treatment regimens and the development of vaccines. These lockdown measures succeeded in slowing the infections and deaths, but had a catastrophic impact on the global economy.

Epidemiologists warn that future virus pandemics could be even more lethal. Most novel diseases arise through contact between humans and nonhuman animals. As civilization expands ever further into formerly wild regions, more new contacts between humans and reservoirs of formerly isolated disease organisms will occur. This is just one of the ways that destruction of habitat for other creatures boomerangs to reduce human well-being.

Meanwhile, bacteria and fungi are evolving immunity to antibiotics.[9] Some of these "super bugs" have emerged from factory farms and feedlots where livestock are routinely fed antibiotics to encourage growth and suppress disease under stressful, overcrowded conditions, and from over-prescription of antibiotics for human patients. As antibiotic-immune microorganisms proliferate, hospitals are losing their primary means of battling infections.

In effect, we humans are exercising interspecies exclusionary power, both purposefully (in the cases of "weed" species, including microbes like the polio virus, which we try to wipe out), and inadvertently (in the case of millions of species that simply get in our way), on a scale likely not seen for the last several hundred million years.

The results are apparent everywhere in declining numbers of species of insects, fish, amphibians, birds, and mammals. It has been estimated that humans—along with our cattle, pigs, and other domesticates—now make up 97 percent of all terrestrial vertebrate biomass. The other three percent are comprised of all the songbirds, deer, foxes, elephants and on and on—all the world's remaining wild land animals. Meanwhile deforestation and other land uses are

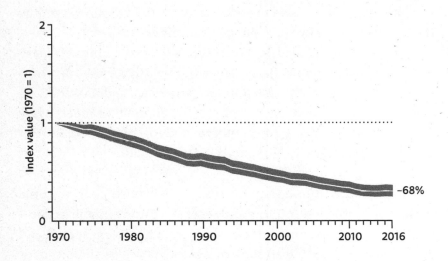

Figure 5.1. Global wildlife decline (Global Living Planet Index with confidence interval). Credit: World Wide Fund for Nature, Living Planet Report 2020.

wreaking devastation on the world's *plant* biodiversity, with one in five plants currently threatened with extinction.

According to the World Wildlife Fund, two-thirds of all wildlife has disappeared in the last 50 years. A third of this loss has resulted from over-exploitation, another third from habitat degradation, and the final third through changes in the environment—which include increasing land use for agriculture and urbanization.[10]

Biological richness is being lost even at the microscopic level. Our use of agricultural chemicals has led to the disappearance from farm soils of bacteria, fungi, nematodes, and other tiny organisms that provide natural fertility. As these microscopic soil communities are destroyed, carbon is released into the atmosphere. Even in the human gut, microscopic biodiversity is on the decline (partly due to the widespread use of antibiotics), leaving us more prone to immune disorders, multiple sclerosis, obesity, and other diseases.

Biologists call this widespread, rapid loss of biodiversity the Sixth Extinction. The geological record tells of five previous events when enormous numbers of species perished; the most severe occurred at the end of the Permian period, 250 million years ago, when 96 percent of all species disappeared. Evidently, we are now in the early stages of another massive die-off of species, though not yet on the same scale as those five previous cataclysmic events.

What does loss of biodiversity mean for people? At the very least, it implies that today's children are set to inherit a world in which many of the animals that filled the lives, dreams, and imaginations of our ancestors, and that provided the metaphors at the root of every human language, will be remembered only in picture books. But biodiversity loss also has enormous practical implications for public health and agriculture.

Among other things, natural systems replenish oxygen in the planetary atmosphere, capture and sequester carbon in soils and forests, pollinate food crops, filter freshwater, buffer storm surges, and break down and recycle all sorts of wastes. As we lose biodiversity, we also lose these ecosystem services—which, if we had to perform them ourselves, would cost us over $30 trillion annually, according to several estimates.

Yet our very tendency to be preoccupied with the *human* impacts of the biodiversity crisis may tell us a great deal about why it is happening in the first place. Many Indigenous cultures believed that other species have as much right to exist as we do. But as cities, houses, cars, and communication media have come to dominate our attention, we have become ever more self-absorbed. We barely notice as the oceans are emptied of life, or as the sound of birdsong disappears from our lives.

Throughout the world, successful programs for biodiversity protection have centered on limiting deforestation, restricting fishing, and paying poor landowners to protect habitat. An international coalition of scientists, conservationists, nonprofits, and public officials has called for setting aside half the planet's land and sea for biodiversity protection.[11] Biologist Edward O. Wilson's book *Half Earth* makes the case for this proposal; Wilson estimates that it would reduce the human-induced extinction rate by 80 percent.[12] It's a bold idea that faces enormous political and economic obstacles. But unless we do something on roughly this scale, wildlife faces an immense threat, one with a distinctly human face.

Land is a fundamental source of wealth, and wealth is a primary form of social power. Therefore, giving up control of land (with the goal of preserving habitat) in effect means ceding power. Naturally, many people prefer not to give up land and the power that it brings; they would prefer to use even more land for mining and other resource extraction, for agriculture, even for renewable energy projects. Further, as long as the human population is growing, increased land use can always be justified for humanitarian purposes. So, how to avoid the trade-off?

Some people think technology could help. Using CRISPR-Cas9 gene-editing technology, it may now be possible to bring some animals and plants back from extinction. Indeed, ecologists at the University of California, Santa Barbara, have already published guidelines for choosing which species to revive if we want to do the most good for our planet's ecosystems.[13] By establishing a genetic library of existing species, we could give future generations the opportunity to bring any organism back from beyond the brink. Doing so could help restore ecosystems that once depended on these species. For example, mammoths trampling across the ancient Arctic helped maintain grasslands by knocking down trees and spreading grass seeds in their dung. When the mammoths disappeared, grasslands gave way to today's mossy tundra and snowy taiga, which are melting and releasing greenhouse gases into the atmosphere. By reviving the mammoth, we could help slow climate change by turning the tundra back into stable grasslands.[14]

Nevertheless, as exciting as it may be to contemplate "Jurassic Park"-like projects reviving long-gone animals like the mammoth or the passenger pigeon, bringing back a few individual plants or animals will be a meaningless exercise if these species have no habitat. Some species cannot reproduce in captivity, and zoo specimens do not perform ecological functions. Many scientists involved in extinct species revival efforts understand the need for habitat, and aim only to revive species that could help restore ecosystems.[15] Still, it's important that we keep our priorities straight: *without* habitat, the revived species themselves are only ornaments. Habitat protection is the real key to reversing biodiversity loss; species revival is just a potentially useful afterthought. And habitat protection means reducing the potential for humans to extract wealth from Earth's limited surface area, even if we find ways to treat land and water in ways friendlier to wildlife. Ultimately, we cannot avoid the trade-off between biodiversity and the increasing power of human beings.

Resource Depletion

Imagine you're at a big party. The host wheels out a tub of ice cream—the giant ones they use at ice cream parlors—and yells, "Help yourselves!" A few people grab scoops, spoons, and bowls from the kitchen and start digging into the frozen top. As they work the tub and it warms up, the ice cream becomes easier to scoop; more people bring spoons and bowls, and ice cream flies out of the tub as fast as people can eat it. When the ice cream is over half gone, it becomes harder to get at; people have to reach in farther to scoop, and they're bumping in to each other. The ice cream isn't coming out as fast anymore, and some people lose interest and turn their attention to the cake. Finally, a small group of the most intrepid scoopers are literally scraping the bottom of the barrel, resorting to small spoons to get the last bits of ice cream out.

That's depletion. The faster you scoop, the sooner you arrive at the point where there is no more left. Like our hypothetical tub of ice cream, Earth's resources are subject to depletion—but usually the process is a little more complicated.

First, material resources come in two kinds: *renewable* and *nonrenewable*.[16] Renewable resources like forests and fisheries replenish themselves over time. Harvesting trees or fish faster than they can replenish will deplete them, and if you do that for too long it may become impossible for the species to recover. We're seeing that happening right now with some global fish stocks.

Nonrenewable resources—like minerals, metals, and fossil fuels—don't grow back at all. Minerals and metals can often be recycled, but that requires energy, and usually the resources gradually degrade as they're cycled repeatedly. When we extract and burn fossil fuels, they are gone forever.

Another complicating issue is *resource quality*. Our hypothetical ice cream is the same from the top of the tub to the bottom. But most nonrenewable resources vary greatly in terms of quality. For example, there are rich natural iron deposits called magnetite in which iron makes up about three-quarters of the material that's mined; at the other end of the spectrum is taconite, of which only about one-quarter is iron. If you're looking to mine iron, guess which ores you'll start with. High-grade resources—the "low-hanging fruit"—tend to be depleted first.

Defining which resources can be considered "high-grade" is also a little complicated. We've just discussed ore grades. But there are also issues of accessibility: how deeply is the resource buried? And location: is it nearby, or under a mile of ocean water, or in a hostile country on the other side of the planet? There's also the issue of contaminants: for example, coal with high sulfur content is much less desirable than low-sulfur coal.

Let's see how issues of resource quality play out in the case of one of the world's most precious resources, crude oil.

Modern oil production started around 1860 in the United States, when deposits of oil in Pennsylvania were found and simple, shallow wells were drilled into them. This petroleum was under great pressure underground, and the wells allowed that oil to escape to the surface. Over time, as the pressure decreased, the remaining oil needed to be pumped out. When no more oil could be pumped, the well was depleted and abandoned.

Exploration geologists soon discovered oil in other places: Oklahoma, Texas, and California; and later in other parts of the world, particularly the Middle East. During the century and a half that we humans have been extracting and burning oil, hundreds of thousands of individual oil wells around the world have been drilled, depleted, and abandoned.

Oil deposits are generally too big to be drained from a single well; many wells drilling into the same underground reservoir are called an oilfield. Oilfields are geological formations where oil has accumulated over millions of years. Some are small, others are huge. Smaller oilfields are fairly numerous. The super-giant ones—like Ghawar in

Saudi Arabia, which in its glory days of the 1990s yielded nearly ten percent of all the world's oil on a daily basis—are extremely rare (there are 131 super-giants in total, out of some 65,000 total oilfields). Like individual oil wells, oilfields also deplete over time.

Today, most of the world's onshore crude oil deposits—often classified as *conventional oil*—have already been discovered and are in the process of being depleted. The oil industry is quickly moving toward several kinds of *unconventional oil*—such as the oil sands in Canada, *deepwater oil* in places like the Gulf of Mexico, and *tight oil* (also known as *shale oil*) produced by fracking primarily in North Dakota and Texas. Compared to conventional oil, unconventional oil resources are either of lower quality, or are more challenging to extract or process, or both; therefore, production costs are significantly higher, and so are the environmental impacts and risks.

As fossil fuels are depleted, their energy profitability (the energy returned on energy invested in producing them, or EROEI) generally declines. That means more and more of society's overall resources have to be invested in producing energy. It also means oil prices are likely to become more volatile. Oil companies have a harder time making a profit, while consumers often find oil less affordable.

During the past century, our transportation systems were built on the assumption that oil prices would remain continually low and supplies would continually grow; meanwhile, the petroleum industry was structured to anticipate low extraction costs. Now that extraction costs are increasing because we're relying more on unconventional resources, there is no longer an oil price that works well for both producers and consumers. Either the oil price is too high, eating into motorists' disposable income, thereby reducing spending on everything else and making the economy trend toward recession; or the oil price is too low, bankrupting oil producers.

The increasing volatility of prices under conditions of depletion suggests that the free market may not be capable of managing nonrenewable resources in a way that meets everyone's needs over the long run—especially those of future generations, who don't have a voice in the discussion. In theory, when a resource that's in high demand becomes scarce, its price rises to discourage consumption and encourage substitutes. But what if poor people need that resource, too? What if the entire economy depends on ever-expanding cheap supplies of that resource? What if a substitute is hard to find, or fails to match the original resource in versatility or price? What happens to producers if costs of extraction are rising rapidly, but a temporary surge in supply or a fall in demand causes prices to plummet far below production costs? In just the last decade, we've seen all of these problems begin to play out in oil markets.

The main alternative to market-based resource extraction and distribution is for governments and communities to collaborate on a program of *resource management*. We've done this with renewable resources by establishing quotas on fishing and by protecting old-growth forests. However, conservation efforts have seldom been undertaken with nonrenewable resources like fossil fuels and minerals. Humanity's collective plan evidently is to extract and use these resources as quickly as possible, and to hope that economically viable substitutes somehow appear in time to avert economic catastrophe.

In agrarian societies, the most crucial depletion issue is usually the disappearance of soil nutrients—nitrogen, phosphorus, and potassium, which together make soil fertile enough to grow crops. In such societies, soil fertility largely determines the size of the population that can be fed. To avert population declines resulting from famine, successful agrarian societies like the ancient Chinese learned to recycle nutrients using animal and human manures (as we saw in Chapter 3). In contrast, modern industrial societies typically get these nutrients from nonrenewable resources: we mine phosphate from large deposits in China, Morocco, and Florida, and ship it around the world. And we produce artificial nitrogen fertilizers from fossil fuels like natural gas. Since both phosphate deposits and fossil fuel deposits will ultimately be depleted, we will have to find ways to break industrial agriculture's dependence on these nonrenewable resources if civilization is to become sustainable over the long term.

Meanwhile, our ability to supply plant nutrients artificially has enabled us to largely ignore the depletion of topsoil itself, of which we are losing over 25 billion tons per year globally through erosion and salinization.

In short, the depletion of renewable and nonrenewable resources is a very real problem, and one that probably contributed to the collapse of societies in the past.[17] Today we consume resources at a far higher rate than any previous civilization. We can do this mainly due to our reliance on a few particularly useful nonrenewable and depleting resources, namely fossil fuels. Energy from fossil fuels enables us to mine, transform, and transport other resources at very high rates; it also yields synthetic fertilizers to make up for our ongoing depletion of natural soil nutrients. This deep dependency on fossil fuels raises the question of what we will do as the depletion of fossil fuels themselves becomes more of an issue.[18]

The depletion of resources results from our pursuit of power in the present, but depletion limits humanity's pursuit of power in the future. We all contribute to resource depletion when we buy goods, use energy, and dispose of trash. The more power we have, the more stuff we can consume. The bigger the scale of our collective human metabolism, and the faster it grows, the more we deplete the environment. We are now using nearly every natural resource faster than nature can regenerate it or technology can substitute for it. And, as a result of exponential growth in extraction rates, half of all the fossil fuels and many other industrial resources that have ever been extracted have been consumed in *just the past half-century*.[19]

If we continue on our current course, we are headed toward a crash from which there can be no industrial recovery, because the raw materials needed for operating an industrial economy will be mostly gone—either burned or dispersed across the planet in forms and amounts that would make collection and reprocessing uneconomic in any realistic future scenario.[20]

Again, we face a trade-off between power and sustainability—in this case, the ability to sustain our access to supplies of resources. It would appear that the only sane basis for an economy would be to use mainly renewable material resources, always at rates equal to or

Figure 5.2. Global extraction/production of resources in 1875, 1945, and 2015 (metric tons).

Year	Iron Ore	Cement	Phosphate Rock	Gypsum	Coal (tons/mtce*)
1875 (British supremacy)	25,000,000	5,000,000	750,000	500,000	265,000,000
1945 (American supremacy)	159,000,000	50,000,000	10,900,000	9,800,000	1,072,000,000
2015 (Chinese supremacy)	2,300,000,000	4,100,000,000	241,000,000	260,000,000	4,895,000,000

* (mtce = million metric tons coal equivalent)
Source: Christopher Clugston, *Blip: Humanity's 300 Year Self-Terminating Experiment with Industrialism*, (St. Petersburg, FL: Book Locker, 2019), p. 232.

Figure 5.3. Power Center import vulnerability ratings of nonrenewable resources.

Resource	Europe (EU)	USA	China	Russia
Aluminium	High	High	Moderate	None
Bauxite	High	EXTREME	High	High
Cement	None	Low	None	None
Coal	Moderate	None	None	Low
Copper	High	High	High	None
Gypsum	High	Low	None	None
Iron Ore	High	None	High	None
Lead	Moderate	Low	Moderate	Negligible
Natural Gas	EXTREME	Low	Moderate	None
Nickel	High	Low	Moderate	None
Oil	EXTREME	High	High	None
Phosphate Rock	EXTREME	Moderate	Low	None
Potash	High	High	Moderate	None
Steel	Low	Low	None	None
Sulfur	None	Low	Moderate	None
Tin	EXTREME	EXTREME	Moderate	High
Zinc	High	High	Moderate	Low

Source: Christopher Clugston, *Blip: Humanity's 300 Year Self-Terminating Experiment with Industrialism*, (St. Petersburg, FL: Book Locker, 2019), pp. 274–275.

below their rates of natural replenishment, and to use nonrenewable resources only as they can be fully recycled.

We are very far from that condition now. And we are generally moving away from it rather than toward it (i.e., rates of nonrenewable resource use are still growing). Efforts to create a "circular economy" (one in which all materials are reused or recycled) represent a step in the right direction, but are destined to be woefully inadequate if we refuse to dial back total levels of consumption.

Soaring Economic Inequality

Economic inequality is typically framed as a problem of distribution of power; however, a little thought reveals that this way of looking at inequality, while valid up to a point, is also misleading.

As more wealth flows through a society, the members of that society will tend to become more unequal, until or unless efforts are undertaken to limit inequality by redistributing wealth, or until a political or social crisis forces redistribution. Once wealth inequalities exist, those with the most wealth will have an advantage which they can use in order to gain even more of an advantage. Anyone who has played the game Monopoly a few times knows that, while a wealth advantage may be hard to establish at first, such an advantage will allow you to rake in more and more income. Then suddenly you own everything. In society, too, wealth begets even more wealth. Since wealth is a key form of social power, this means inequality is at least partly a result of too much power.

Sometimes wealth accumulation by elites ends in revolution; other times it simply leads eventually to the failure of the overall economy. Either way, it seldom ends well. Walter Scheidel's *The Great Leveler* (2017) is one of the most thorough studies to date of economic inequality in world history. Sadly, Scheidel finds that inequality has seldom been substantially reduced through peaceful means. Rather, the most sweeping leveling events resulted from four different kinds of powerful shocks to society. Scheidel calls these the "Four Horsemen" of leveling—mass-mobilization warfare (which we touched on in Chapter 4), transformative revolution, state failure, and lethal pandemics.

Economic historians (including Scheidel, Thomas Piketty, and Peter Turchin) have begun tracking inequality in terms of the proportion of income and wealth concentrated in the top one percent of society. US wealth inequality, measured this way, reached a high level at the turn of the 20th century; Turchin attributes this to elites' disproportionate capture of increasing overall national wealth flowing from industrialization and trade.[21] Then, following the Progressive-era reforms, as well as New Deal programs (including more steeply progressive taxation) created during the Great Depression, inequality declined until the mid-1970s. Scheidel argues that this period of economic leveling was largely due to the two World Wars, which drove governmental redistributive reforms. Turchin and Scheidel agree that, from the 1970s until the present, inequality has reasserted itself—partly due to Reagan-era tax cuts for the wealthy, and partly due to changes in the financial system (which we'll discuss shortly).[22]

SIDEBAR 22

Inequality in Economic and Political Power in the US, and Its Consequences

Increasing the inequality of social power within a society tends to lead to worsening outcomes with regard to health, longevity, and social solidarity, while broader distribution of power tends to lead to better outcomes. This observation can be illustrated by comparing the United States with other industrial democracies.

During recent decades, political power has become more concentrated in the US, as documented the nation's declining score on the annual Democracy Index.[23] Relatively few people participate in elections, compared to other countries, and government exerts little effort to increase participation (indeed, the Republican party actively seeks to make voting more difficult). The share of Americans in unions has plummeted from 35 percent in the mid-1950s to about ten percent today (union membership is higher in other industrialized countries), and the decline has come about largely because government policy has made it easier for employers to suppress unionization.

Incarceration rates are higher in the US than in any other wealthy country, and the prison population is disproportionately Black and Latino. Time in prison leads to lingering health problems and to difficulty in finding well-paying work. And in nearly all states, felons are prohibited from voting, either for a specified period or for life.

These expressions of political power imbalance are clearly tied to greater concentration of economic power. While the ratio of executives' salaries to workers' salaries has risen in most countries in recent decades, the trend is most extreme in the US. This is partly because the US has a lower minimum wage. In addition, the racial wealth gap in the country is enormous and growing. Black families in 2016 had two percent of the wealth of white families.[24]

Tax policy in the US also increasingly favors the wealthy, when compared with policies in other industrialized nations, and income tax rates for high earners have fallen sharply in recent decades.

The United States has done less than other nations to discourage corporate concentration. Large companies are better able to fend off workers' demands for better compensation and working conditions.

Finally, due to their mostly privatized healthcare system, Americans pay more for drugs, medical procedures, and doctor visits.

The outcomes for the US have been grim: in nearly every high-income country, people have both become richer over the last three decades and have been able to enjoy substantially longer lifespans. But the United States is an exception. While average incomes have risen, most of the gains have gone to the wealthy, and life expectancy has risen in only three years since 1990. No other industrialized country has seen a similar lag in growing lifespan. Meanwhile, political polarization and economic inequality are leading to dwindling life satisfaction for most Americans, as measured by international surveys. All of these trends probably contributed to the nation's failure to meet the challenge of the coronavirus pandemic.

Did worsening US economic inequality cause an increase in vertical political power—or vice versa? An argument could be made either way. Perhaps the process of power concentration, once set in motion, simply proved to be yet another self-reinforcing feedback.[25]

The *Gini index* shows income distribution within a nation; the higher the Gini number, the greater the inequality. The Gini numbers for many nations, including the United States and China, have been increasing in recent years (during this period, the world's poorest peoples and nations have seen only marginal improvements, at best, in per capita wealth and income). Most political and social scientists say this is a dangerous situation because very high levels of inequality erode the legitimacy of formal and informal governance—including both elected leaders and economic institutions. As wealth and income inequality grow, imbalances in political power tend to follow. For example, a recent study by sociologists Martin Gilens and Benjamin Page found that the policy preferences of the bottom 90 percent of wealth holders in the US are consistently ignored by lawmakers in deference to the desires of the top ten percent.[26]

Throughout the past century, industrial nations have used progressive taxation—as well as public spending on health care, pensions, and unemployment insurance—as ways of restraining the tendency for wealth to become concentrated in ever-fewer hands (we'll discuss these strategies at more length in Chapter 6). These policies have certainly helped, but they clearly weren't full solutions to the inequality dilemma, in light of the recent trend toward greater concentration of wealth. In the US, new data shows that the income of the top one percent of earners has grown 100 times faster than the bottom 50 percent since 1970. Today, just eight men together enjoy as great a share of the world's wealth as the poorer half of humanity— over 3.5 billion people. It is almost certain that this degree of inequality has never existed before.[27]

To gain a historical and systemic perspective on this issue, it's helpful to understand wealth inequality in terms of the *commons*— the cultural and natural resources that are accessible to all members of a society, and not privately owned. In most pre-industrial economies, the commons included sources of food as well as natural materials for making tools and building shelters. People with little or no money could still subsist on commonly held and managed resources, and everyone who used common lands had a stake in preserving them for the next generation. During and especially after the Middle Ages in Britain and then the rest of Europe, common lands were gradually enclosed with fences and claimed as private property by people who were powerful enough to be able to defend this exclusionary appropriation, using laws and force of arms. The result was that people who would otherwise have been able to get by without money now had to buy or rent access to basic necessities. Again, the rich got richer, while the poor fell further behind.

The school of economics known as *neoliberalism* prescribes privatization (i.e., further exclusion) as the cure to nearly every economic ill. Thus, the past half-century—during which neoliberal economic ideas have prevailed throughout most of the world—has seen massive privatization of natural resources, industries, and institutions in the United States, in the United Kingdom, in the former Soviet Union and its Eastern Bloc of nations, and in poor, less-industrialized nations that supply resources to industrial economies.

Another cause of the recent increase in world (and US) wealth and income inequality is a process that's come to be known as *financialization*. If the charging of interest on loans tends to pump money upward within the social pyramid, then it stands to reason that a significant expansion of total debt, kinds of debt, and ways of creating and managing debt would serve to accelerate the wealth pump. In recent decades, networked computers and global communications have made it possible to make money on trading and insuring many new kinds of financial investments. As a result, those who spend their lives chasing financial profit have seen their wealth and income rise much faster, on average, than those who work for an hourly wage or yearly salary.

For the past few generations, most people have been living under the assumption that, if we just increase the size of the pie (that is, increase overall wealth as measured by GDP), then inequality will be less of a problem: each person will have a bigger slice, even if a few get portions so huge that no individual could possibly consume so much. But that assumption has been shown to be false: we're reaching limits to the expansion of the pie (i.e., economic growth is stagnating or reversing), and inequality is burgeoning to levels such that increasing numbers of people are questioning the legitimacy of a system that benefits elites so much and so unfairly.

If wealth is power, and if inequality is a problem of too much power, then there is really only one peaceful solution to the problem of growing inequality: reduce the wealth of those who have the most. Ultimately that means reducing the wealth-power not just of billionaires, but also of rich countries relative to poor ones. We must all aim for a relatively simple life that's attainable by everyone, and that's able to be sustained without harming ecosystems.

History, in Walter Scheidel's view, is not encouraging in this regard. While he acknowledges the leveling effects of unionization, progressive taxation, and government redistributive programs, he sees these as historically dwarfed by the impacts of the Four Horsemen mentioned above. Nevertheless, policy-based leveling helps at the margins, and we would do well to *anticipate* the next major leveling event that's beyond our control—which, as we'll see later in this chapter, could take the form of a debt-deflation financial crisis.

Figure 5.4. United States top one percent income share (pre-tax).

Credit: World Inequality Database.

Pollution

In nature, waste from one organism is food for another. However, that synergy sometimes breaks down and waste becomes poison. For example, about 2.5 billion years ago cyanobacteria began to proliferate; because they performed photosynthesis, they gave off oxygen—far more of it than the environment could absorb. This huge, rapid influx of oxygen to the atmosphere was toxic to anaerobic organisms, and the result was a mass extinction event. Clearly, humans aren't the only possible source of pollution.

However, these days the vast majority of environmental pollution does come from human activities. That's because we humans are able to use energy and tools to extract, transform, use, and discard ever-larger quantities of a growing array of natural substances, producing wastes of many kinds and in enormous quantities.

The social power issues related to pollution can often be untangled simply by identifying who is polluting, and who is on the receiving

end of pollution. The powerful tend to benefit directly from economic processes that produce waste, while insulating themselves
from noxious byproducts; while the powerless share much less in the
economic benefits and tend to be exposed to a much greater extent to
toxins. Polluters often count on society as a whole to clean up their
messes, while they keep the profits for themselves. Thus, the ability
to pollute, without the requirement to fully account for that pollution
and its impacts, is itself a form of power.

Nobody sets out to pollute. Pollution is always an unintended
byproduct of some process that uses energy and materials for an
advantageous purpose. Humans were already causing pollution in
pre-industrial times—for example, when mining tin or lead, or when
tanning leather near streams or rivers. Today, however, with much
higher population levels, and with much higher per capita rates of
resource usage, examples of environmental pollution are far more
numerous, extensive, and ruinous.

Arguably, the modern environmental movement began with the
publication of Rachel Carson's *Silent Spring* (1962), which warned the
public about the invisible dangers of new synthetic pesticides, such
as DDT. These chemicals were hailed as giving humanity the power
to control insect pests, which regularly decimated crops, and disease microbes that spread death and misery via illnesses like chol-
era and malaria. The benefits of pesticides were undeniable—but so
were the unintended side effects. DDT accumulated in animal tissues
and worked its way up the food chain; it caused cancer in humans
and was even more harmful to bird populations. Carson's book led
to the banning of DDT in the United States and changed public attitudes about pesticides generally. Many other pesticide regulations
followed. Yet, despite these successes, the overall global load of pol-
lutants has continued to worsen, and new pesticides—often inade-
quately tested—have replaced DDT.

The pesticides most widely used today are collectively known as
neonicotinoids; they are being blamed both for widespread crashes
in the populations of bees and other insects, and for serious impacts
to aquatic ecosystems worldwide.[28] Multiple studies have linked

bee colony collapse disorder—which has decimated bee numbers in North America and Europe—with neonicotinoids; some of the latest studies suggest the pesticides kill colonies slowly over time, with queen bees being particularly affected.[29]

Like pesticides, artificial ammonia-based fertilizers were welcomed as a humanitarian boon, dramatically increasing crop yields and staving off famine. But fertilizer runoff from modern farming creates "dead zones" thousands of square miles in extent around the mouths of many rivers. Fertilizer acts as a nutrient to algae, which proliferate and then sink and decompose in the water. The decomposition process consumes oxygen, depleting the amount available to fish and other marine life.

Air pollution from burning coal in China is so thick and hazardous that it causes nearly 5,000 deaths per year (millions more face shortened lives), and the haze sometimes drifts as far as the West Coast of the United States. A combination of firewood, biomass, and coal burning, compounded by population growth, has similarly resulted in deadly and worsening air quality in India. A study based on 2016 data showed that at least 140 million people there breathe air that is polluted ten or more times the World Health Organization's safe limit. India is home to 13 of the world's 20 cities with the worst air pollution.

Nuclear power plants supply electricity reliably; but, in rare instances where something goes terribly wrong, persistent radioactive pollution can result. Soil in areas around the melted-down nuclear reactors in Fukushima in Japan and Chernobyl in Ukraine will be radioactive for centuries or millennia. Attempts to stem the Fukushima disaster have produced more than a million tons of contaminated water stored in tanks at the site. As more water accumulates, cleanup managers see no solution but to release radioactive water into the sea.

Plastic packaging offers affordable convenience and protection, especially in the food industry, where it prevents the spoilage of thousands of tons of food annually. However, plastic particles (a large proportion of which originated as food packaging) in the Pacific Ocean have formed giant floating gyres, and it's been projected that by 2050 the amount of plastic in the oceans will outweigh all the remaining fish.[30] What's more, plastic packaging leaches small amounts of organic chemicals, some known to cause cancer, into the foods they protect. Many of these chemicals are known to mimic the action of hormones, and are believed to contribute to diabetes, obesity, and falling male sperm counts as well as other fertility problems, both in humans and other animals.

Another endocrine system-disrupting class of chemicals is known as PFAS (polyfluoroalkyl substances), used, for example, in making Teflon for nonstick cookware.[31] These chemicals have been spread globally (they've been detected, for example, in Arctic polar bears and fish in South Carolina), and are estimated to be in the blood of over 98 percent of Americans.

A 2013 report by the UN Environment Programme and World Health Organization noted that endocrine diseases and disorders are on the rise globally, including cancers, obesity, genital malformations, premature babies, and neurobehavioral problems in children. Total incidences of human cancers are increasing at between one and three percent annually worldwide, according to some estimates.[32]

Immune system impacts, reproductive system impacts, and impacts on wildlife are all tied to chemicals that disperse in the environment without breaking down. At least 40,000 industrially produced chemicals are now present in the environment. Few are specifically regulated, and of those that are, regulation started long after their introduction. The initial burden is on those impacted to prove chemicals harmful, not on industry to prove them safe.

It is possible in some instances to reduce pollution to relatively harmless levels without having to forgo economic benefits, just by substituting safer alternatives for existing chemicals, fuels, and industrial processes. For example, many companies are developing and using bioplastics to replace plastics made from petrochemicals. However, it will take decades to identify and develop those safer alternatives in all instances; and, in many cases, alternatives may be more expensive or less functional. In the meantime, the only way to ensure human and ecosystem health is simply to stop using polluting chemicals, fuels, and processes. That means giving up the power to control our environment in many ways we currently do.

Overpopulation and Overconsumption

It's often said that there is safety in numbers. But there is also power in numbers: a larger population size can sometimes translate to a bigger economy, a larger voter base, or a more formidable army—all of which confer advantages to nations or sub-national groups. In pursuit of such advantages, societies, and groups within societies, have long sought ways to increase their numbers.

As we saw in Chapter 3, the adoption of agriculture was likely both a result of crowding, and a facilitator of further population growth. But overall that growth was typically slow and uneven, punctuated by famines and plagues.

Big God religions, discussed in Chapter 3, based part of their success on their ability to convince their adherents to have more children, thereby out-competing rival groups. For example, the Bible enjoins believers to "be fruitful and multiply," and census records today show that Christians, Muslims, Hindus, and Jews still outbreed the religiously unaffiliated.[33] The Mormon Church likewise urges believers to have large families, and in just 170 years it has grown from a few hundred followers to 15 million globally. While missionary activities and conversion played a significant role in this expansion, high fertility has been responsible for a persistent, underlying growth trend for Mormonism. "It is no accident," writes Ara Norenzayan in *Big Gods*, "that religious conservative attitudes on women's rights, contraception, abortion, and sexual orientation are conducive to maintaining high fertility levels."[34]

While agriculture enabled larger population sizes, and Big God religions promoted higher fertility among followers, for the past couple of millennia actual rates of world population growth were typically a tiny fraction of one percent annually. Still, simple arithmetic shows that even slow rates of increase, if compounded, can eventually lead to very large numbers. At the start of the 19th century, English cleric and scholar Thomas Malthus wrote *An Essay on the Principle of Population*, in which he hypothesized that any improvements in food production, rather than resulting in higher living standards, would instead ultimately be outpaced by population growth, since "the power of population is indefinitely greater than the power in the earth to produce subsistence for man."[35] Malthus, a deeply conservative thinker, wrote his essay to counter the utopian expectations of writers like William Godwin, who believed that progress would lead to a future human society characterized by freedom and abundance.

Throughout the two centuries since Malthus, evidence has consistently contradicted his dire warnings, and his name is now evoked to dismiss any and all long-range pessimism regarding population growth and resource availability. As we've seen, guano deposits in South America provided fertilizer for farmers in Europe and the Americas in the 19th century; and just as those deposits were exhausted, new synthetic fertilizers made from fossil fuels became available. Together with tractors, new farming techniques, other soil amendments, and new synthetic pesticides, ammonia fertilizers enabled a dramatic increase in farm yields. Millions of acres of forest were cleared to make room for more farms. At the same time, better sanitation and new disease-fighting drugs lowered death rates. As a result, while population has grown roughly eight-fold since the publication of Malthus's book, food production has more than kept pace.

Still, the realistic prospect of continuing to expand agricultural yields at the historically recent rate is widely disputed. New farmland can be brought into production only by taking more habitat away from other creatures, thereby exacerbating the biodiversity crisis. Simply applying more fertilizer won't help much, as the limits of its effectiveness have nearly been reached, and environmental impacts from excessive fertilizer use are soaring.[36] Meanwhile current farming practices are resulting in the annual erosion or salinization of tens of billions of tons of topsoil annually worldwide.[37] The difficulty of continuing to defeat famine with current methods is at least implied by the recent proliferation of calls for completely new food production methods that don't require soil or agricultural land.[38]

World population growth rates in the mid-20th century were probably the highest in all of human history. Between 1955 and 1975, population increased at over 1.8 percent per year, with a peak rate of growth of 2.1 percent occurring between 1965 and 1970. For perspective, recall from Chapter 1 that an annual growth rate of one percent results in a doubling of population size in 70 years; two percent growth doubles population in just 35 years.

The percentage rate of increase is one way to understand population growth; another is to track the actual numbers of humans added each year (births minus deaths). This annual increase peaked at 88 million in 1989, then slowly declined to 73.9 million in 2003, then rose again to 83 million as of 2017—even though the percentage rate of growth had declined to about 1.2 percent. A declining rate of growth can still produce a larger annual increase in human numbers because it's a smaller percentage of a much bigger number (this is one of the devilish consequences of exponential growth). We're currently adding about a billion new humans every 12 years.

Generally, wealthier nations have seen a decline in their population growth rates in recent decades, but annual growth rates remain above two percent in many countries of the Middle East, sub-Saharan Africa, South Asia, Southeast Asia, and Latin America. The world growth rate is projected to decline further in the course of the 21st century. However, the United Nations projects that world population will reach more than 11 billion by 2100—with most of that growth occurring in poorer countries where it will be difficult for people to obtain the additional housing and food they will need.

If human numbers were to continue growing at just one percent per year, population would increase to over 115 *trillion* during the next thousand years. Of course, that's physically impossible on our small planet. So, one way or another, our population growth will end at some point. The question is, *how*—through plan or crisis?

Population growth makes every environmental crisis harder to address,[39] and contributes to political and economic problems as well (for example, nations with high population growth rates are more likely to have authoritarian governments).[40] Reproduction is a private decision, a human right. Yet, quite simply, the more children we have now, the more human suffering is likely to occur later this century. If we don't find ways to reduce our numbers humanely, nature will find other ways.

Demographics is the statistical study of population, and demographers speak of a "demographic transition," which describes the tendency for population growth rates to decrease as nations become wealthier. Most policy makers hope to avoid the problems associated with population growth simply by pursuing more economic growth. However, as we have already seen, further economic growth is incompatible with action to halt climate change and resource depletion.

Ecological footprint analysis measures the human impact on Earth's ecosystems. Our ecological footprint is calculated in terms of the amount of land and sea that would be needed to sustainably yield the energy and materials we consume. According to the Global Footprint Network, at our current numbers and current rate of consumption we humans would need 1.5 Earths' worth of resources to sustainably supply our current appetites. At the US average amount of consumption, we would need the equivalent of at least *four* Earths to sustain us. Of course, we don't have four or even one-and-a-half planets at our disposal—yet we are still in effect using *more* than one Earth by drawing down resources faster than they can regenerate, thereby reducing the long-term productivity of ecosystems that would otherwise be available to support future generations.

Human impact on the environment results not just from population size, and not just from the per capita rate of consumption, but from both factored together (with further variation in impact depending on the technologies applied to productive and consumptive purposes). Clearly, different countries' per capita rates of consumption vary greatly: the ecological footprint of the average American is almost 11.5 times as big as that of the average Bangladeshi. On the other hand, Bangladesh has a population density of 1,116 persons per square kilometer, as compared with the US, with a population density of 35 people per square kilometer. Within both nations, there is also a great deal of economic inequality—and thus a vast variation in levels of consumption. Overpopulation is tied to other problems, and solving it won't automatically solve these others; but ignoring it tends to make related problems worse.

There are many ways to avoid recognizing and confronting the problems of overpopulation and overconsumption. Pointing fingers and insisting that other nations are to blame is one; another is to hold firmly to the illusion that economies can somehow continue to grow without using more energy and materials, thereby providing for more people without further depleting or damaging nature.[41] But, once again, a problem ignored is still a problem.

Giving up population growth means giving up a certain kind of power—one that many groups within society currently prize. But, for humanity as a whole, the power of population growth is ultimately self-defeating, and we are approaching or beyond natural limits. If we do not choose to rein in our power of population growth, we will face a series of snowballing crises that will disempower us in ways none of us would choose. (We'll return to the discussion of population growth and what to do about it in Chapter 7.)

Global Debt Bubble

As we saw in Chapter 3, debt and money are inseparable, and money can be understood as quantifiable, storable, and transferrable social power. Debt has the further characteristic of time-shifting consumption: with it, we can consume (thus enjoying the fruits of power) now, but pay later. This can create problems when "later" finally arrives.

Many individuals, and whole societies, have overindulged in debt throughout history as a way of making life easier for the moment, with only a vague hope of paying it off in the future; or as a way of investing capital with the expectation of earning interest or dividends, but without adequate understanding of the risks involved. As societies pass through periods of optimism and expansion, there is a tendency for people and businesses to take on larger loads of debt. Such episodes of widespread borrowing and lending are sometimes called inflationary bubbles, and they can be based on irrational group psychology. As manias crash on the shoals of reality, debts can be defaulted upon in large numbers, initiating a deflationary depression. Such boom-and-bust cycles can be traced back to the first societies to use money and debt.[42]

In Chapter 4, we saw how decades of access to cheap, abundant fossil fuels led to an unprecedented expansion of population, production, and consumption—the Great Acceleration. That astonishing period of expansion was marred by temporary financial setbacks—repeated recessions, panics, and depressions (for example, in the US, the recessions of 1882–85 and 1913–14; the panics of 1893, 1896, 1907, and 1910; and the depressions 1873–79 and 1920–21, capped of course by the Great Depression of 1929–41). The last of these, especially, along with the two World Wars, led to a series of general overhauls of the global monetary and financial systems.

For many centuries, gold and silver had served as anchors to national monetary systems and international trade: while paper deposit receipts for precious metals could be issued in excess of actual metal holdings (usually by banks or governments) and used as money, the knowledge that an eventual accounting or reckoning could occur, requiring the surrendering of actual gold or silver, usually kept such inflationary temptations in check. Instances of hyperinflation were relatively rare (one notable exception was the French hyperinflation of 1789–1796).

However, by the late 19th century, it was clear that the world had entered a new era of abundant energy, technological innovation, expanding industrial production, scientific discovery, and financial opportunity. We've already seen (in Chapter 4) how this led to a crisis of overproduction, which was largely solved with expansions of advertising and consumer credit. But the Great Acceleration also posed a monetary problem: there was simply not enough gold and silver money in existence to facilitate all the investment and spending that were otherwise possible. The solution consisted of a gradual retreat from the use of precious metals and an increasing reliance on fiat currencies (i.e., currencies not backed by any physical objects or substances) that could be called into existence by banks. The rules for the creation of money were set forth by new central banks (including the US Federal Reserve Bank, created in 1913), which were themselves authorized by national governments. Governments could also create money by issuing bonds (i.e., certificates promising to repay borrowed money at a fixed rate of interest at a specified time)—a means of deficit spending.

Anticipating growth in tax revenues, and wishing to fund public services and military expansion, governments tended to overspend and thereby to become over-indebted. Nevertheless, as economies snowballed during the Great Acceleration, government debt came to be normalized. Consumer credit and business debt also ballooned, and were also accepted as innocuous facilitators of growth.

The result was, of course, a steep increase in overall debt. This could be measured in several ways. First was simply in terms of the number of units of a given national currency. But large increases in debt and spending tended to lead to inflation, diluting the actual purchasing power of the currency. Debt could also be translated to its purchasing-power equivalent. Finally, government debt could be measured in terms of the debt-to-GDP ratio (too much debt per unit of GDP has historically led to crises in which governments have been unable to make payments on debt; that in turn has led to financial panics in which large numbers of people suddenly wished to sell stocks and bonds, thus dramatically lowering their value). This ratio has fluctuated during the modern industrial era, for the US and the world as a whole, but in recent years the ratio has grown significantly, even while GDP has been increasing.

The discipline of economics was mostly formulated during the Great Acceleration (as described in Chapter 4). Economists observed the economic expansion that was occurring around them and adopted the absurd assumption that growth is normal and desirable, and can continue forever. Economists became, in effect, the brain (such as it is) of the Superorganism. While they were able to collect and analyze economic statistics at ever increasing levels of detail, they generally failed to appreciate environmental limits. Oddly, conventional economic theory also proposed simplistic or misleading explanations for the workings of money and debt. Partly as a result, economists tended to be surprised by each new recession or panic, and searched for ways to explain it after the fact.[43]

For the highly industrialized nations of North America and Western Europe, the rate of economic growth peaked in the first few decades after World War II. Especially in the United States, GDP growth began to slow in the 1970s, while total debt started to grow even faster than before—perhaps partly as a result of government and central bank efforts to continue expanding consumption, employment, and tax revenues. Another reason for the debt takeoff was the start of the process of financialization, mentioned previously in our discussion of economic inequality. The deregulation of the financial industry, and the introduction of new kinds of debt instruments, including new forms of financial derivatives (which are, in effect, bets on the future values of stocks, bonds, and other financial instruments), starting in the 1980s, led to a growth of the financial industry relative to the size of the rest of the economy, as measured by GDP.

As economists Carmen Reinhart and Kenneth Rogoff have documented, the over-extension of debt during prosperous times can lead to a debt-default crisis when general conditions worsen (i.e., when business profits fall, unemployment increases, and purchases slow), as defaults cause the economy to decline much faster than would otherwise be the case.[44] But history has little precedent for today's situation—in which the entire networked global economy is leveraged to the hilt with debt and derivatives, all based on unrealistic expectations of future growth. Climate change and resource depletion virtually ensure the eventual end of this temporary era of abundant energy, rapid population growth, and industrial expansion. The prospect of the end of growth poses a host of problems for governments and for the populace as a whole. But add to that political conundrum the possibility of the collapse of our global financial house of cards, and the prospects become truly daunting.

I was among those who thought the world might have reached the end of growth back in 2008, when all hell was breaking loose in world financial markets. However, government and central bank efforts to forestall collapse succeeded by piling even more debt onto a system already in danger of drowning in debt (today's world total debt level, at over $180 trillion, is higher than before the 2008 crisis, and is approximately 300 percent of world GDP—a ratio typically seen in economies on the verge of financial collapse).[45]

The years after the 2008 crash saw some semblance of recovery, in that growth, as conventionally measured, was revived. Unemployment levels in the United States were at historic lows prior to the COVID-19 pandemic. But the effort required to achieve this recovery was truly astonishing. Trillions of dollars, euros, and yuan were created and spent by central banks in the early 2010s, with much of this money directed toward purchases of government bonds. More trillions were called into existence through government deficit spending. Some analysts point out that, in the US at least, during the decade after 2008 the dollar amount of cumulative government deficit spending exceeded the dollar amount of GDP growth.[46] By November 2019, US federal government deficit spending had increased to over one trillion dollars per year. As we are about to see, economic mayhem caused by the COVID-19 pandemic required even more deficit spending by governments, along with the injection of even more capital into the financial systems by central banks.

During the post-2009 "recovery," central bank bond buys skewed the entire financial system. Business strategist and financial consultant Graham Summers called the situation circa 2019 the "Everything Bubble" (the title of his recent book), because when government bonds, which serve as the foundation of our current financial system, are in a bubble, then all risk assets (i.e., equities, commodities, high-yield bonds, real estate, and currencies) are in a bubble too. Thus, efforts contributing to the "recovery" did not solve our underlying economic problems, but only hid them. There had been no significant reforms to the financial system or efforts to reduce society's reliance on unsustainable debt. Moreover, injecting money into financial markets to stem losses just increased society's inequality of wealth and income, since nearly all the new money was funneled to investors.[47]

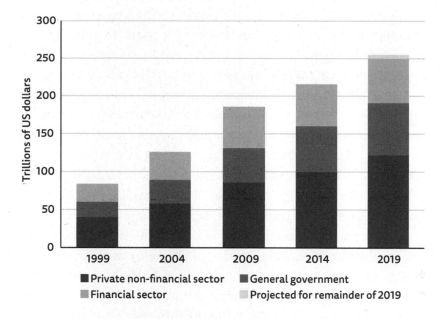

Figure 5.5. Growth in total global debt by sector, 1999–2019.

Credit: Enda Curran, "The Way Out for a World Economy Hooked On Debt? More Debt,"
Bloomberg, December 2, 2019.

The 2009–2019 "recovery" proved to be merely a temporary reprieve, giving way to a different sort of economic crash in the immediate wake of the coronavirus pandemic. As nations implemented lockdowns, unemployment levels soared, reaching in the US 14.4 percent in May, 2020. Central banks once again undertook defensive measures, in many cases on an even larger scale than in 2008–2009, with the US Federal Reserve bailing out investors with large-scale purchases of corporate bonds. Meanwhile, the US Congress quickly passed a relief package totaling $2.2 trillion (a second, smaller relief bill followed at the end of 2020). As of this writing, the long-term economic consequences are unclear, though a deep and extended global recession or even depression appears possible.

Some imaginative economists have proposed Modern Monetary Theory as a way out of our collective debt trap.[48] The basic idea is that governments can, and should, create all the interest-free money they want, and use it to fight climate change with a Green New Deal or to reduce economic inequality by way of a Universal Basic Income (UBI).[49] Such proposals are useful up to a point: doing away with interest on government debt would free up needed capital for projects and programs that benefit the public, and UBI could reduce misery for the poorest. However, we can print as much money as we desire, and add zeros to the end of the number denoting the average salary, but if raw materials and energy aren't growing as fast as the money supply, the likely eventual result will be runaway inflation, which could be catastrophic for everyone.[50] In fairness, Modern Monetary Theorists have given considerable thought to the problem of inflation, and have come up with ways of limiting it—such as by levying deficit-reducing taxes during times of full employment, to reduce aggregate demand. Realistically, however, we are facing a future with declining raw materials and energy, so the best we can hope for is a fair sharing of the burden of economic contraction.

Any long-term solution to the global debt bubble must eventually include massive debt forgiveness, which would require investors and banks to accept sizable financial losses—effectively giving up the power that their financial holdings confer on them. Such a program of debt forgiveness would have to be accompanied by a major overhaul of the financial system, so that the latter would no longer rely on the continual overall growth of the economy, if we are to avoid simply creating conditions for yet another bubble in the future.

Weapons of Mass Destruction

As we saw in chapters 3 and 4, warfare has been a significant driver of cultural evolution since the beginning of complex societies thousands of years ago. Warfare has helped turn us into the most cooperative species on Earth, and has led to the development of a wide range of technologies, from the horse cart to the internet. Warfare has made us formidable and powerful—but also extremely vulnerable: the next total war could spell the end of the human experiment.

We humans have long sought improved weapons to make killing easier, safer, and more effective from the point of view of those wielding the weapons. Since hand-to-hand weapons have always entailed the most risks, special effort was invested in developing projectile weapons, from slings to bows and arrows, catapults, and guns. Only in the last couple of thousand years, and especially the last century, have societies developed weapons of mass destruction (WMDs), which are typically defined as weapons that can kill or bring significant harm to many people at once, or cause great damage to human-made structures (e.g., buildings), natural resources (farmland, oilfields), or the biosphere. The following are the main WMD categories.

Chemical weapons were likely first developed around 1000 BC by the Chinese, who used arsenic smoke to incapacitate or kill enemies on the battlefield. Mustard gas, which blisters the lungs, was employed extensively in World War I; its sporadic use continues nearly to the present (it was deployed by ISIS against the Syrian Army during the battle in Deir ez-Zor in 2016). In 1995, sarin gas was used to attack subway trains in Tokyo, and ricin has been used in various terrorist plots up to the present.[51] Types of chemical weapons include blister agents, blood agents, choking agents, nerve agents, tear gas, vomiting agents, and psychiatric compounds. Chemical weapons are generally not as dangerous as biological weapons due to the inefficiency of delivery systems and the need for significant quantities of the chemical. However, unlike bioweapons, chemical weapons take effect immediately, and many are relatively easy to manufacture using existing industrial chemical infrastructure. The 1925 Geneva Protocol banned the use but not the possession of chemical weapons, while the Chemical Weapons Convention, adopted in 1997, outlaws the production, stockpiling, and use of chemical weapons and their precursors.

Biological weapons were first used in North America in 1763, when British officers distributed blankets containing smallpox to Native Americans (Australian and US soldiers later used this tactic in mass-murder campaigns against native peoples). Since then, however, biological weapons have been used primarily by individuals, rather than groups. Types of biological weapons include bacteria (plague, anthrax), viruses (smallpox, hepatitis, avian influenza), and toxins (botulism). Biological weapons work by contact with the skin, through ingestion, or via the lungs. Bioweapons could theoretically be genetically engineered to attack specific genetically similar populations, or even certain food crops. The biggest drawback of bioweapons, from the standpoint of potential users, is that impacts are difficult to control, and the attacker's population could be vulnerable to the unleashed disease organism. The Convention on the Prohibition of the Development, Production and Stockpiling of Bacteriological and Toxin Weapons and on their Destruction, signed in 1972, was the first multilateral disarmament treaty banning the production of an entire category of weapons, and it remains in effect.

Cyberweapons are designed to damage computers or information networks through, for example, computer viruses or denial-of-service attacks. Cyberwarfare may not meet the typical definition of the term *war*, as the objective is not to kill soldiers or civilians outright, but to undermine or incapacitate the basic systems and functions of an organization or society. Many nations, including the US, UK, Russia, China, India, Israel, Iran, and North Korea have active and rapidly growing cyber programs for offensive and defensive purposes. Cyberwarfare can be used in propaganda operations, as it was in the 2016 US presidential election, for the purpose of undermining governance and social cohesion. Cyberattacks can also target infrastructure, including electricity grids, water systems, financial systems, and communications systems, all of which are typically computerized and connected to the internet in some fashion. It is now possible to cripple an enemy country without either firing a shot, or *directly* killing or injuring a single person (though indirect mortality could be very substantial). It may still be true that power grows from the barrel of a gun (as Mao Zedong famously said), but now it also springs from a computer keyboard.

Robotic weapons are not considered WMDs, but are becoming increasingly common and deadly. Military robots were introduced in World War II (German Goliath tracked mines—essentially bombs in small, remote-controlled tanks) and the Cold War (Soviet radio-guided "teletanks"), but were little used until fairly recently. Today, the US employs drones such as the Pioneer and Predator to gather intelligence or to deliver air-to-ground missiles. These are remotely operated from a command center, sometimes on the other side of the planet. Since 2000, the Central Intelligence Agency and Joint Special Operations Command at the Pentagon have carried out hundreds of drone strikes in Yemen, Pakistan, and Somalia. Advocates say the drone program has saved American lives and reduced the need for deadly ground operations. But the program has also killed hundreds, if not thousands of civilians, according to data collected by the NGO Airwars.[52] Antiterrorism experts warn that such civilian killings often have a radicalizing effect on the societies on the receiving end. Moreover, human rights organizations (such as Human Rights Watch) have criticized drone "targeted killing" operations as "clear violations of humanitarian law."[53] Robotic snipers and sentries, intended for urban combat, are under development, as are robotic bombers and fighter aircraft. However, no international robotic weapons treaties are currently in effect.

Nuclear weapons were developed and used by the United States during World War II; the Soviet Union soon began its own nuclear program, and both nations competed throughout the Cold War to build more and bigger warheads, and more robust guided missile and bomber delivery systems. Currently nuclear weapons are believed to be in the arsenals of nine countries. By 1986 the number of active nuclear weapons had climbed to 70,300; today, decades after the collapse of the Soviet empire and the end of the Cold War, there are about 9,500 nuclear warheads in service and nearly 14,000 total nuclear warheads in the world (many decommissioned weapons are stored or only partially dismantled).[54] Individual warheads vary in firepower: the most powerful US weapon in active service is the B83 nuclear bomb, with a 1.2 megaton yield (i.e., it produces a blast equivalent to 1.2 million tons of TNT), while tactical weapons intended for use in the battlefield range down to 0.02 kiloton.[55] The largest nuclear weapons in service are capable of destroying entire cities, with radiological effects spreading hundreds of miles in all directions. A full-scale nuclear war could effectively destroy nations within minutes, killing hundreds of millions outright, and could raise enough radioactive dust and smoke high enough into the atmosphere so as to alter the global climate, possibly leading to a "nuclear winter" during which no crops would grow anywhere on the planet.[56] New generations of nuclear weapons are currently being designed, with the objective of making them smaller and more readily usable in foreseeable conflicts. Hypersonic delivery systems (Mach 5 and above) are being engineered for the purpose of thwarting anti-missile defense batteries.[57] Nuclear weapons could conceivably be deployed by terrorist organizations, though the latter are more likely to attempt to build a radiological weapon, which combines highly radioactive materials with conventional explosives. The Nuclear Proliferation Treaty entered into force in 1970, and since then 190 nations have signed on. But four nations with nuclear weapons (India, Pakistan, Israel, and North Korea) are not signatories.

Over all, trends in weaponry during the Great Acceleration have favored the development of weapons that are easier and faster to deploy (hence offering less opportunity for moral reflection prior to their use), and that are often more lethal. While weapons confer power, including the power of self-defense, a case can persuasively be made that humanity has far too much killing power, in that we now face the possibility of a future world war that civilization, perhaps humanity itself, and much of the biosphere, might not survive.

The only real solution to the problem of weapons of mass destruction is to ban and dismantle at least the deadliest of our weapons— which means giving up power. As we have seen, steps along these lines have already been taken with chemical, biological, and nuclear weapons treaties. Yet we clearly have not done enough, as the risk of annihilation remains, and, in the view of some weapons watchdogs, is still worsening.

SIDEBAR 23

Guns: The Power to Kill Cheaply and Easily at a Distance

America's TV and movie crime dramas tend to glorify guns. The good guy (usually a guy) wields a gun, a tool of lethal power, using it to overcome evil. It's no wonder that many young boys want a toy gun at some point, so that they can feel more powerful; and many grown-up American men maintain their youthful fantasies into adulthood, building up sizeable arsenals. However, for any society, the cost of having more guns is always more gun violence and bloodshed.

The very first guns were invented in China around the year 1000; however, it was Europeans who developed portable and reliable firearms in the late 14th and early 15th centuries, just in time for them to be used in the conquest of the Americas. Not only did Europeans use guns to kill and capture Native Americans; they also traded guns to Indians for pelts. Then Native tribes used guns on each other and against the settlers.

In his book *Thundersticks: Firearms and the Violent Transformation of Native America* (2016), David J. Silverman details the role guns played in the deaths of hundreds of thousands of Native Americans from roughly 1600 to 1900. In the Southeast, colonists exchanged guns for Indian slaves, depriving as many as 50,000 people of their freedom and killing many more in the process. By the early 1700s, the Floridian peninsula had been virtually emptied of its Indigenous population.

In the Northeast, the Iroquois League nations began trading for munitions as soon as Dutch flintlock muskets arrived in the 1630s. By the mid-17th century, the Iroquois had become the preeminent military power as far west as the Mississippi River. The downfall of the Iroquois ensued when enemy tribes gained guns. This pattern would be repeated across North America for more than two centuries. Silverman writes:

> The results were terrible, with intertribal wars and related outbreaks of epidemic diseases dramatically reducing the population of nearly every Native group.... Some groups were completely wiped out. In the long term, however, the growing balance of power, and recognition of the high cost of gun warfare, produced something of a détente. By the end of the [18th] century, people who expected their young men to prove themselves as warriors would have to look outside the region for victims among the poorly armed tribes of the continental interior. As they did, the gun frontier spread with them, leaving a trail of devastation that was becoming a signature of colonialism in Indigenous North America.

The brutal transformation of societies by guns continues up to the present in Afghanistan, the Middle East, Africa, and wherever conflict and societal breakdown lead men to take up arms.

The gun is usually not considered a weapon of mass destruction, since it typically kills only one victim at a time. However, the widespread proliferation of guns has translated to fatalities on a massive scale. In 2016 alone, an estimated 250,000 people died as a result of firearms worldwide (about 64 percent of gun deaths were homicides, about 27 percent suicides, and 9 percent resulted from accidental shootings). That year, a negligible number died from the use of weapons of mass destruction.

Most nations make efforts to control guns by licensing them, by banning certain kinds of weapons, by restricting sales, and by restricting certain groups (such as children and convicted felons) from purchasing them. In some countries, such measures are so successful that gun deaths are rare and even police seldom carry guns.

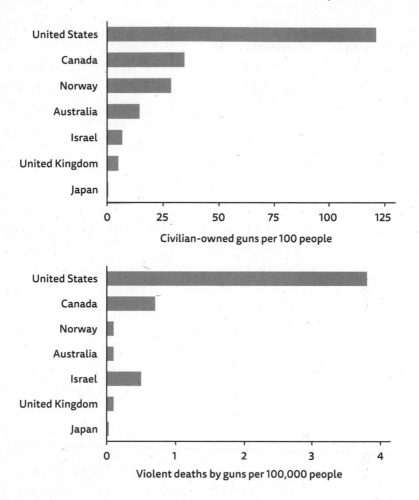

Figure 5.6. Gun ownership and violent deaths.

Credit: Small Arms Survey. Gun ownership data from 2017, at smallarmssurvey.org. Violent deaths data from 2018, at smallarmssurvey.org/tools/interactive-map-charts-on-armed-violence.html.

As we saw in Chapter 4, humans have developed a unique set of specializations (primarily language and toolmaking) that enable us to outperform virtually any other organism that has evolved an extraordinary ability. That is, with cooperation and the appropriate prostheses, we can fly higher than any vulture and faster than any falcon; see further than any eagle; smell more keenly than any bear; and lift heavier burdens than any elephant. But, as we saw in Chapter 1, organisms nearly always pay a price for extreme specialization. So do we, with our tool-making and linguistic specializations. Indeed, we humans have reached the point where the risks posed by our evolutionary successes are beginning to outweigh their advantages. We have taken over too much land from other species, we use too much energy, we have become too numerous, we've created too much debt and inequality, and we have weapons that are too deadly.

The only realistic solutions to the crises we have surveyed in this chapter involve giving up some of our power. Philosophers have warned us for centuries about the perils of power. "Power corrupts," Lord Acton famously admonished. But most such cautions typically concerned the abuse of social power; few philosophers could foresee the possibility that human beings—by harnessing millions of years of stored ancient sunlight—might, with their physical power, one day overwhelm the natural systems that make life possible.

There is a profound irony in all of this: most people who understand that inequality is increasing, that birds and insects are disappearing, and that future generations may not be able to survive on this heating planet, feel *powerless* to change the situation. One can protest, one can write letters. But the momentum of decades of over-empowerment seems unstoppable. We must somehow overcome our power-seeking habit, which has been present for all of human history, *plus* the over-powered state we've more recently attained as a result of our adoption of fossil fuels. We will examine this conundrum in more depth in Chapter 7.

In the next chapter we will explore the crucial question: is it even possible for our species to voluntarily and deliberately rein in its own power?

OPTIMUM POWER

Sustaining Our Power Over Time

It always has been so. The grievances of those who have got power command a great deal of attention; but the wrongs and the grievances of those people who have no power at all are apt to be absolutely ignored. That is the history of humanity right from the beginning.

— VARIOUS, *The Suffragettes*

He who reigns within himself, and rules passions, desires, and fears, is more than a king.

— JOHN MILTON

IN 1964, SOVIET ASTRONOMER NIKOLAI KARDASHEV PROPOSED a scale for measuring a civilization's level of technological advancement based on the amount of power it can command. The scale has three levels:

- Type I civilization, also called planetary civilization, can use all of the power reaching its planet from its closest star. Estimated to be in the range of 10^{16} to 10^{17} watts, this amount of power would be three or four orders of magnitude more than humans are currently able to wield (2×10^{13} watts).
- Type II civilization, also called stellar civilization, can control power at the scale of its solar system (the Sun's power is about 4×10^{26} watts).
- Type III civilization, or galactic civilization, can control power at the scale of its entire galaxy (the Milky Way galaxy shines at 4×10^{37} watts).

Theoretical physicist and futurist Michio Kaku has suggested that humans may attain Type I status in a century or two, if our technological advancement continues apace and we don't blow ourselves up or destroy our planet in the meantime. Achieving Type II status would take us a few thousand years, and Type III status up to a million years.[1]

If the Kardashev scale presents a visionary extrapolation of evolution and astrophysics, the Fermi Paradox offers a more pessimistic take on our future. Proposed by Italian-American physicist Enrico Fermi in 1950 during conversations with fellow physicists, this paradox consists of the contradiction between the high statistical likelihood that many other technologically advanced civilizations exist in the universe, and the lack of any reliable observations to indicate the presence of even one such civilization. Fermi's reported question was, "Where is everybody?"[2]

Scores of scientists from various disciplines have spent significant effort attempting to explain the paradox. Wikipedia lists 23 solutions that have been published by one or another scientist or team of scientists. The solution (among those 23) that fits best with the story of human power developed in this book is summarized by the dismal sentence, "It is the nature of intelligent life to destroy itself." If there are other intelligent species that rise to planetary dominance by way of fossil fuels (which was the only realistic pathway available to *Homo sapiens*), they may do essentially what we are doing—they will undermine their planetary climate and deplete their fossil energy resources. There's no one else out there because other technologically advanced species have either self-destructed, or have lost their ability to maintain advanced technology before they could use it to send out interstellar signals saying, "Hello! We're here! Is anyone out there?"[3]

But there is another solution to the Fermi Paradox, one that's not mentioned in the Wikipedia summary. Perhaps there are indeed at least some intelligent extraterrestrial species in the universe, but, rather than flaming out after seeking to maximize their power, they have instead found ways to limit their own power voluntarily. They've kept their power within bounds that would permit the long-term flourishing of healthy, diverse ecosystems on their home planets, so that they and the other beings they share their planets with can continue to thrive for millions of years. They have expressed their intelligence not by dominating ever-greater cosmic domains, but by developing their aesthetic creativity and spirituality, and they spend their days simply enjoying and caring for their small but beautiful worlds.

I'd argue that careful observation of nature suggests that these are the two most likely solutions to the Fermi paradox, and that they point to the most likely futures for our own species.[4] Further, the two solutions are not mutually exclusive; the first could be a brief pathway to the second.

This chapter will make the case that we humans are capable of controlling our thirst for power, and we have a long history of doing so. That capacity is rooted in similar behavior expressed throughout nature. Indeed, we need a new concept, as significant in its way as the maximum power principle, to describe this universal tendency. I call it the *optimum power principle*, defined as the tendency of natural and human systems to sacrifice some measure of power in the present so as to maximize power *over a longer period of time*. It doesn't contradict the maximum power principle; it merely adds the element of duration.

As we are about to see, power can be curbed by involuntary means: an organism or system of organisms can collide with natural limits and either be extinguished completely, or reorganize itself into a lower-power state that can be sustained. On the other hand, power can also be curbed voluntarily: an organism or system of organisms can, at least under some conditions, learn to anticipate limits and adapt its behavior to stay within them.

Let's explore the involuntary curbing of power first.

Involuntary Power Limits: Death, Extinction, Collapse

Every organism faces an ultimate existential limit to all its powers, in the form of death. The very idea of death can be frightening and depressing. Upon death, an entire universe of perceptions, feelings, and actions coming to focus in a particular individual vanishes forever. Why would something so awful to contemplate be inevitable?

Evolution must have a good reason for death, and it's not hard to see. In principle, there is no reason organisms couldn't have evolved to be immortal. Actually, a few come close. The bacterium *Deinococcus radiodurans* can survive intense radiation, extreme cold, and corrosive acids. And tardigrades (known colloquially as water bears), a phylum of tiny water-dwelling eight-legged segmented microanimals, seem immune to dehydration, high heat, and even the vacuum of space. They've survived all five mass extinctions.

However, in order to be immortal (or nearly so), organisms have to settle for extreme limits on their powers of motion and perception, and they probably need to remain tiny. Larger organisms and more sophisticated organs inevitably accumulate injury to their tissues over time. DNA sustains damage from natural (or human-manufactured) chemicals in the environment, or from cosmic radiation or copying errors. Cells sometimes divide incorrectly, proteins can misfold, and organisms can succumb to disease or injury. Nature's strategy is to let cells and organisms eventually die, and thus to cede the opportunity of existence to their replacements. After all, if all organisms were immortal but still capable of reproduction, they would proliferate and accumulate to the point where all possible food sources would be consumed and there would be no space for anyone new. There's just no getting around the necessity of death.[5]

Nonhuman organisms appear not to be aware of the inevitability of their own death, so they don't have to cope with that awareness. A few intelligent animals (including crows and elephants) take note of the deaths of their comrades and appear to mourn them, but we don't know if they are able to contemplate their own mortality. For us humans, though, usually beginning in late childhood, language and rational thought ensure that we inescapably know that everyone will die sooner or later, ourselves included. Thus, death is a big deal long before the event itself. A field of psychology known as Terror Management Theory addresses the psychological conflict between our self-preservation instinct and our knowledge of our own eventual demise, and thereby seeks to explain a wide array of cultural institutions that appear to promise immortality—including, but not limited to, religious beliefs and rituals.[6] (We will return to Terror Management Theory later in this chapter, when we discuss the psychological basis of climate denial.)

The extinction of a species is a form of collective death, the passing of a whole way of being. Some species manage to hang around for a very long time: cyanobacteria have been here for 2.8 billion years, and horseshoe crabs for a respectable 450 million years. But the average mammalian species persists just one to ten million years. It's likely that extinctions—especially mass extinctions—clear the way

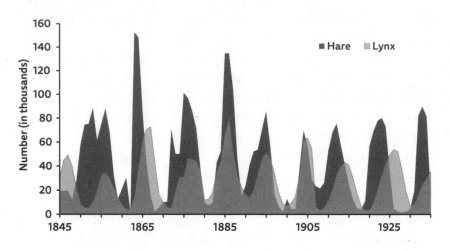

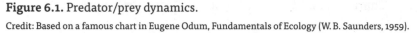

Figure 6.1. Predator/prey dynamics.

Credit: Based on a famous chart in Eugene Odum, Fundamentals of Ecology (W. B. Saunders, 1959).

for the evolution of new life forms, but the precise role of extinction in evolution is still under investigation. All we can say for sure is that warm-blooded species like ours tend to persist for only a very few multiples of the timespan we humans have already been here.

Another natural power limit derives from predation. Nature keeps the population levels of organisms in check via predator-prey relationships, which form the warp and weft of the food web. Predators—including micropredators, in the forms of viruses and bacteria—limit the population of prey species, but a decline in the population of prey species (due to any cause, including overpredation) can lead to a fall in the population of predators. Typically, the abundance of prey and predators is characterized by cycles, with the population peaks of predators lagging those of prey.

As a way of introducing a few concepts in population ecology that we'll find useful in a moment, let's consider one example—the field mouse, or vole. Its numbers in any given area vary according to the relative abundance of its food (typically small plants), which in turn depends on climate and weather. The local vole population size also depends on the numbers of its predators—which include foxes, raccoons, hawks, and snakes. A wet year can result in abundant plant growth, which temporarily increases the land's *carrying capacity* for voles, allowing the vole population to grow. This growth trend is likely to *overshoot* the vole population level that can be sustained in succeeding years of normal rainfall; the result is an eventual partial *die-off* of voles. Meanwhile, during the period that the population of voles is larger, the population of predators—say, foxes—increases to take advantage of this expanded food source. But as voles start to disappear, the increased population of foxes can no longer be supported. Over time, the populations of voles and foxes can be described in terms of *overshoot and die-off cycles*, again tied to external factors like longer-term patterns of rainfall and temperature.

We humans killed off most of our macropredators a long time ago. Occasionally someone still gets munched by a crocodile, alligator, mountain lion, bear, or tiger, but that's an exceedingly unusual way to go. Micropredators are a different story. Throughout human history, epidemics of infectious diseases were frequent and lethal. The Black Death of the Middle Ages temporarily reversed the growth trend of the global human population, and the influenza outbreak of 1918 killed between three to five percent of the total population at that time. As of this writing, the world is still attempting to cope with a coronavirus pandemic. And public health professionals are concerned that today we could be creating conditions for even worse pandemics in the future.[7] (See sidebar 21, "Rising Risk of Disease and Pandemic" in Chapter 5.)

The principles of population ecology apply as much to people as to other organisms: we have managed to increase Earth's carrying capacity for humans through agriculture and the application of fossil fuels to food production (via tractors and nitrogen fertilizers) and transportation (moving resources from where they are abundant to where they are scarce, so that more people can be supported overall). Today we are evidently in a condition of overshoot, in view of the fact that fossil fuels are finite resources and will be difficult to substitute, and also the fact that we are depleting topsoil, fresh water, and other essential natural substances.[8] If we are indeed in an overshoot phase, we must do what we can to avert or minimize a die-off event.

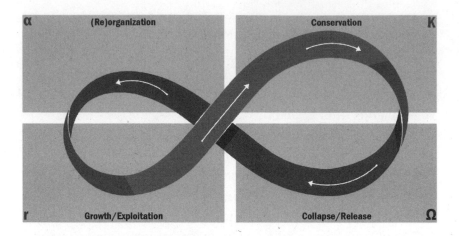

Figure 6.2. The adaptive cycle.

Credit: Based on the adaptive cycle as developed by Buzz Holling and others. See "Adaptive Cycle," Resilience Alliance, resalliance.org/adaptive-cycle.

The study of predator-prey cycles and other population dynamics has led ecologists, including path-breaking resilience theorist Buzz Holling, to develop the concept of the *adaptive cycle*. The cycle encompasses four phases: exploitation, conservation, release, and reorganization.[9]

Imagine, for example, a Ponderosa pine forest. Following a disturbance such as a fire (in which stored carbon is *released* into the environment), hardy and adaptable "pioneer" species of low-growing plants and small animals fill in open niches and reproduce rapidly. This *reorganization* phase of the cycle soon transitions to an *exploitation* phase, in which slower-growing species that can most effectively exploit available resources over the long term start to dominate. This transition makes the system more stable, but at the expense of diversity. During the *conservation* phase, resources like nutrients, water, and sunlight are so taken up by the dominant species that the system as a whole eventually loses its flexibility to deal with changing conditions. These trends lead to a point where the system is susceptible to a crash—another *release* phase. This may come in the form of a wildfire in which many trees die, dispersing their nutrients, opening the forest canopy to let more light in, and providing habitat for shrubs and small animals. The cycle starts over.

In effect, the adaptive cycle is a description of the process whereby communities of organisms collide with limits and rebound. It's only natural to want to apply this template to human communities as well, and doing so yields some useful insights.

There have been thousands of human cultures, defined by unique languages and sets of customs, but only 24 or so complex societies (or civilizations), defined by the presence of cities, writing, and full-time division of labor.[10] In their early days, complex societies are populated with generalist pioneers (people who do lots of things reasonably well) living in an environment with abundant resources ready to be exploited. These people develop tools to enable them to exploit their resources more effectively. Division of labor and trade with distant regions also aids in more thorough resource exploitation. Trading and administrative centers appear and expand. Money is increasingly used to facilitate trade, while debt enables a transfer of consumption from the future to the present. Specialists in violence, armed with improved weaponry, conquer surrounding peoples.

Complexity (more kinds of tools, more social classes, more specialization) solves problems and enables the accumulation of wealth, leading to a conservation phase during which an empire may be built and great achievements made in the arts and sciences. However, as time goes on, the costs of complexity accumulate and the resilience of the society declines. Tax burdens become unbearable, natural resources become depleted, environments become polluted, and conquered peoples become restless. Aspirants to the elite classes become more numerous and begin to compete more strenuously with one another for a limited number of official positions within institutions. At its height, each civilization appears stable and invincible. Yet it is just at this moment of triumph that it is most vulnerable to external enemies and internal discord. Debt can no longer be repaid. Conquered people revolt. A natural disaster may then break open the façade of stability and control.

Collapse often comes relatively swiftly, leaving ruin in its wake. But at least some of the components that made the civilization great (including tools and elements of practical knowledge) persist, and the natural environment has an opportunity to regenerate and recover, eventually enabling reorganization and a new exploitation phase—that is, the rise of yet another civilization.

As noted in Chapter 3, Peter Turchin and his colleagues have analyzed quantifiable data from hundreds of agrarian societies (most were nations within a larger civilization, as France and Norway are nations within the European Union and today's global industrial civilization). Turchin found clear evidence of cyclical behavior, with an average periodicity of about 300 years. Societies grew in wealth, geographic extent, and other measures of power, then shrank and simplified, at least temporarily and to some extent.[11] Sometimes the retrenchment was so catastrophic that it warrants the term *collapse*. No complex society has ever just gotten bigger and bigger without eventually experiencing crisis and contraction. Thus contraction, even collapse, should be considered normal features of societal evolution.

In summary, death, extinction, and societal shrinkage or collapse are inevitable phases of individual and collective existence. But these are usually imposed by circumstance; except in the case of suicide (which is almost uniquely human, and is tied to our awareness of our mortality), they're not deliberately chosen. Learning more about them helps us only so much in thinking about ways humanity could voluntarily reduce its own power in order to avert the crises discussed in Chapter 5. However, observation of nature suggests that *self-limitation* is also possible.

Self-Limitation in Natural and Human-Engineered Systems

At first thought, human beings might seem to be the only organisms capable of the kinds of conscious calculation, judgment, and planning that would be required for deliberate self-limitation of power. But other creatures engage in self-limiting behavior as well.

Power regulation occurs first of all *within* organisms via homeostasis. Organisms exhibit dynamic equilibrium, in which temporary imbalances in internal temperature, fluids, or pH trigger rebalancing. For example, in humans, when internal or external temperature exceeds a pre-set limit, we perspire, thereby cooling our skin until its temperature is back within an acceptable range. Each vital variable is controlled by one or more homeostatic mechanisms, which together maintain life.

Homeostasis is an example of what systems theorists call negative (or self-limiting) feedback, where the output of a system is fed back to the system in a way that tends to reduce fluctuations in key system parameters, whatever the cause. The cruise control system in a car, for example, can be set for a target speed. The car, in this example, is the system. One of its subsystems (the speedometer) produces output information about the car's speed. That output is fed back as an input to the cruise control, which adjusts the accelerator, increasing or decreasing fuel flow to the engine so as to maintain the target speed even when the car is going uphill or downhill.

Much of our modern technology simply wouldn't work without mechanical or electronic ways of mimicking homeostasis. The thermostats in our homes keep them from getting too hot or cold. The governor or speed limiter of an engine measures and regulates the engine speed (a classic example is the centrifugal fly-ball governor on a reciprocating steam engine, which uses the effect of inertial force on rotating weights driven by the machine output shaft to regulate its speed: when the shaft spins faster, the balls connected to it fly further apart, thereby closing a valve and limiting the fuel input into the engine). *Cybernetics* is a scientific field studying all such technological control systems. The principle, whether in biology or engineering, is the same: in order for a system's operation to proceed in a sustained fashion, internal conditions, including power in various forms, have to be continually managed and controlled through balancing feedback.[12] (Predator-prey relationships, discussed above, are an example of balancing feedback in the context of ecosystems.)

Sometimes the output of a system feeds back to the system in a way that continually amplifies an aspect of the system's behavior. This is positive, or self-reinforcing feedback—which is a confusing term for most people when they first hear it in introductory courses on systems theory. We all appreciate what is colloquially termed "positive feedback," in the forms of compliments and favorable job reviews. However, positive feedbacks in systems are almost always signs or causes of trouble. Self-reinforcing feedback occurs when the output of a system creates a cause-and-effect loop. For example, when a microphone and a loudspeaker are both connected to a PA system and the mic gets too close to the speaker, or the amplifier volume is turned too high, sound from the speaker is fed back through the mic and amplified continually, resulting in a loud squeal or screech. Jimi Hendrix was a master at controlling electric guitar feedback so as to create thrilling musical effects, but uncontrolled feedback is not so thrilling.

Many of the crises discussed in Chapter 5 can be understood in terms of the failure of balancing feedbacks, or the ramping up of self-reinforcing feedbacks. The global climate offers many instructive examples. Here is an instance of balancing feedback: long before humans started burning fossil fuels, large amounts of carbon dioxide were already venting into the atmosphere from the respiration and decomposition of organisms, wildfires, and volcanic eruptions. Carbon was (and is) absorbed by green plants, by carbonate rocks, and by the oceans. When CO_2 levels increased steeply following major volcanic eruptions roughly 93 million years ago, the rebalancing process took a long time: green plants flourished, more carbonate rocks formed, and CO_2 levels eventually settled back to their long-term normal range.

Today, as a result of very high and growing carbon emissions from fossil fuels, we seem to be setting off a series of self-reinforcing climate feedbacks. Higher air temperatures are melting the north polar ice cap, which exposes dark water. The exposed water absorbs more heat from sunlight than does white snow or ice, thus heating the atmosphere further, thereby melting more ice. Higher temperatures also melt permafrost soils in northern latitudes, causing the decomposition of buried organic matter as the permafrost thaws; this releases CO_2, which creates higher atmospheric temperatures, which melt more permafrost.

In several respects, the Great Acceleration was itself an example of positive, self-reinforcing feedback. As humans gained access to more energy from fossil fuels, they increased their economic activity, which increased their demand for energy. More energy, plus artificial fertilizers, enabled more food production, which enabled more population growth, which again led to more energy demand, as well as more demand for food. Growth created the conditions for more growth, the same way a wildfire—at least in its initial stages—creates the conditions for its own spreading, as it heats and dries the vegetation around it and generates winds that broadcast embers.

The point of this discussion of systems, homeostasis, and feedback deserves to be underscored: self-limitation is inherent in all systems; they can't function without it. Moreover, the failure of crucial self-limiting behaviors or mechanisms is always cause for concern and can be catastrophic.

SIDEBAR 24

The 2,000-Watt Society

An organization associated with the Swiss Department of Energy has suggested that a "sustainable" level of power for each human might be 2,000 watts.[13] The calculations used to derive this figure appear to be simple and straightforward. From the organization's website:

> The 2,000-Watt Society represents a sustainable and socially just society. For every person on earth, 2,000 watts of continuous power (primary energy) are available. This is enough to ensure prosperity and a high quality of life. Today, the primary energy consumption per capita worldwide is on average 2,500 watts—with enormous country-specific differences.

> The proposal for a 2,000-watt society is a useful attempt to quantify the amount of power that each human could use fairly and sustainably. What lifestyle could 2,000 watts support?

That depends on several factors, including the efficiency of technologies for transportation, manufacturing, and communication. As we saw in the sidebar, "How Much Power?," in Chapter 1, a typical human in an early agrarian society could wield and control 718 watts on a continuous basis, while a typical American today has nearly 10,000 watts of continuous power at her fingertips. If every human ran their life at 2,000 watts, that would imply a 20 percent reduction in overall energy usage for humanity, and an 80 percent reduction for the average American. For a glimpse of a 2,000-watt society, go to any country whose GDP per capita is roughly $15,000, such as Serbia, Brazil, the Dominican Republic, or China and spend some time with people living on the national median income.

The 2,000-Watt Society site offers no discussion of the trade-off between consumption and population implied in any attempt to achieve sustainability: if our global population were half what it is (four billion instead of eight billion), everyone could enjoy 4,000 watts of power and be just as sustainable as eight billion at 2,000 watts. On the other hand, if population continues to grow, a reduction in per capita power from the current 2,500 watts to 2,000 watts would at some point (depending on the rate of population growth) yield no ongoing reduction in total global energy usage.

Is a 20 percent reduction in overall energy usage enough to achieve sustainability? That would imply that we humans are in environmental overshoot by 20 percent. But according to analysis by the Global Footprint Network, our degree of overshoot is currently greater than 40 percent. Even if all eight billion of us were operating at 2,000 watts each, we would still be drawing down nonrenewable resources like minerals and metals, overusing renewable resources such as fish and forests, and taking habitat away from other creatures. Perhaps we should aim instead for a 1,500-watt society—or less!—at least until global human population begins to decline.

A self-limiting organism or colony of organisms can restrict its own growth in many ways. A single organism may have a genetically determined maximum size, while a colony of organisms may release waste that is toxic to the colony once it exceeds a certain size. In the case of parasites, self-limitation of the colony's size may be advantageous to continued survival: if the number of parasites becomes too high, they will kill the host, and hence themselves.

In some prey species, self-limitation of numbers keeps predators away by ensuring that there isn't enough food available for them to bother with. This is a frequent survival strategy for rare species (a group of organisms that are very uncommon, scarce, or infrequently encountered). Ecologist Glenda Yenni, in her published work based on observations and modeling of desert plants and animals, argues that "strongly self-limiting rare species are common."[14] Theory predicts that rare species should quickly go extinct. But observation shows that they don't; instead, they can persist for very long periods, and any given environment may include many species that are each represented by very few individuals. Yenni concludes that rare species "are rare because they are more self-limiting."

> ...[S]elf-limitation occurs when a species is more negatively affected by other members of the same species than it is by members of other species. The stronger this self-limitation is, the more a species is negatively affected when its numbers get too high. While this can prevent these species from becoming abundant, it also means that a species with strong self-limitation is more positively affected when its numbers are very low, i.e., it can rebound quickly when its population becomes small.[15]

Some species limit themselves by specializing on a rare resource, whose scarcity limits the species' reproduction and population growth. Others are limited by frequency-dependent predation, wherein a species may be very susceptible to a predator when its population grows, but undetectable to that same predator when it is very rare.[16]

Take, for example, the American pika (*Ochotona princeps*), a small relative of the rabbit. It specializes on a rare habitat type, talus fields in high alpine meadows. The pika's resting body temperature is only about 3°C lower than lethal body temperature (due to a high metabolic rate and low thermal conductance), so it is confined to places where it can easily thermoregulate. It can survive without drinking liquid water, getting most of its water from the vegetation it eats. Pikas build up body fat all summer so as to make it through the winters, which they spend in their burrows. Though they are widespread in the American West, at each location they are relatively scarce compared to other montane mammals (marmots, squirrels, chipmunks, and rabbits) due to their choice to subsist in this extremely harsh environment.[17] They have limited their population size by specializing so much, but they have gained relative stability.

Rare species that are self-limiting trade away the opportunity for momentary abundance, but they gain resilience against the likelihood of extinction, which explains why there are so many such species around to observe. "The conclusion using either abundance or energy use," writes Yenni, "is that though strong self-limitation seems at first a counter-intuitive candidate to explain the persistence of rare species, it arises as a relatively prevalent pattern across many types of ecological communities."

Abundant species, which more often are generalists, follow a different survival strategy. Rather than restricting their own population through their choice of food or habitat, they are more likely to be population-constrained by external factors such as predators and variations in food supply. Humans have increasingly taken this latter pathway. In doing so, we have also found ways to push against external limits, such as through the elimination of our predators (including disease organisms), and through intensified food production. This strategy has worked spectacularly well for us, in that we have been able dramatically and quickly to grow our population and consumption levels. But now those expanded external limits are starting to snap back, or threatening to do so. Perhaps our only way out is to learn or relearn voluntary self-limiting behaviors. Fortunately, we have some history to fall back upon.

Taboos, Souls, and Enlightenment

If other organisms have the capacity for self-limitation, humans certainly do, too. Indeed, self-limiting behavior is extolled and supported in the traditions of cultures around the globe.

In hunter-gatherer communities, as we've seen, authority was situational and nearly everything was shared. Bullies were eliminated through ostracism or capital punishment. There was little opportunity for the development of extreme inequality of any kind, and power relations were almost entirely horizontal rather than vertical. Children were taught to be humble and self-effacing so as to maintain solidarity within the group. Anthropologist Richard Lee, who studied the !Kung people of southern Africa, noted that when a hunter brought back a prized animal to share with the band, he always talked about how skinny and worthless it was. If he failed to do so, others would complain about the meat and make fun of him. When Lee asked about this, he was told: "When a young man kills much meat, he comes to think of himself as a big man, and he thinks of the rest of us as his inferiors. We can't accept this. We refuse one who boasts, for someday his pride will make him kill somebody. So we always speak of his meat as worthless. In this way we cool his heart and make him gentle."[18]

Taboos against overhunting were traditional methods of self-restraint and ecological stewardship. One example: the Bayaka of the Congo placed leaf cones on paths that led into parts of the forest where hunting had been unsuccessful, thus warning others to avoid it, and giving game populations time to recover.[19] Such practices were widespread and varied. Tribal taboos regulating the harvest of vulnerable species took at least six forms, according to anthropologists Colding and Folke.[20] These included "segment taboos," which forbade individuals of a certain age, sex, or social class from harvesting a resource; "temporal taboos," which banned the use of a subsistence resource during certain days, weeks, or seasons; "method taboos," which restricted overly efficient harvesting techniques that might deplete the stock of a resource; "life-history taboos," that forbade the harvesting of a species during vulnerable periods of its life history such as spawning or nesting; "specific-species taboos," which protected a species at all times; and "habitat taboos," which forbade human exploitation of species within particular reefs or forests that served as biological reserves or sanctuaries. Given the evidence that ancient peoples, as they migrated into new territories, often hunted abundant prey species to the point of extinction, it seems probable that Indigenous conservation practices were learned over a long time, through trial and error.[21] As Clark Monson points out in his thorough review of the subject, Indigenous resource management is now being studied widely as a model for modern practice.[22]

In horticultural societies, social power took the form of prestige, but the Big Man was only able to gain this prestige through his generosity and encouragement of others; he lacked the ability to coerce anyone else in the group into doing anything. Periodic potlatch (giveaway) feasts kept material inequality to a minimum and ensured that everyone in the group shared in whatever surplus was produced. While such traditions did not limit overall resource usage, they did keep economic inequality to a minimum.

As cooperation in ever-larger groups made vertical social power possible, checks on the accumulation of authoritarian power often fell away. Still, in early state societies, extreme inequality of wealth was somewhat blunted by periodic debt Jubilees, in which families were reunited and regained access to land. Also, kings always eventually died, and uprisings sometimes toppled governments.

Even in early state societies, the peasantry usually still had access to common resources of various kinds (as we saw in Chapter 5). Over the past couple of centuries, economists have debated whether communal management of land and other resources limits or encourages abuse and overuse. In an 1833 essay, British economist William Forster Lloyd used a hypothetical example of the dire effects of unregulated grazing on common land to argue that privatization of land leads to superior management. In 1968, in a widely discussed essay titled "The Tragedy of the Commons," ecologist Garrett Hardin made the same point, suggesting that common resources such as the atmosphere and oceans are inherently prone to being polluted and overused by members of society who thereby personally gain, while leaving society as a whole to deal with negative impacts.[23] However, more recently economist Elinor Ostrom has shown that, in most instances, Indigenous societies managed common resources responsibly. Basing her argument on her own field studies of pasture management in Africa and irrigation systems management in villages of Nepal, as well as numerous carefully designed experiments with test subjects, Ostrom found that societies frequently manage common resources successfully through mutual self-limitation.[24]

While early kingdoms exemplified vertical social power in the extreme, the pendulum of history was set to swing back toward mutualism and horizontal power. During the Axial Age, new Big God religions proclaimed the holiness of voluntary poverty and service to others (as discussed in Chapter 3). Even members of royal families were expected to at least pay lip service to these new ideals. In China, India, and the Mediterranean region, prophets, philosophers, and sages proclaimed the holiness of self-limitation. Here are just a few representative quotations from ancient texts (many more are collected in the anthology *Less Is More*, by Goldian Vandenbroeck, from which these are borrowed), giving a taste of the ethic common to the new religions and philosophies of the Axial Age:

- Epicurus: "Poverty, brought into conformity with the law of Nature, is great wealth."
- Socrates (via Plato): "...I do nothing but go about persuading you, old and young alike, not to take thought for your person and your properties, but first and chiefly to care about the greatest improvement of the soul."
- Matthew 6:20: "Lay up for yourselves treasures in heaven, where neither moth nor dust doth corrupt, and where thieves do not break through nor steal."
- Matthew 6:28–29: "And why take ye thought for raiment? Consider the lilies of the field, how they grow; they toil not, neither do they spin: And yet I say unto you, that even Solomon in all his glory was not arrayed like one of these."
- Matthew 5:5: "Blessed are the meek, for they shall inherit the earth."
- Mohammed: "Poverty is my pride."

- I Ching: "Limitation must be carried out in the right way if it is to be effective. If we seek to impose restrictions on others only, while evading them ourselves, these restrictions will always be resented and will provoke resistance. If, however, a man in a leading position applies the limitation first to himself, demanding little from those associated with him, and with modest means manages to achieve something, good fortune is the result."
- Tao Te Ching: "He who knows he has enough is rich."
- Confucius: "The superior man understands what is right. The inferior man understands what will sell. The superior man loves his soul. The inferior man loves his property."
- Appollonius of Tyana (1st century, writing of his travels in India): "I saw Brahmans living upon the earth and yet not on it, and fortified without fortifications, and possessing nothing, yet having the riches of all men."[25]

Axial Age religions focused on the idea of the soul—an inner essence of each individual which grows as a result of prayer, contemplation, and selfless good works, but atrophies when the person's behavior is selfish. The worst behavior of all is that which has a tarnishing effect on the souls of others—as when a king leads his people astray morally. The idea of the soul was, and is, partly a means of denying the reality of death. But it has also served to encourage self-limiting, prosocial behavior at all levels of the social pyramid.

The uplifting or purification of the soul was likened to a journey whose ultimate objective was a transcendent state of being. In Eastern traditions, this goal was described as enlightenment; in Western traditions, as sainthood, or simply as wisdom. Enlightenment or saintliness was to be achieved by self-limiting behavior (voluntary poverty, fasting, silent contemplation or meditation, and withdrawal from worldly concerns), and good works on behalf of others—especially on behalf of the souls of others. By setting aside the pursuit of worldly power, one could develop an intangible inner power.

Meanwhile, the Big God was thought of as a Higher Power, capable of creating or destroying the universe and yet interested in the affairs of every person. Compared to this colossal potency, the power of any human individual, even an emperor, was vanishingly insignificant; yet the Higher Power was accessible to the lowest of the low via prayer and meditation. We should, it was believed, maintain an attitude of humility before this unseen Almighty, and make daily decisions on the basis of our considered assessment of the actions He/She/It would find most pleasing.[26]

The attitudes engendered by these new religions and philosophies worked their way throughout society, beginning with child-rearing. Hunter-gatherer parents taught their children the values of sharing, thrift, and modesty. In the Axial Age, these ancient and universal values were revived and supercharged with the belief that the very souls of one's children were at stake.

Secular social developments also increasingly underscored the righteousness of horizontal power and mutual self-limitation. Repeatedly throughout post-Axial Age history, vertical social power was checked by people acting together. Even though kings and emperors often still reigned, the notion of legal rights of citizens came to be discussed, disputed, and codified. In Greece, democracy emerged as an alternative to the rule of kings—though the opportunity to vote was limited to free, property-owning men. Centuries later, conflict between the English King John and a rebel group of barons led to the creation of the Magna Carta, a charter signed in 1215 protecting church rights, and guaranteeing the barons freedom from illegal imprisonment, access to swift justice, and limitations on feudal payments to the crown, to be implemented through a council of 25 barons. In subsequent centuries, that charter would serve as the template for constitutions guaranteeing citizen rights and circumscribing the powers of officials. As Russian scientist and anarchist philosopher Peter Kropotkin documented in his remarkable book *Mutual Aid: A Factor of Evolution* (1902), European free cities (which were self-ruling constitutional entities) and self-governing guilds of artisans exemplified horizontal social power throughout the medieval period. In many instances, secular checks on vertical power were fortified by religious or spiritual beliefs: kings and emperors were seen as unjustly usurping the authority of God, while popular movements for the "leveling" of society proclaimed the holiness of poverty, modesty, and care for others' needs.

Throughout history, individuals have given up wealth and other forms of social power for ethical reasons. Gautama the Buddha is described in scriptures as having been a prince who renounced his hereditary advantages to work ceaselessly for the enlightenment and uplifting of all sentient beings. Mohandas Gandhi, a prosperous lawyer in South Africa, became the Mahatma ("Great Soul") by giving away his worldly possessions, taking on a lifestyle of conspicuous voluntary poverty, and dedicating his life to freeing India from the yoke of British colonialism. Leo Tolstoy, a wealthy Russian count, became a Christian anarchist and pacifist, adopted peasant garb, and opposed private property. Today's philanthropists such as Bill Gates and Warren Buffett dimly echo the sacrifices and achievements of such historical predecessors. It could, of course, be argued that those renunciates were simply trading one form of power for another—the power to compel for the power to inspire. Nevertheless, their example reminds us that the refusal of power is not only possible; it can change history.

Today it's difficult, maybe even impossible, to imagine a politician proclaiming, "Vote for me and together we will reduce our wealth and power so as to tackle global existential issues like climate change and inequality." Yet, for many centuries, Buddhist, Christian, and Muslim spiritual leaders have said to their followers, in effect, "Follow me and give up wealth and other forms of worldly power, and you will be happier."[27] And untold millions did just that.

Of course, not everyone responded to such appeals. Wars still raged, kings still sought the greater status of emperor, merchants still sought riches. Class structures, inequality, and ecological abuse persisted because elites *generally* don't voluntarily relinquish their power and privilege (they tend to do the opposite). Agrarian cultures still tended to grow too big and complex for their inhabitants to understand the ecological damage they were doing. And in patriarchal societies lacking awareness of population problems and benign forms of birth control, familial reproductive pressures to have children still overwhelmed any concern about long-range, collective demographic consequences.

Nevertheless, a precedent had been established. To assume that people simply won't voluntarily sacrifice power in order to serve the common good, even on a large scale, is simply incorrect.

Taxes, Regulations, Activism, and Rationing: Power Restraint in the Modern World

As we saw in Chapter 4, fossil fuels have greatly increased the size and complexity of modern societies. While the kingdoms and empires of the past had to balance the powers of the royalty and aristocracy with those of merchants, the church, and the military, today's industrial nations feature additional power centers, including corporations, banks and other financial institutions, political parties, various communication media, arms manufacturers, unions, and nonprofit advocacy groups. In addition, global power is contested by nations and alliances of nations, using trade, espionage, and propaganda as weapons even when no shooting war exists. Contests for power have become so complicated that a thorough analysis would require hundreds of pages of text. However, our purpose here will be simply to explore how physical power and social power are restrained in the modern world.

As we saw in Chapter 5, the impacts from having accumulated too much power are now legion—including climate change, the proliferation of highly lethal weapons (notably nuclear weapons), pollution, habitat destruction, propaganda, extreme inequality, population growth, and resource depletion. We customarily limit these powers or impacts of power through regulation and treaty. Social movements counter power from above with power from below via popular organizing, nonviolent protest, and even revolution. The greater the concentration of power, the greater the variety and intensity of efforts needed to rein it in—and the greater the likelihood that, in some instances at least, those efforts will themselves become corrupted by abuses of power.

Laws and constitutions have evolved to limit the dangerous accumulation of social power even in the absence of religious commandments and prohibitions. The goal of architects of governmental institutions, beginning with Plato, has been a kind of social homeostasis in which checks and balances prevent power from accumulating dangerously in any one sector of society through a self-reinforcing feedback process. The most decisive element of this social homeostasis is representative democracy, which, over the past two centuries, has taken root in roughly half of the world's nations. In a constitutional democracy, institutions of government are intended to provide balancing feedback to one another—as when (in the US) Congress investigates and impeaches a President, or courts strike down laws passed by Congress because they violate the Constitution.

The most radical stance in power-sharing political theory was adopted by anarchist philosophers like Mikhail Bakunin (1814–1876). They argued that the state and all forms of hierarchy are inherently evil, and that authority should flow instead from individuals' talents and labor, with all productive property and factory machinery owned in common by the people. Bakunin differed from Karl Marx (1818–1883), who advocated that the transformation of society to the ideal of perfect distribution ("from each according to their abilities, to each according to their needs") would require an intermediate stage— a dictatorship of the proletariat (Bakunin thought this was both an ideological and tactical mistake).[28] In the US, Eugene Debs (1855–1926), who ran for president five times and was imprisoned for his labor organizing efforts, preached democratic socialism—the notion that elected government should own all industries and divide profits among the workers. During the 20th century, socialist and communist ideas were put into practice to varying degrees in many nations, while anarchism helped inspire thousands of cooperatives and labor unions. In most industrial democracies, the results included better access to health care and other amenities, better working conditions, and a reduction in economic inequality. Unfortunately, the Soviet experiment with Marxism seemed merely to shift authoritarian power from one sector of society (the capitalists) to another (the Party and

its functionaries), rather than altogether doing away with vertical social power, which was its ostensible goal.

Prior to the Great Acceleration, religious institutions managed
wealth inequality by demanding tithes from the royalty, aristocracy,
and merchants, and distributing alms to the poor. Today, however,
inequality is managed to a greater degree through graduated taxation and various government redistributive programs (though religious charities still persist). In Britain, Prime Minister William Pitt
the Younger introduced the first modern income tax in 1798 to pay
for his nation's involvement in the French Revolutionary War. Pitt's
tax was graduated or progressive, in that low earners paid less: tax
rates ranged from less than one percent up to ten percent of income.
In the United States, the first progressive income tax was established
by President Lincoln in 1862, but repealed in 1872. The Sixteenth
Amendment to the US Constitution, adopted in 1913, empowered
Congress to begin collecting income taxes to fund the government,
and, by the mid-20th century, most other countries had likewise implemented some form of graduated income tax, serving both to fund
governments (and their various programs) and to reduce income inequality. In the US, tax rates on the wealthy peaked in the 1950s, when
earners in the highest tax bracket were taxed 91 percent of their income. Taxes on capital gains (which apply only to investors) were introduced in the early 20th century. In recent decades, US tax rates on
the wealthy have fallen sharply, especially since the Reagan administration, partly as a result of political lobbying by the rich, who have
also found a multitude of ways of evading taxation. Unsurprisingly,
levels of wealth inequality have rebounded upward as a result.[29]

Government programs aimed at reducing economic inequality
have come to include transfer payments (welfare, financial aid, and
Social Security) and social safety nets (unemployment benefits,
government-run or subsidized healthcare systems, free education,
rights to housing, legal aid, funds for pensioners and veterans,
consumer protections, and subsidized services such as public transport). Some nations have more robust public spending programs
than others: Europe and Central Asia currently spend the most,

averaging 2.2 percent of GDP; the Middle East, North Africa, and South Asia spend the least, at about 1.0 percent. Unfortunately, nations with generous social programs coincidentally tend to have high per-capita greenhouse gas emissions.[30] In the US, government redistributive programs have become the subject of much political controversy, with right-leaning politicians seeking to reduce or eliminate programs, and their left-leaning colleagues proposing to expand existing programs or create new ones, such as "Medicare for all" or a guaranteed basic income.

Walter Scheidel, in his book *The Great Leveler*, argues that these deliberate efforts to manage inequality have been unusual in historical terms, and only partly effectual. He documents, as we saw in Chapter 4, that most of the economic leveling that occurred in the mid-20th century resulted directly or indirectly from the two World Wars (his larger point, as we noted then, is that increased economic equality has usually come about as a result of the Four Horsemen of mass mobilization warfare, transformative revolution, state failure, and lethal pandemics). Timothy Mitchell, in *Carbon Democracy*, further argues that violent labor disputes during the coal era did much more to promote economic equality than the peaceable tinkering with economic policy that occurred in the post-WWII Oil Age. Nevertheless, as both authors acknowledge, government policy (however arrived at) does impact equity, and so there is every reason for citizens to demand more progressive taxation, higher inheritance taxes, financial transactions taxes, and other economic policies that would make for more equitable distribution of income and wealth.

In the end, all such leveling efforts and influences must push against the inherent tendency of the structures of industrial society (involving energy production, manufacturing, distribution, information flow, and investment) to concentrate power in the hands of various elites—including politicians, financiers, corporate managers, and media gatekeepers. The very nature of complex societies ensures that such power concentrations will emerge; the question is only how, and to what degree, those concentrations will be contested or limited.

The limiting of coercive social power in the modern world is perhaps epitomized in the abolition of slavery. In Chapter 4, we saw how industrial machinery, powered at first by flowing water and increasingly by coal, undermined the institution of slavery in the United States. However, it would be wrong to ignore the role of abolitionists and slaves themselves in that process. Slaves revolted, escaped, and helped others escape; and some former slaves, such as Frederick Douglass, wrote and spoke tirelessly about the horrors of the institution. Abolitionists wrote tracts and pamphlets, ran for office, organized public lectures and demonstrations, and deliberately and often publicly broke laws, thereby risking their freedom and lives, in order to further their cause. They argued the immorality and cruelty of the trade in human persons, forced labor, the separation of families, and the denial of basic human dignity. In doing so, abolitionists set a tactical template for nearly all subsequent human rights and environmental advocacy campaigns.[31] Not only did the struggles to abolish the Atlantic slave trade and the practice of slave ownership in the US and other nations succeed, but similar efforts resulted in the banning of child labor and unsafe working conditions, and the creation of regulations instituting the eight-hour workday and the minimum wage.

Internationally, activism partly inspired by the abolitionists was directed toward the ending of colonialism. In countries in Africa, South and Central America, South Asia, East Asia, the Pacific Islands, and the Caribbean, anti-colonial uprisings, boycotts, demonstrations, and wars continued through the mid-20th century. The power of colonizing nations to directly and brutally commandeer labor and resources from other peoples was eventually mostly terminated (though subtler means of exploitation continue to this day). In India, the anti-colonial struggle was led by Mahatma Gandhi, whose theory and method of nonviolent resistance would be studied and emulated by human rights, antiwar, and environmental protection campaigners everywhere.

Also drawing upon the moral impetus and the successful tactics of the slavery abolition movement, the suffragists of the 19th and early

20th century sought to extend full civil rights—beginning with the right to vote—to women. Some polities (including Sweden and the Dutch province of Friesland) had permitted women to vote as early as the 17th and 18th centuries, but typically only in certain districts or if women owned land. New Zealand gave all female citizens the power of voting rights in 1893. Australia's states began granting women the right to vote (though not to run for office) in the 1890s; by 1902 a national law granted women (except "aboriginal natives" of Australia, Africa, Asia, and the Pacific Islands) not only voting rights but the right to run for federal Parliament. Women in Britain gained voting rights via two laws, in 1918 and 1928; while in the US, women's right to vote was codified in the 19th Amendment to the Constitution, which became law in 1920.

Majorities often have the power to oppress or marginalize minorities. This power, piggybacking on the religious Big God impulse to promote fertility and population growth, has historically led to prohibitions against same-sex sexual behavior. Countering this, activism by gay rights advocates has led to the decriminalization of homosexuality, which has occurred piecemeal in nations and US states over the past half-century (gay people are still legally persecuted in many African and Middle Eastern countries, and Russia maintains laws restricting freedom of expression and association for LGBTQ people). The struggle for gay marriage and other equal rights for people of all sexual orientations is ongoing.

The environmental movement began in the late 19th century with efforts to protect public lands from exploitation; it has since taken on a widening array of issues, including the protection of threatened species, the ending of various forms of pollution, curbing the growth of human population, and the halting of climate change. Tactics borrowed from the abolition movement and from Gandhi's nonviolent anti-colonial campaign have led to a long series of victories, including (in the US) the establishment of the Environmental Protection Agency, and the passing of the Clean Air Act, Clean Water Act, and Endangered Species Act.

However, as Chapter 5 hopefully made clear, these measures have been insufficient to halt climate change and many other snowballing environmental problems. This perceived failure has led some members of the environmental movement to question its tactics. Rather than warning of impending crisis and calling for sacrifice (by reducing consumption and birthrates), as many early environmentalists did, some self-described "bright greens" now argue that only good news, and promises of more economic growth and jobs from clean technology industries, can turn the tide.[32] However, others argue that most of the failures of the environmental movement did not stem from some flaw in the essential message of first-generation environmentalists, who were correct in saying that humanity will have to rein in its powers of production, consumption, and reproduction if it is to avert ecological ruin. Instead, they would say, the larger failure of environmentalism can mostly be chalked up to psychological denial among the general populace and the momentum of economic growth (as we'll discuss later in this chapter). It's true that positive and encouraging messages are helpful, but they need to take the form of believable stories about how we can all thrive together by using less, and in the absence of growth.[33] Meanwhile, worsening news about climate change impacts and species declines has ignited a new phase of environmental radicalism epitomized by Extinction Rebellion, a global movement with the stated aim of using nonviolent civil disobedience to compel government action to avoid tipping points in the climate system, biodiversity loss, and the risk of social and ecological collapse (more on that in the next chapter).[34]

The specter of annihilation from nuclear war led to calls to "ban the bomb" starting in 1957, with the creation of the Campaign for Nuclear Disarmament in Britain. Meanwhile, as the civilian nuclear power industry grew, fears of risk of accidents and concern over the links between nuclear power and the production of fissile material for nuclear weapons led to environmentalists' engagement with the nuclear issue.[35] New nuclear power plants were discouraged through massive ongoing protests (for example, in Sonoma County, California), and protests also contributed to the closure of power plants and a related weapons facility in Hanford, Washington.

In normal times, market economies ration goods by price, so that the people with the most money can consume the most of any good they choose. However, during wartime, in the case of extreme scarcity of essential goods, or when consumption of a good needs to be controlled for some other reason, modern governments have instituted quota rationing, a collective form of self-limitation. Britain and Germany instituted rationing during World War I and again in World War II as a way to meet the basic needs of ordinary citizens while directing large quantities of resources to the military. Overall economic inequality declined as a result. During the Second World War, the United States likewise issued ration coupons—for fuel, food, clothing, and tires, among other things. In his book *Any Way You Slice It*, author Stan Cox recounts how Americans willingly, even enthusiastically participated in the program; one woman was quoted at the time as saying that "rationing is good democracy."[36] During the first three years of the British rationing program, "overall consumer spending dropped 15 percent and shifted sharply toward less resource-intensive goods," according to Cox.[37] Britain continued its rationing program well after the end of the war, and surveys showed that, during the period of rationing, Britons were generally better fed and healthier than either before or after.

Food rationing began long ago, in ancient Mesopotamia and Egypt, and subsidized food rationing is still commonly practiced—though less often as a way to conserve scarce commodities than as a way to ensure that people with low incomes have access to essential nourishment. In the US, the Supplemental Nutritional Assistance Program (SNAP, formerly known as the Food Stamp Program) served 40 million Americans in 2018, roughly nine percent of the population. Food rationing programs have been implemented in the past, or are currently in effect, in nations as diverse as Argentina, Bangladesh, Brazil, Chile, China, Colombia, Cuba, Egypt, India, Iran, Iraq, Israel, Mexico, Morocco, Pakistan, the Philippines, the Soviet Union, Sri Lanka, Sudan, Thailand, Venezuela, and Zambia. Cox points out that, in the future, societies may face increasing scarcity of water, food, and energy, and may need to find ways to fairly reduce carbon emissions; rationing could play a role in each instance. If economies need to be deliberately shrunk in order to reduce energy usage, quota rationing could provide a means of degrowing them fairly, and with minimal pain and sacrifice.

Surveys suggest that high levels of willing participation in rationing programs during wartime depended on three primary factors: a shared sense of immediate crisis; a common belief that the crisis would pass, so that rationing would be only a temporary inconvenience; and a sense that sacrifices are being shared fairly. It may be a challenge to design future rationing programs in the context of scarcity that is ongoing, and crises that are difficult for many people to understand. Nevertheless, precedent shows that there are effective alternatives to price rationing when markets fail.

The homeostatic mechanisms of modern societies, which take the form of mutually self-limiting institutions whose purpose is to solve human problems (laws, police, courts, and governmental redistributive programs), have somewhat reduced the requirement for Big Gods as motivators of pro-social behavior. Religious affiliation has declined in industrial democracies, and especially in those that provide robust social services—notably the Scandinavian countries. Denmark, for example, is a majority atheist nation, yet manages to remain highly cooperative and peaceful. Ara Norenzayan suggests this is because effective big government can be a replacement for religion, and vice versa.[38] The chain of causation is unclear. Is it that, when people believe that government will take care of them and punish anti-social cheaters, their need for a Big God recedes, or does a decline in religious faith lead people to vest more confidence in the problem-solving ability of government? Either way, if and when big governments lose their ability to solve problems, people may yearn for older forms of social control.

The good news is that, in the modern world, we've gotten somewhat better than we used to be at managing social power (though reversals are common). Social evolution has proceeded ever more rapidly, creating institutions and strategies for rebalancing when trends get out of hand—when a leader grows too arrogant, when the wealth disparity between rich and poor becomes unbearable, or when the daily operations of society threaten the integrity of natural systems.

The bad news is that our expanding power-management strategies haven't always kept up with our even more rapid development of powers of fossil-fueled population growth, resource extraction, and industrial production. Systems put in place to prevent social and environmental harm are being overwhelmed (we'll explore some of the reasons for this in a moment). And despite social services and progressive taxation, wealth has become more concentrated in fewer hands.

Further, many of the new rights and freedoms we've gained in the last couple of centuries arose in the context of a society of abundance. What happens if and when the current period of abundance shifts to one of scarcity? Will those freedoms and rights persist? Or will we turn back toward starker forms of vertical social power that prevailed in earlier eras? Ominously, as the fossil-fuel era grinds to a close, we are seeing a trend toward the rejection of liberalism and democracy in at least some nations. This is a subject to which we will return in Chapter 7.

Games, Disarmament, and Degrowth

Before addressing the question of why sophisticated modern power-limiting efforts have been insufficient, it's worth briefly exploring mathematicians' contribution to the discussion of mutual power self-limitation. In the last few decades, game theory—typically defined as the study of mathematical models of strategic interaction among rational decision-makers—has become integral to economics, philosophy, international relations, business, and evolutionary biology.[39] While many early game theorists chose to address the real-world problem of nuclear disarmament, their findings apply to any situation in which power holders must negotiate a stand-down, including international climate negotiations. Game theory has also been used to explain the evolution of cooperation within and among species. The relevant question that game theory addresses is: If I give up some of my power, how can I be sure that you or someone else will not take up that power and use it against me?

The game that has been studied most extensively is *the prisoner's dilemma*, which presents a situation where two parties, separated and unable to communicate, must each choose between cooperating and competing. The most desirable outcome for both will be realized if they cooperate, but there is no way for either to know at the outset that this is the case. Here's a simple, frequently cited example of the game: suppose two members of a gang of bank robbers, Bill and Joe, have been arrested and are being interrogated in separate rooms. The authorities have no other witnesses, and can only prove the case against them if at least one betrays his accomplice and testifies to the crime. Each must choose to cooperate with his accomplice and remain silent, or defect and testify. If each cooperates with the other and remains silent, then the authorities will only be able to convict them on a lesser charge, which will mean one year in jail each (1 year for Bill + 1 year for Joe = 2 years total jail time). If one testifies and the other does not, then the one who testifies will go free and the other will get three years (0 years for the one who defects + 3 for the one convicted = 3 years total). However, if each testifies against the other, each will get two years in jail for being partly responsible for the robbery (2 years for Bill + 2 years for Joe = 4 years total jail time).

Each robber always has an incentive to defect, regardless of the other's choice. From Bill's perspective, if Joe remains silent, then Bill can either cooperate with Joe and do a year in jail, or defect and go free. He would be better off betraying his comrade. On the other hand, if Joe defects and testifies against Bill, then Bill's choice becomes either to remain silent and do three years or to talk and do two years in jail. Again, he would be better off testifying, thus getting two years instead of three.

The paradox of the prisoner's dilemma is essentially this: both robbers can minimize the total jail time that the two of them will face only if they both cooperate with one another (in which case they get two years total), but each separately faces incentives that drive him to defect; and if they both defect, they will end up doing the maximum total jail time (four years).

Solutions to the prisoner's dilemma are hard to find if the situation is narrowly defined; but real-world prisoner's dilemmas are typically more complex and offer many opportunities for solution. The tragedy of the commons, mentioned earlier, is an example of a prisoner's dilemma: if each shepherd grazing her herd on a common green chooses to maximize the size of her herd, the green will be overgrazed and all the sheep will starve. Yet each shepherd is individually incentivized to do just that. Nevertheless, in reality, shepherds have faced this dilemma for several thousand years in innumerable locations, and have found ways to cooperate to the benefit of all.

A prisoner's dilemma game is often played only once; but, in the real world, many types of competitive and cooperative human interaction play out over a long period and are repeated many times. As people see the results of their choices, they recalibrate and find ways to reward cooperation or punish defection among fellow players. They can create formal rules and institutions that alter the incentives faced by individual decision makers. There are also informal incentives to cooperate, such as the desire to have a favorable reputation among one's peers, as well as disincentives such as social opprobrium.

Given time, groups of people tend to develop psychological and behavioral biases toward increased trust in one another, long-term future orientation, and inclination toward reinforcement of cooperative behavior and discouragement of anti-cooperative behavior. While these biases can be temporarily reversed during periods of intense social conflict, they otherwise tend to evolve through a selection process within a society, or by group selection across different competing societies. Over all, they lead individuals often to engage in personally "irrational" choices that lead to the most beneficial outcome for the group as a whole.

Disarmament is frequently described as a prisoner's dilemma: if one nation gives up military assets, it may leave itself vulnerable to attack by another country that hasn't disarmed. Nevertheless, many arms treaties have been negotiated during the past century, limiting certain kinds of weapons (such as nuclear warheads) or whole classes of weapons (such as biological and chemical weapons). Typically, success is achieved through a sequence of carefully designed and monitored stages, so that no country is unacceptably exposed at any one stage.[40]

One nation, Costa Rica, has unilaterally disarmed completely: it decommissioned its armed forces in 1948. Similarly, South Africa unilaterally gave up its nuclear weapons in 1989 in an effort to create greater regional stability and to foster respect within the international community.

In 1979, game theorists began using computer programs to run the prisoner's dilemma and similar games in tournaments; the winner was often a simple "tit-for-tat" program that cooperates on the first step, then, on subsequent steps, does whatever its opponent did on the previous step. Such programs, with only slight modification, have also been used to model natural selection and the evolution of cooperative behavior.

Game theory has obvious implications for society's ability to reduce greenhouse gas emissions and halt climate change. Peter John Wood of Australian National University has published an overview of research along these lines.[41] The best solutions, he finds, involve the use of carrots and sticks—rewards for compliance and punishments for non-compliance—as well as linkage with other issues, such as trade, so that nations will find it in their best interest to participate.

As we saw in Chapter 5, global problems such as climate change, pollution, resource depletion, and species extinctions can probably ultimately be addressed only by shrinking the total human enterprise. But degrowth presents yet another prisoner's dilemma: if one politician or political party proposes to shrink the economy and reduce population over time, thus calling for collective belt-tightening and sacrifice, another politician or party is likely to respond by saying that there is no need for such effort: just vote for me and we can all enjoy more growth and prosperity! Similarly, if one nation degrows its economy but others grow so that total production and consumption remain unchanged or even increase, nothing has been gained. Ecological economists have been puzzling over this particular prisoner's dilemma for many years; the best solution they've found so far is to call for increased societal focus on happiness and well-being: perhaps we can increase the social factors that create feelings of life satisfaction while we engage in the otherwise contentious work of reducing population and consumption.[42] I'll return to this point in the next chapter.

Unfortunately, game theory tends to gloss over the historically constituted, real-world power relations between participants and assumes that people (and nations) are entirely rational. Even within this idealized theoretical framework, the more actors that are involved and the fewer the mechanisms for monitoring and sanctioning free riders, the lower the chances of cooperation. On the whole, our current international system, driven by profit and power, is not altogether conducive to building cooperation and discouraging free-riding.[43]

Nevertheless, life is full of prisoner's dilemmas. And, in principle, we are perfectly capable of playing these games so that everyone wins. It all comes down to trust—the basis of social capital and horizontal power. Trust isn't always rational, and time and effort are required to build it. In difficult times, trust is far more valuable than the material wealth to which we often aspire or cling.

Denial, Optimism Bias, and Irrational Exuberance

In this chapter we have seen that the ability to limit power is rooted in biology and has a long history in all human societies; further, engineers routinely find ways to incorporate power-limiting mechanisms in technologies of all sorts. I have proposed the optimum power principle—the observation that organisms often curb their power in the present as a way to maximize power over the long run—as an addendum to the maximum power principle. Even prisoner's dilemmas have solutions. If the existential crises we surveyed in Chapter 5 are indeed the result of too much power, there is no reason (in principle, at least) why humanity cannot power down to solve them. So, given the capacity and innumerable opportunities for power moderation, why do we humans still appear to be running headlong toward a global catastrophe driven by power excesses and abuses?

The adaptive cycle tells us that temporary imbalances in nature and human societies are natural and inevitable; over time, rebalancing occurs. However, those temporary imbalances are sometimes particularly large; that is, the growth phase of the cycle can sometimes swing to extremes. Humanity's access to fossil fuels has propelled us on growth and conservation phases that are utterly unprecedented in terms of the levels of power and population size that we have achieved. With so many opportunities for growth, we've tended to set aside ancient cultural attitudes and practices promoting self-restraint, even if doing so blinds us to the overwhelming likelihood of a cyclical collapse/release phase of unprecedented magnitude, and makes it harder for us to do things that would reduce the scale of the impending calamity.

But there's more. We are inherently subject to a set of collective and individual psychological mechanisms that make it easy for a large power imbalance to appear, but difficult for us to recognize and minimize that imbalance before serious problems occur. These mechanisms take the forms of denial, optimism bias, our tendency to lie to ourselves and others, the Overton Window, our genetically-based pursuit of status, our addiction to novelty, our tendency to discount the future, and the lottery winner's syndrome. Let's look briefly at each of these, and see how and why they all tend to keep us from limiting our power excesses.

The phrase "climate denial" may trigger thoughts of efforts by fossil fuel companies to sow public doubt about the reality of global warming. These efforts are certainly real, but they have been successful largely because denial itself is a deeply entrenched human capacity. In their book *Denial: Self-Deception, False Beliefs, and the Origin of the Human Mind*, Ajit Varki and Danny Brower suggest that the awareness of our own mortality (which arose along with the development of language sometime in the Pleistocene) might have stopped human evolution in its tracks. That is, our expectation of personal extinction would have made us so depressed and so cautious that we probably wouldn't have been able to compete successfully with other species, or other members of our own species who were not so burdened, if not for the simultaneous appearance of a fortuitous adaptation—our ability to deny death and other unpleasant realities.[44] As we became aware of the inevitability of our own death, we quickly learned to deny that awareness so we could get on with our day-to-day business. Denial thus served an evolutionary function as an essential tool of terror management. Over time, our denial muscle probably strengthened—and it has arguably done so especially in recent decades.

If individual mortality is terror-inducing, coming to terms emotionally with collective death is utterly beyond us. Scientists have been aware of species extinctions for the past couple of hundred years, since the beginnings of modern biology and paleontology.[45] We have therefore also become aware of the possibility of our own species' extinction. That awareness has become more acute during the past 70 years or so, since the start of the global nuclear arms race. But human extinction is a subject few people wish to consider, let alone bring up in polite conversation (although apocalyptic novels and movies are becoming increasingly popular). While we know that each of us will eventually die, we implicitly count on the persistence of future generations, and the survival of human culture, in order to maintain our psychological equilibrium. The thought that the entire human enterprise, supporting all our collective dreams and accomplishments, could disappear in a cloud of smoke or an endless stream of carbon emissions is unbearable. So, we psychically bury the prospect of human extinction, even as we go about creating the means for its occurrence.

Denial of climate change is therefore more than just a political tool for maintaining corporate profits (though this it certainly is). It is also a collective coping mechanism.

The most common form of denial—whether of death or climate change—is mental compartmenting. We create a mental compartment for death and another for climate change; but there are also compartments for favorite old movies, recipes, opinions about politics, and so on. We aren't literally denying the reality of death or climate change; it's right there, in its compartment. Every so often we look in that compartment and feel fearful. But the amount of time we spend looking in any particular compartment is typically proportional not to its actual importance, but to its ability to satisfy our interests and emotional needs. Sooner or later an event (perhaps a visit to the doctor's office) forces us to look into the death compartment; but by then the damage from years of smoking or other unhealthful habits is already done. Much the same will likely be true with climate change.

The individual and cultural coping mechanism of denial has a flipside—an optimism bias that again leads us to believe that we are less likely to experience a negative event than we actually are. Neuroscientist Tali Sharot, in her book *The Optimism Bias: A Tour of the Irrationally Positive Brain*, cites surveys and experiments showing that the phenomenon is real and pervasive.[46] Its mirror image, pessimism bias, affects people suffering from clinical depression, but is otherwise rare. This natural tendency toward optimism has served an evolutionary purpose—it encourages us to take risks in order to reap rewards. But it also steeply increases our vulnerability and hobbles our ability to respond in the era of climate change and other converging crises.

Sometimes collective optimism bias feeds back on itself, resulting in mania, bubbles, and booms. As Scottish journalist Charles MacKay put it in his still-relevant 1841 book *Extraordinary Popular Delusions and the Madness of Crowds*, "Men...think in herds; it will be seen that they go mad in herds, while they only recover their senses slowly, and one by one." Federal Reserve Board chairman Alan Greenspan used the phrase "irrational exuberance" to describe the dot-com stock market bubble of the 1990s, but many bubbles preceded that one, and more bubbles have followed, including the fracking frenzy of the 2010s. Collective manias spread and intensify because no one wants to miss out on the "next big thing." We may rationally know that the boom can't last forever, but we don't want to be the one who gets left behind. In a sense, the Great Acceleration can be thought of as the grandest popular delusion in the history of our species: from the outset it was obvious that fossil fuels would eventually deplete, but we treated them as though they would last forever.

Both denial and optimism bias depend on deception, including self-deception. Yuval Noah Harari, in his popular book *Sapiens*, makes the point that our development of language back in the Pleistocene gave us the capacity to create myths and useful fictions. Language enabled us to talk about things that don't exist—which is an essential ability if you want to design a new machine from scratch or create a new company. But we have gotten so good at creating fictional entities that the real world has become easy to deny and ignore. As Harari puts it:

> Ever since the Cognitive Revolution, Sapiens have thus been living in a dual reality. On the one hand, the objective reality of rivers, trees and lions; and on the other hand, the imagined reality of gods, nations and corporations. As time went by, the imagined reality became ever more powerful, so that today the very survival of rivers, trees and lions depends on the grace of imagined entities such as the United States and Google.[47]

Collective denial and optimism bias make it difficult to talk to friends and relatives about the looming climate crisis, resource depletion, and other trends that imperil our future. We are sensitive to one another's subtle cues, and change the subject when it's clear that the discussion has touched a nerve. The same is true on a national level: there are some things we just don't want to talk about. The range of acceptable public discourse is called the Overton Window, named for Joseph P. Overton, who stated that an idea's viability depends mainly on whether it falls within this range of acceptability, not on its inherent truth or usefulness. According to Overton, this window frames the range of policies that a politician can recommend, or ideas she can talk about, without appearing too extreme to gain or keep public office.

As a result of decades of sustained collective effort, the scientific community has brought climate change within the Overton Window—at least some of the time, and in most nations. However, one important and necessary response to climate change is still well beyond the window—a deliberate policy of degrowth. As discussed in Chapter 5, continued annual growth in population and consumption makes it ever harder for the world's nations to reduce greenhouse gas emissions in order to minimize climate change. An obvious solution would be to reduce population and consumption. That would, of course, pose a challenge in a world that has come to see growth as essential to the economic health of nations. Nevertheless, the logical necessity of degrowth is inescapable, and a few economists have proposed ways of dealing with the difficulties it would pose.[48] As a gauge of degrowth's proximity to the window, consider this fact: the International Panel of Climate Change (IPCC) of the United Nations periodically produces hundreds of models to show how government policies of various kinds would impact emissions and global warming. But degrowth has never been included among those policies.

Incidentally, the truth-telling ability of Greta Thunberg, the young Swedish climate activist who was *Time* magazine's Person of the Year in 2019, is (in her view, and that of some psychologists as well) partly due to her autism—which makes her less aware of social cues and hence less prone to hypocrisy. If no one but Thunberg has had the courage to tell world leaders to their faces, "We are in the beginning of mass extinction, and all you can talk about is money and fairy tales of eternal economic growth," that's largely because people with autism are less aware of the Overton Window.[49]

Another psychological mechanism making it difficult for us to rein in our powers so as to avert global crisis is the pursuit of status. Throughout the evolution of complex organisms, notably vertebrates, status has served as a way of minimizing the costs of competition. Animals compete for mates and food, but competition carries costs. Signals of status within a species establish which individuals are more or less likely to be successfully challenged, so overall there is less energy wasted in competition. Tendencies among modern humans to acquire status symbols—expensive cars, clothes, and houses—are therefore deeply rooted in evolution.[50] If we're told that big, powerful automobiles and jet vacations are sealing the fate of future generations, that message has to overcome the lure of status in order to get our attention and change our behavior.

We humans are also wired to respond to novelty—to notice anything in our environment that is out of place or unexpected and that might signal a potential threat or reward. Most types of reward increase the level of the neurotransmitter dopamine within the brain. Experiments have found that if an animal's dopamine receptor genes are removed, it explores less and take fewer risks—and without some exploration and risk taking, individuals have reduced chances of survival. But the brain's dopamine reward system, which evolved to serve this practical function, can be hijacked by addictive substances and behaviors. This is especially problematic in a culture full of novel stimuli specifically designed to attract our interest—such as the hundreds of advertising messages the average child sees each day.

Addictions to shopping or to acquiring status symbols are hard to overcome because they are reinforced by our innate brain chemistry. They can be as hard to defeat as a drug dependency. If our environment is filled with potential dopamine reward system hijackers (which it is, primarily due to cheap energy and profit-maximizing consumer capitalism, magnified by the reward systems built into social media), then it stands to reason that more of us are likely to end up spending much of our lives chasing after momentary feel-good experiences that soon turn sour. That's why our society is overwhelmed with high levels of drug, gambling, sugar, television, social media, and pornography addiction.

As we've seen in this chapter, human societies have learned to tame biologically rooted reward-seeking behavior with culturally learned behaviors geared toward self-restraint and compassion for others. Prudence, thrift, and the willingness to sacrifice on behalf of the community are functions of the neocortex—the part of the brain unique to mammals; and even though they are rooted in evolutionary imperatives, they are also at least partly learned by way of example. Most pre-industrial human societies expended a great deal of effort to provide moral guidance, often through myths and stories, to foster pro-social behavior. When a culture ceases providing this needed educational effort—either because self-restraint and empathy are no longer seen as important, or because the society is so overcome with basic survival challenges that it simply doesn't have the resources to devote toward educating the next generation—then these values can become seriously eroded.

Consumerism, the economic system that was invented to solve the problem of overproduction, hijacked our brains' reward pathways for status and novelty, and it has also deliberately eroded our learned social adaptations for restraint and compassion. It reduced the perceived social value of thrift and sacrifice on behalf of community in order to promote the ideal of individual gratification through consumption. Again, this system was put in place with what industrial and governmental leaders regarded as the best of intentions—that is, with the hope of expanding markets, creating jobs, maximizing profits, and increasing tax revenues so that governments could provide more services. But consumerism makes it harder for us to address converging global crises.

We also have an innate tendency, when making decisions, to give more weight to present threats and opportunities than to future ones. This is called *discounting the future*—and it makes it hard to sacrifice *now* to overcome an enormous *future* risk such as climate change. The immediate reward of vacationing in another country, for example, is likely to overwhelm our concern about the greenhouse gas footprint of our airline flight. Multiply that tendency by billions of individual decisions with climate repercussions, and supercharge it with a systemic drive to maximize each company's quarterly profit margins, and you can see why it's difficult to actually reduce total greenhouse gas emissions.

To make matters even worse, many of us in wealthy nations suffer from lottery winner's syndrome.[51] Sociological studies of lottery winners show that many actually experience a *reduction* in happiness and well-being: they're overwhelmed by choice and excess, their relationships become discolored by jealousy and suspicion, and they often become more socially isolated and feel less empathy toward others. Some end up gambling their money away, divorcing, or turning to drugs or alcohol. In a sense, the people who benefitted most from the fossil-fueled Great Acceleration (i.e., middle- and upper-class citizens) are like lottery winners: they have collectively experienced a vast and rapid increase in wealth. They have been encouraged to think that they must somehow deserve this level of wealth, and their sense of empathy toward poorer communities—both domestically and globally—has shriveled. They may also feel more isolated, and are more likely to pursue high-risk behaviors with a high potential reward so as to extend and repeat the initial high they got when they realized they had the winning ticket.

Two final barriers to collective self-limitation in the modern world need to be mentioned, and they are perhaps the most obdurate of all; they are less psychological in nature, more structural. First: nationalism and patriotism tend to hide our common humanity behind divisive notions of global rivalry and national chauvinism. Hence governments and peoples insist on mistrusting each other, and on refusing to cooperate in rationally facing our collective overshoot predicament.

Second: any collective effort to degrow the economy in order to halt climate change, or for any other reason, must confront the relentless logic of capital expansion. Interest must be paid on existing debt to avoid default; and expectations of higher incomes must be met if politicians are to maintain their approval ratings. In short, the maintenance and reproduction of the system require ever more accumulation of capital and social power.

These two factors support and reinforce each other (even if they occasionally come into conflict). In tandem, they create a cultural matrix that encourages feelings, attitudes, urges, beliefs, actions, and ideas that promote accumulation and growth, while discouraging those that undermine accumulation and growth. Anyone who wants to rein in the depredations of a system that has already grown too big to be maintained over the long term must swim upstream against this powerful current.

Of course, not everyone is in denial, not everyone suffers from optimism bias, and not everyone discounts the future and suffers from the lottery winner's syndrome. Some of us, like Greta Thunberg, are even immune to the Overton Window. But in all, we have formidable barriers to overcome if we are collectively to understand and respond to the global crises that our way of life is provoking.

◆   ◆   ◆

The maximum power principle would seem to suggest there's little a species like ours can do if it faces problems created by the accumulation of too much power. It's in our genes, after all, to gain and apply as much power as we can. If we don't do it, the next organism (or person, or country, or company) will, and we will fall by the wayside in life's evolutionary struggle.

As we've seen in this chapter, self-limitation is in fact widespread and essential in nature. The maximum power principle is a true and useful concept, but it requires a supplement—the optimum power principle—which adds the element of time. Organisms routinely limit power in the present so as to have and use more of it in the long run. There is plenty of precedent for self-regulation in nature and human history.

Nevertheless, as a result of the fossil-fueled Great Acceleration, we have gotten so used to growth in population and consumption (two self-reinforcing feedbacks) that we think such growth is normal and essential. Potential checks on power, which could stop or reverse not only climate change but economic inequality, resource depletion, and other deepening crises, have a lot of momentum to overcome.

Given our current levels of denial, and the vested interests of our elites, the overwhelming likelihood is that humanity is in for a release phase (in terms of the adaptive cycle) that will be unprecedented in its severity. However, that could ultimately help clear the way for a different way of being—the second solution to the Fermi Paradox discussed at the start of this chapter. We could learn to focus our attention on beauty and happiness rather than acquisition of material wealth. We could excel in self-control, rather than seeking to further control nature and other people.

Will we flame out or learn to live within limits? In the next and final chapter, we'll look to the future of humanity, and the future of power.

THE FUTURE OF POWER

Learning to Live Happily Within Limits

*Knowing others is intelligence, knowing yourself
is true wisdom. Mastering others is strength,
mastering yourself is true power.*

— LAO-TZU

*If only no one were ever to acquire material power
over others! But to the human being who has faith
in some force that holds dominion over all of us, and
who is therefore conscious of his own limitations,
power is not necessarily fatal. For those, however,
who are unaware of any higher sphere, it is a deadly
poison. For them there is no antidote.*

— ALEKSANDR SOLZHENITSYN

FORECASTING THE FUTURE IS A FOOL'S GAME. THE WORLD, after all, is a chaotic, nonlinear system. We cannot know how global events will play out because there are just too many variables—including too many sources of power, and too many actors seeking to use, abuse, increase, or limit that power. But it is sometimes possible to map constraints and opportunities, as well as likely routes of branching cause-and-effect chains. In this book so far, we have surveyed enough of the biological and cultural factors at play leading up to the present to attempt to do just that.

The range of realistic possible outcomes that could unfold by the end of this century is likely bounded by the two solutions to the Fermi Paradox discussed at the beginning of Chapter 6. At one end of the spectrum of outcomes lies collapse and possible human extinction; at the other end, systematic self-restraint regarding per-capita consumption levels and human numbers, with collapse largely averted. I'll not waste time describing a scenario in which consumption growth and population growth continue to the end of the century, because such a scenario is of vanishingly low probability, in my view.[1] There just aren't enough natural resources and waste sinks to allow it to happen.

There's a lot of territory between the extremes of human self-annihilation and sufficient self-restraint, and this final chapter explores that territory. We will begin by considering the future trajectory that leads to the dismal first solution to the Fermi Paradox, in which intelligent life simply destroys itself. But we'll not dwell there too long. Instead, we'll spend most of the chapter looking at what will be required to achieve the second solution (in which intelligent beings learn to live happily and beautifully within limits), examining the tradeoffs between population, consumption, and efficiency necessary to achieve that outcome, as well as adaptations needed both in society at large and in our own daily existence.

We will finish with an exploration of life goals in a future society of self-restraint. If humans are competing less with one another for wealth and social power, to what might they aspire, and what might give them satisfaction? As we'll see, the answer may lie with the development of internal forms of self-control via the arts and spirituality.

All Against All

Let's first assume that humanity attempts to continue its unbridled pursuit of increasing rates of energy usage, increasing power over nature, increasing social power, increasing military power, and denial of the likely consequences. How might that scenario play out? Again, my goal is not to offer a specific forecast of the future, just to propose a general scenario by extrapolating the current trends surveyed in Chapter 5.

If we follow the path of denial, the results will not be pretty. By the end of the century, the planet's carrying capacity for humans could shrink significantly. Due to climate change, soil depletion, water shortages, and consequent loss of agricultural productivity, it's possible that, in the latter half of the century, significantly more people will be dying by starvation, pestilence, natural disasters, or violence, rather than from old age. It's also possible that deaths will far outnumber births, leading to uncontrolled population decline. There is no point in trying to imagine the specific circumstances leading to such high mortality rates; there are many possible routes. In any case, the psychological terrors unleashed would be profound and intergenerational.

One might think that, as dire events began to unfold, everyone would finally awaken to the reality of climate change, species extinctions, resource depletion, and overpopulation, and do whatever was necessary to keep society from further descending the chute of collapse. However, once the impacts of these problems really started showing up unmistakably, nearly everyone's attention would likely be fixated on effects rather than causes, or on imagined causes rather than real ones. And some problems, like climate change, are slow to develop and hard to reverse once they've reached certain thresholds. Over time, an increasing proportion of the population would be forced to spend most of its effort directly scrounging for living space and food, with little means left over for the maintenance of larger social systems. The descent would then become a series of self-reinforcing feedbacks that would be difficult to halt.

One of the most crucial of these feedbacks has to do with trust. At the end of Chapter 6, I described trust as social capital. It's trust, after all, that enables humans to cooperate in large groups. These days pundits tell us that worsening political polarization in many countries (such as the US) is making democratic governance less and less workable.[2] Polarization is both a symptom and driver of diminishing trust. Once trust is gone, time and effort are required to rebuild it. And without increasing (rather than diminishing) levels of trust, we cannot address global problems like climate change, overconsumption, overpopulation, and the spread of nuclear weapons. As we saw in Chapter 6, these problems are prisoner's dilemmas. They can only be solved with concerted, global effort based on a deliberate process of trust building.

As trust erodes, so does cooperation. Large groups splinter into smaller ones. Ironically, within their smaller subgroups, people might feel more trust and cooperativeness than they did within the larger group as its bonds weakened. So, social breakdown might actually yield a giddy temporary satisfaction for some people. Further, within smaller groups, people might tend to blame rival groups for worsening conditions. And if conditions continue to worsen because nothing is being done to improve them (since small groups acting alone won't be able to solve global problems), then unraveling may simply fuel more unraveling. Under such circumstances, when violence appears it can turn into an endless round of seemingly justified reprisals serving largely to vent and intensify emotion.

The breakdown of trust will cleave existing fault lines between nations, regions, and communities, and between economic classes (for example, between investors versus corporate managers versus salaried professionals versus hourly wage laborers versus the unemployed), as well as between age strata, ethnicities, and religions. As economic pressures mount, each group will seek to insulate itself from suffering as much as possible, while blaming other groups for the worsening crisis.

A particularly nasty set of self-reinforcing feedbacks may arise from the inherent dynamics of our capitalist economic system. Recall our discussion of the adaptive cycle in Chapter 6, in which we saw that complex human societies, like ecosystems, tend to pass through phases of growth/exploitation, conservation, collapse/release, and reorganization. In the current two-century, fossil-fueled running of the cycle, enormous fortunes were made during the growth/exploitation phase via resource extraction and manufacturing. As energy quality began to decline (during the last half-century) due to depletion of the easiest-to-get fossil fuels, capitalists in wealthy countries began to make less of their profit from conventional manufacturing and more from high tech and finance. Debt began to grow much faster than GDP. This likely marked the commencement of the conservation phase of the cycle.

During the collapse/release phase of the cycle, capitalism will be transformed once again. The impetus will be irresistible: profits must still be made; but, with energy supplies declining rather than growing, and debt in all forms being defaulted upon, new sources of profit will have to be found. Capitalists will, out of necessity, seek to profit from the collapse of society and from conflict between regions. The making and selling of weapons, kidnapping for ransom, the organization of political and religious cults for the fleecing of the faithful, the manipulation of opinion, and the theft of warehoused supplies and hoarding of necessities in order to raise prices—these have always existed as perverse paths of profit-seeking at the fringes of capitalist society, but they could become significant new growth industries. Needless to say, such efforts to benefit from others' misery would only further rupture whatever social trust remained.[3]

As ecosystems break down, as agriculture becomes problematic due to climate change, as wildfires and droughts rage, and as water and food become scarce, more people may be forced to flee their homelands. Trickles of refugees and migrants may turn to rivers, rivers to swirling floods. Anti-migrant fervor, which has already taken hold in many European nations and the United States, might spread and deepen.

Governments may be overwhelmed. As governments' ability to provide services, guarantee rights, and punish cheaters breaks down, many people could move away from faith in Big Government and return to faith in Big Gods, perhaps grasping at new forms of those religions that claim to represent "true" or "original" teachings.[4] In addition, entirely new religions may quickly emerge and flourish. Meanwhile, as trust continues to fail, conflict between religions and sects is likely to increase, as well as conflict between believers and nonbelievers (most likely in the form of persecution of nonbelievers by believers).

Global trade depends on flows of fuel and credit, along with general adherence to anti-piracy laws (which prevent goods in shipment from being hijacked, branded products from being cheaply imitated, and proprietary technologies from being stolen). All of these, again, depend on trust. If and when global trust erodes, trade will become more problematic. But we have never been more dependent on global trade. Today it is nearly impossible to manufacture a computer or smartphone in just one country, using local raw materials (this is likewise true for many life-saving pharmaceuticals). Increasingly, as more of our machines and infrastructure become computerized, this means no country can maintain business as usual on its own. If global trade were cut off, then over time people in many nations would find workarounds. In many cases they would simply do without certain products; in others, they would come up with locally produced alternatives, which would likely be more expensive or less functional. Alternatively, they might find ways to keep various machines functioning for years or even decades by cannibalizing other similar equipment (as the Cubans already do with their pre-Castro-era cars). The United States has a great variety of raw-material resources available domestically; even if it could not access rare earths from China or copper from South America, it could in theory manage to maintain a relatively high level of technology and infrastructure (think 1960s, but with a few residual high-tech bells and whistles). Nevertheless, even in the best case, severe ongoing adjustments would be required.

If economic inequality continues to worsen, a backlash may eventually come. Occupy Wall Street, the Tea Party, Black Lives Matter, and the Arab Spring were merely faint, early indicators of the sorts of uprisings that are possible. Many of those earlier protests fizzled, were co-opted, or were violently put down (that doesn't seem to have been the case with BLM), and future rebellions may meet similar ends—but not before overturning governments in some cases, and provoking even more political polarization in others. Eventually the fury of those left behind during the Great Acceleration may be unquenchable—especially when the dispossessed realize that their moment in the Sun is never coming, because the resources that enabled history's grandest banquet are mostly gone.

The explosion of resentment of the poor against the rich may incite a corresponding rage among the young against the old (look out, Boomer and Gen Xer). Climate activism may take a violent turn as young people realize they have been consigned to a future of crisis and deprivation.

In 20th-century geopolitics, Anglo-American economic and military power dominated most of the world. The main European challenger to this hegemony, Germany, was twice defeated in war. Another power center emerged in Eurasia in the form of the Soviet Union, but was ultimately undermined economically; a tipping point came when the United States, acting with Saudi allies to lower the price of oil (the USSR's main source of revenue), pushed the Soviet empire toward collapse.[5] In the latter years of the 20th century, some international observers were already forecasting an end to Anglo-American dominance. In the early 21st century, the tide has turned somewhat: a former KGB intelligence officer, Vladimir Putin, appears to have successfully contributed to the political destabilization of the United States, Britain, the European Union, and NATO using information warfare (though it's worth remembering that the US has a long history of interfering in other nations' politics). Some geopolitical commentators are of the opinion that, while the US won the Cold War, Russia is seeking to win a Long Cold War, or at least to get even with the US for its past meddling in Russia's internal affairs.[6]

Meanwhile, China is on track to become the world's largest economy in just a few years as a result of its low-cost manufacturing capability. Several historians have noted the parallels with the circumstances leading up to World War I: a global hegemonic power (then Britain, now the US) is facing a rapidly industrializing challenger (then Germany, now China) that is seeking to expand its sphere of influence (both then and now: Africa and the Middle East).[7] Hence strategists in Washington and Beijing are preparing for war.[8]

Old alliances are increasingly in ruins. But it would be premature to declare the dawn of a China-centric era of global supremacy. That's because current geopolitical shifts are occurring at the precise moment that climate change and other global crises are beginning to bite. There may be many geopolitical losers in the decades ahead, but there may be no real winner.

During the coronavirus pandemic that began in 2020, some nations acted quickly and competently to reduce infection rates. Other nations, notably the US, dithered and denied for months, leading to hundreds of thousands of preventable deaths. It seems highly likely that, as the economic and social consequences of the pandemic continue to reverberate, nations that responded competently (including small nations such as New Zealand, Finland, Norway, Denmark, and Taiwan) will enjoy more social cohesion and better economic prospects, while those that mounted a poor pandemic response may be hit by continuing economic hardship. Meanwhile, public health experts warn more new pandemics could arise in the years and decades ahead, possibly ones with far higher fatality rates than COVID-19. (See sidebar 21, "Rising Risk of Disease and Pandemic," in Chapter 5.)

In this topsy-turvy world, established assumptions are being stood on their heads. In older democracies like the US and UK, political parties' constituencies are being scrambled, with those that formerly represented working-class interests now supporting views espoused by globalist elites (favoring free trade, open doors to immigration, and religious and cultural pluralism), while those that have long represented the wealthy now also focus on populist cultural issues (opposing both immigration and women's rights, favoring specific ethnic groups and religions). The confusion is likely only to grow.

Geography and climate will largely determine how the future unfolds for any given human population. These factors were pivotal throughout millennia of human history, but during the last couple of centuries we have become accustomed to an unusual situation in which climate zones and geographic boundaries such as mountain ranges and rivers could be largely ignored; after all, we could easily fly above such natural obstacles, creating new political boundaries merely by drawing lines on maps. As we lose easy fossil-fueled mobility, waterways and mountain passes will again become key routes for trade, and nature will reassert itself as a limiting factor for habitation.

The Middle East, the area of the world that exports the largest amounts of oil and gas, is increasingly volatile—the consequence of decades of meddling by world powers eager to control the planetary fuel spigot. The region is also highly susceptible to climate impacts, primarily intense and long-lasting droughts leading to declining food production (a factor in the Syrian civil war).[9] Global warming may make much of the Middle East simply uninhabitable. Geopolitical theorists have long held that if global conflict erupts, the point of ignition will likely be somewhere in this culturally ancient and religiously pivotal part of the world.

The very worst scenario for the remainder of this century is one in which ecological and social trends provoke a nuclear war. In that case, human extinction becomes a possible consequence; if there were survivors, they would (as Nikita Khrushchev reportedly said) envy the dead.

Even without nuclear Armageddon, the picture is grim indeed—though not necessarily without the prospect for some relief. In past instances of civilizational collapse, the process took up to three centuries and there were periods of partial recovery.[10] After the Black Death in 14th-century Europe, a time of economic leveling and relative political calm followed, simply because so many people had died that there were now more goods to go around on a per capita basis. Labor shortages ensued, and peasants demanded better terms from their lords. We may see analogous moments of reversal and partial recovery late in this century or in the next.

However, that is a faint hope in the face of all that stands to be lost if we continue down the path of denial. If we're to avoid such a dark road, or if we hope to minimize the casualties along the way, then we will have to engage in some forms of collective self-limitation. What might those look like?

In the six sections that follow, I will use the pronoun "we" to mean "we humans," as I explore what must be done to restrain our powers of consumption and reproduction in order to prevent the scenarios described above. However, in many respects "we" as individuals are not all in the same boat: various groups of us benefit differently from the status quo based on our gender, ethnicity, nation of residence, and so on. And if even some of the social fragmentation I have just described ensues, then dissimilarities of privilege and impact may worsen. Power struggles are inevitable over the short run, and, if humanity is to avert an all-against-all future, groups adhering more to communitarian and ecological principles must prevail over the forces of continued capital accumulation and vertical power. How such an outcome can be furthered constitutes a vital strategic discussion—one that, in some respects, might seem best placed at this point in the narrative arc of the chapter. However, it's essential first to describe the general direction in which human cultures will necessarily be mutating. I'll discuss the coming power struggle near the end of the chapter, in a section titled "Fighting Power with Power."

Trade-Offs Along the Path of Self-Restraint

How far will we humans have to pull back the reins of population and consumption in order to halt the process of social and ecological unraveling? How many people can Earth sustainably support? What levels of technology and consumption can be maintained over the long haul?

There are no simple answers to these questions. Probably the most straightforward rule of thumb would be to keep pulling back on the reins of population and consumption until the problems resulting from too many people using too much stuff are resolved. One researcher figures we will ultimately return to hunting and gathering, with a global human population numbering in the millions rather than billions.[11] Others say we might be able to sustain a global population of up to three billion at a comfy industrial scale of production and consumption.[12] But those are really just guesses. Any surer grasp of the way ahead depends on coming to grips with trade-offs implicit in the project of restraining aspects of human power that are currently running amok. Those trade-offs mainly involve factors not only of population and per-capita consumption levels, but efficiency as well.

The trade-off between population and consumption is fairly obvious. If we don't somehow humanely reduce population, then overall consumption cuts would have to be drastic in order for us to live once again within sustainable planetary resource budgets (of course, consumption cuts would at first have to be borne primarily by the wealthy, but the global middle class would also have to pare back significantly—a prospect that is politically fraught). On the other hand, if we do gradually and humanely reduce our population size through fewer births, then per-capita consumption levels wouldn't have to drop quite so much in order for us to achieve a sustainable society.

Efficiency, the third variable, is a bit of a catch-all term, in that it can be used to refer to a wide array of strategies such as making longer-lasting products, managing wastes better, substituting renewable materials for nonrenewables in manufacturing and building, replacing fossil fuels with low-carbon energy sources, adopting more ecologically sound agricultural practices, and decoupling economic growth from resource consumption and energy use. So far, policy makers have used hopes for efficiency as an excuse for nearly complete inaction on overconsumption and inadequate action with regard to overpopulation. That's entirely understandable. Efficiency gains are brought about through research and the development of new technologies—and most policy makers love the notion of research leading to technological advances; after all, that has been a significant source of increased economic power and growth during the past couple of centuries. Further, leaning heavily on efficiency means we don't have to pay as much attention to population and consumption, and attempts to rein in either of these is nearly always met with social, political, and even religious pushback.

But we have to be realistic about how much we can do with efficiency alone. We'd all probably be happy with more durable products—though doing away with planned obsolescence might soon cut into sales of new products and thereby inhibit economic growth and profit. Better waste management and more recycling would make for a cleaner environment, something we'd all appreciate, but it would cost more money and has its own limits. Substituting renewable raw materials for nonrenewable ones is easy in some cases (bioplastic compostable forks and spoons work just as well as petroplastic disposable ones), but in other instances there are trade-offs in terms of performance, cost, and profitability.[13] That's why nobody is proposing to make cars or airliners entirely from renewable materials. In addition, there is the very likely prospect of using renewable materials like wood or agricultural waste faster than they can regrow: indeed, global ecological footprint analysis tells us we're already significantly overusing Earth's biocapacity.[14] We've already discussed (in Chapter 4) some of the challenges of substituting renewable or nuclear energy for energy from fossil fuels: it can be done, but it would be foolish of us to expect to continue using as much energy as we currently do, during and after the transition away from coal, oil, and natural gas. Agroecologists have for decades proposed improvements to standard agricultural practices that would build topsoil, conserve water, protect biodiversity, and yield healthier food. Agricultural reform sounds like a no-brainer, but the large majority of farmers, agribusinesses, and government ag bureaus have stuck with methods that are cheaper and more profitable over the short run, even if they result in system-crushing long-term costs.

The decoupling of economic growth from increased consumption of raw materials and energy is really the holy grail of efficiency. Decoupling comes in two strengths: mild-strength (or *relative*) decoupling, which implies using relatively less energy and stuff for each unit of economic growth; and high-strength (or *absolute*) decoupling, which implies reducing the total use of resources even as the economy continues to grow. Almost all economists believe that relative and absolute decoupling will be inevitable features of further technological innovation, and that the benefits will be ongoing and cumulative. Indeed, decoupling is (for these folks) the real key to banishing the contradiction inherent in trying to avert snowballing environmental impacts while failing to deal with population and consumption issues.

Unfortunately, it turns out that decoupling has been oversold. A 2015 paper in *Proceedings of the National Academy of Sciences* showed that even the relative decoupling that most economists believe industrial nations have already achieved is largely the result of false accounting.[15] And two recent metastudies by a team of scientists in Finland have confirmed and elaborated on that conclusion.[16] When Americans buy Chinese-made products, that usually results in usage of less domestic energy and materials than if the products were manufactured in the US. But the energy and materials still need to be accounted for, even if they have been used elsewhere. Some efficiency investments (such as a switch to renewable energy sources, which would do away with the inefficiency of converting energy from fossil fuels into electrical power) would achieve a measure of decoupling, but that would be a one-off accounting benefit. Once those measures were in place, we'd be back to a nearly exact correlation between economic growth and increased energy and materials usage. Further, renewable energy sources would introduce a different set of inefficiencies: due to the intermittency of wind and sunlight, society will have to invest heavily in redundancies in energy production capacity (i.e., building far more solar panels and wind turbines than are needed at optimal times of sunlight and wind availability) and energy storage.

In reality, averting collapse would require all three strategies—managed population decline through fewer births, reduction in per-capita consumption of energy and materials, and a thorough efficiency overhaul of just about everything we do. I've already suggested that these efforts would likely provoke resistance. Since citizens of wealthy industrial nations have the highest per-capita consumption rates, it would be up to them to shoulder the bulk of the consumption cuts, so some of the resistance would come from those quarters. The wealthiest people in all nations, including wealthy people who live in relatively poor countries, would have to make the steepest cuts of all—and these are powerful people who are used to having their way. The well-off would bristle at having to tamp down their consumption levels, while also having to make hefty efficiency investments. The world's poorer nations, many of which have high rates of population growth, would likely resent having to direct significant effort toward changing cultural and economic narratives that currently favor large families. Religious conservatives would be altogether suspicious of the goal of managing population levels. Neoliberal economists (not to mention the entire profit-driven system of corporate capitalism and all of its institutional actors—media, banks, transnational corporations, governments, military establishments, etc.) would howl at the very thought of limiting consumption—the engine of the consumer economy. There are ways around these pitfalls, but all those pathways would require trust, persuasion, and sustained, shared work, along with a dramatic increase in the political power of those doing the persuading relative to those needing to be persuaded.

Another major war would squander societies' cooperative powers in a spasm of violence just when those powers are needed for the project of collective self-restraint. Avoiding war altogether might require reducing and eventually eliminating not just nuclear weapons, but most other military hardware as well. Game theory (discussed in Chapter 6) shows that disarmament can be most successfully achieved through planned stages, with verification at each juncture. The point is to build trust cumulatively, stage by stage. One approach might be for nations to agree to reduce military spending as a percentage of GDP in verifiable increments—though it is difficult to imagine such an initiative taking hold without widespread and persistent demand from the public in many nations.

Trust-building across all scales of organization—from neighborhoods to communities to nations to regions—that enables us to reduce population, taper off unnecessary consumption, and improve economic efficiency is needed to avert collapse. And much of this effort needs to come to fruition not in a century or two, but in a couple of decades if we are to halt the process of ecological collapse. Will we succeed? Recent history suggests the path that world leaders will most likely take is something like the following: modest efforts toward efficiency, zero effort at proactively reducing consumption, only limited efforts at reducing population growth rates in a few nations, and half-hearted efforts (at best) at weapons reduction. Unfortunately, merely muddling through on these terms is likely to lead to crisis after crisis. Again, there is a range of possible outcomes. The more we cooperatively reel in our out-of-control powers, the better our prospects.

SIDEBAR 25

Viktor Frankl and the Will to Meaning

With our collapsing democracies, an imploding biosphere, and reverberations from the recent pandemic, it's no wonder that people despair. The Austrian psychoanalyst and Holocaust survivor Viktor Frankl presciently discussed despair and its resolution in his book *Man's Search for Meaning* (1946).

Already a successful psychotherapist before he was sent to Auschwitz and then Dachau, Frankl was part of what's known as the "third wave" of Viennese psychoanalysis. Reacting against both Sigmund Freud and Alfred Adler, Frankl rejected the first's theories concerning the the primacy of the "will to pleasure" and the latter's "will to power." In contrast, Frankl wrote that: "Man's search for meaning is the primary motivation in his life and not a 'secondary rationalization' of instinctual drives."

Of course, meaning is an attribute of words; therefore non-linguistic (i.e., nonhuman) animals presumably don't engage in a search for meaning. Nevertheless, for language-making people, Frankl argued that literature, art, religion and other cultural phenomena that place meaning at their core are the very basis for finding purpose in life. In his private practice, Frankl developed a methodology he called "logotherapy"—from *logos*, Greek for "reason"—describing it as arising from the fact that "striving to find a meaning in one's life is the primary motivational force in man."

While in the camps, he informally worked as a physician and therapist, finding that acting as analyst to his fellow prisoners gave him purpose and satisfaction. In those discussions with patients, he came to conclusions that became foundational for humanistic psychology.

One such conclusion was that the "prisoner who had lost faith in the future—his future—was doomed." Frankl recounts how even in the camps, where suicide was endemic, the prisoners who seemed to have the best chance of survival were not necessarily the strongest or physically healthiest, but those somehow capable of directing their thoughts toward a sense of meaning. A few prisoners were "able to retreat from their terrible surroundings to a life of inner riches and spiritual freedom," and in the imagining of such a space there was the potential for survival.

Frankl wrote that he "grasped the meaning of the greatest secret that human poetry and human thought and belief have to impart: *The salvation of man is through love and in love.*"[17]

The Fate of the Superorganism

There's a paradox hidden in the project of global human self-restraint. As we've just seen, global cooperation is needed to address global problems. Yet, as we restrain certain of our powers, societies will need to become more localized (i.e., deglobalized), and may do so by default even if that is not the intention.

Globalization and urbanization are trends with a shelf life. Both depend on high rates of energy usage—especially for transportation, communication, and data processing. Current petroleum-based transport fuels are depleting and polluting. The project of substituting electric transport for liquid-fueled transport has begun, but is going much too slowly to avert crisis, and will be limited in scope by energy transition trade-offs and barriers already discussed.[18] As a result, we probably face a future of less and more costly transportation. That means less global trade. It also translates to a less mechanized, more labor-intensive, and more localized food system, which entails more people growing food in the countryside and fewer living in cities.[19]

A more locally-organized and more rural society may serve its members' psychological health better than one that is globalized and highly urbanized: we evolved living in small groups and function best in contexts where we know one another face-to-face. And living in more direct contact with nature encourages us to maintain a healthy relationship with it. Crucially: when political and social entities grow in size, social power increases, leading to the possibility of inequality in power and the abuse of power on a larger scale. One way to keep power abuses and inequalities manageable is to keep the scale of social organization small. (On the other hand, a reversion to a form of social organization characterized by competing agrarian states would likely bring with it a rigid class structure and sharper distinctions in gender power. This is the main reason I see the general adoption of a decentralized horticultural way of life as a more desirable path, at least in theory.)

Ecologists have been telling us that "small is beautiful" since the 1960s, but trends have gone in just the opposite direction, resulting in the flourishing of the Superorganism.[20] Nobody designed this vast, intricate web of global interconnectedness, and no one can control it. It is an emergent phenomenon—something that could not have been predicted on the basis of a thorough knowledge of its constituent parts. We may be able partially to control various subsystems of the Superorganism, but the entity itself has its own inherent priorities and imperatives. It favors anything that leads to its growth; it discourages anything that impedes expansion.[21]

SIDEBAR 26

Dethroning GDP: Key to Limiting the Power of the Superorganism

GDP, or Gross Domestic Product, was adopted by the US and other nations during the period when the consumer economy was being designed and implemented (i.e., shortly after the end of World War II). It is essentially a measure of the amount of money flowing through a national economy, and, indirectly, of the materials and energy being used. An expanding population also helps GDP to grow. GDP is, in effect, the speedometer of consumerism. And the Superorganism, which is in consumerism's metaphorical driver's seat, usually wants to go faster—rarely slower.

While GDP has become the most universally used indicator of the health of national economies, it is spectacularly unsuited for that role, as has been well documented.[22] GDP prioritizes the monetization of everything: if we become more self-reliant or start sharing more instead of paying for goods and services, that reduces the GDP. Moreover, GDP doesn't distinguish between economic activities that hurt the environment and those that heal; nor does it tell us if inequality is increasing or decreasing. Indeed, in recent decades an increase in GDP has reliably correlated with increasing carbon emissions and worsening inequality.

During the last couple of decades, many nations such as the US and Japan were finding GDP growth harder to achieve; now, since the advent of COVID-19, even many formerly accelerating economies seem to be in a persistent funk. The world appears to be colliding with the limits to growth that systems scientists have long warned about. If GDP growth will soon no longer be possible, and if there are good reasons to limit growth in population and consumption anyway, then relying on alternative economic indicators rather than GDP seems both desirable and inevitable. We tend to get what we value and aim for, and GDP values and aims for the wrong things. Therefore, ditching GDP may be a pivotal way for society to limit destructive power.

Alternative economic measures exist, some of which are already in wide use. Unemployment levels and measures of economic inequality (including comparisons of the amount of national income captured by the "one percent" versus everyone else) are better than GDP at showing how well an economy is satisfying most people's needs, and measuring greenhouse gas emissions gives a general sense of environmental harm being inflicted. Gross National Happiness (GNH) has been proposed as an alternative to GDP, as has the Genuine Progress Indicator (GPI); both aggregate information about human and environmental well-being, though in somewhat different ways.[23]

For example, millions of individuals, as well as thousands of nonprofit environmental organizations and even many for-profit corporations are deeply concerned about climate change. Yet, as we have seen, humanity as a whole is still increasing its greenhouse gas emissions, cutting forests, and doing all the other things that undermine climate stability. Why are sustainability efforts failing? It's easy enough to blame greedy oil executives, but doing so merely shifts attention away from our own collective insistence on maintaining overall economic growth. This growth imperative—which at its core is a drive toward power and capital accumulation—washes up and down, left and right throughout the institutions (political parties, governments, and businesses) and levels of organization within society. Even if, individually, we can see that human society is exceeding natural limits, we collectively elect leaders who promise more resource extraction, manufacturing, and trade. The Superorganism has its way.

In a sense, the Superorganism is just a large-scale expression of certain innate human traits. But it is also something genuinely new. It is all of humanity acting together, largely unwittingly, to form the most powerful living entity in Earth history. While humans have been cooperating in groups for a very long time, the global Superorganism is only decades old.

Reversing the trends leading toward global crisis and collapse might seem to require the breakup or disempowerment of the Superorganism. Yet, since we need global communication and cooperation to address global problems, it might be preferable for the Superorganism to survive and mature—to become aware of itself and of natural limits. Humanity became ultrasocial through its development of language, cooperation, technology, and hypersociality. Our survival may depend not on repudiating these, but on developing them to a higher level.

The Superorganism is still an infant, and it has not had much opportunity to test its limits. Individual humans can recognize limits and learn how to live within them by trial and error. Even groups of humans are capable of understanding when their collective behavior threatens long-term sustainability. Unfortunately, the global Superorganism currently has the intelligence of an amoeba. It adheres strictly to the maximum power principle, so it knows only how to grow and amass ever more power. This puts it entirely at odds with an environment characterized by natural limits and hidden feedbacks.

A mature global Superorganism governed by the optimum power principle is for now a purely theoretical entity. Perhaps it could emerge through a bottom-up process of building trust and understanding, and through the crafting of cooperative institutions and formal agreements for human self-limitation at ever-higher governmental levels (local, regional, national, and global). Arguably, the annual United Nations climate conferences, along with other government-led international environmental and arms-control gatherings, constitute efforts to raise the Superorganism's level of self-awareness. We can only speculate whether, under conditions of declining global trade, worldwide communications could be maintained in such a way as to facilitate the expansion of such efforts so as to promote not just the survival, but the moral flourishing (in terms of declining violence and inequality) of our human hive.

If we can't do those things, then the Superorganism may be short-lived. That doesn't necessarily mean that humanity will go extinct, simply that the global economy and global communications as we know them will cease to function. And we will have to address global ecological problems without the benefit of international coordination, through bottom-up efforts in many places simultaneously. Such efforts are already sprouting up in towns and villages around the world, but they tend to get much less attention than international climate treaties, largely because climate change is a global problem presumably requiring global solutions.

Frankly, when looking at the factors arrayed before us, it appears to me that the gradual or piecemeal breakdown of global integration is highly likely. This isn't a conclusion I particularly welcome, as it contradicts some values I have long held. We may be facing a less cosmopolitan, more tribal future, like it or not.

SIDEBAR 27

Energy and Human Values

In his book *Foragers, Farmers, and Fossil Fuels* (2015), historian Ian Morris argues that human values have evolved to fit societies' changing modes of energy capture. This argument represents, in many ways, an updating of anthropologist Marvin Harris's ideas on cultural materialism.[24] Morris's book originated as a set of lectures delivered in 2012, and includes essay-responses from classicist Richard Seaford, historian Jonathan Spence, philosopher Christine Korsgaard, and novelist Margaret Atwood.

According to Morris, hunter-gatherers have politically and economically egalitarian values. Gender inequality is variable, but generally low. Levels of interpersonal violence are high.

People in agrarian state societies ardently believe that political and economic hierarchies are justified. Gender inequality is extreme. Levels of interpersonal violence are lower.

Fossil-fuel societies have politically and economically egalitarian values, gender inequality is low, and interpersonal violence is lower still.

In terms of human values, history describes an arc that nearly completes a full circle: the values of hunter-gatherers and fossil fuelers have much in common. Agrarian societies were the outlier, because processes of energy capture favored rigid political hierarchies and division of labor by gender.

Morris doesn't discuss the unique value-sets of horticultural, herding, fishing, and trading societies, which is a pity because doing so would have added richness to his argument (Marvin Harris did explore these modes of energy capture and their cultural ramifications). Nevertheless, Morris succeeds in adequately illustrating his central point that human values are only partially universal; and that where they differ, the difference is usually explainable by the cultural impacts of society's means of capturing and harnessing energy.

That raises the question: as we move away from fossil fuels, what values will be most favored? Unfortunately, Morris offers little guidance as to the likely nature of our post-fossil-fuel energy regime. However, Margaret Atwood contributes an entertainingly written essay to the book, titled "When the Lights Go Out: Human Values after the Collapse of Civilization," in which she offers this germane, if sobering, opinion:

You might think that [when civilization collapses] those of us who are left would go back one step—from fossil-fuel values to agricultural ones—but in conditions of widespread societal breakdown, we'd more likely switch to early foraging values almost immediately, with the accompanying interpersonal violence. Short form: when the lights go off and the police network fails, the looters will be out looting within twenty-four hours. Agriculturalists have land to defend and therefore borders to protect, but urban dwellers minus their usual occupations are nomads, dependent not on what they can grow—that's a long seed-to-harvest cycle anyway—but on what they can scrounge, filch, or kill.[25]

Questioning Technology

For most of human history, technological innovation was slow. Each new generation expected to do things (grow food, build houses, make clothing, fight, and communicate) essentially the same ways their parents did. Even in the relatively fast-paced Italian Renaissance, the appearance of a disruptive new communication medium, weapon, or farming tool was a once-in-a-lifetime phenomenon.

In contrast, we've gotten used to living in a world where the pace of technological change exceeds our ability to adapt to it, much less to assess and control it. Grandparents and parents have difficulty communicating with children because the latter live in a different technological universe. Worldwide, people are systematically empowered or disempowered based on their access to, and mastery of, new technologies.

Every technology has unintended consequences, some better or worse than others. By the time the drawbacks of a new pesticide or social media app have become apparent, it's often too late to put the genie back in its bottle. The market has adjusted to this shiny new thing and livelihoods depend on it.

The pace of technological change is driven by at least five factors:

- *Energy growth*: with more energy available, we are likely to find more tools for applying that energy to solving our problems and expanding our options.
- *Population growth*: with more people working at solving problems, there is greater likelihood of producing innovations.
- *A culture of investment*: with investors on the lookout for the "next big thing," and a system of ownership (via patent rights) in place, inventors are more likely to find backing for their efforts.
- *A culture of research and development*: with universities turning out large numbers of skilled scientists and engineers, there will be more technical wherewithal to imagine and realize new inventions.
- *Inventions available to be discovered*: the number of possible inventions may be infinite, but within any given sphere of human interest there is a hierarchy of importance in inventions. The most important inventions, the proverbial low-hanging fruit, are likely to appear, at least in rudimentary form, in the early phases of a complex society's take-off; later, each new invention is likely to be an extension or refinement of an earlier one, with each successive improvement having a higher incremental cost.

Each of these factors introduces potential technological limits, for reasons we have in some cases already discussed. Available energy and human population cannot continue growing at recent rates for much longer. In addition, investments in research and development have been reaping diminishing returns for the past couple of decades: while new technologies continue to appear, our basic technologies of transportation, manufacturing, and agriculture have been in place for some time, and at this point are largely subject merely to refinement.[26] In a possible future of declining energy, declining population, debt deflation, and social turmoil, it's likely that the pace of technological change will slow to a crawl and many existing technologies may no longer be supportable.

In contrast, most futurists foresee technological change continuing to accelerate along two paths: either toward increasing computerization and automation, or toward redesign of systems to minimize carbon dioxide output and incorporate captured atmospheric carbon into products and materials. While advances in the latter direction would be welcome, low-carbon innovation will have to show significant profit potential to attract much investment in research and development. In reality, meanwhile, most R&D funding is going toward computerization—i.e., a mere extension of current profit-seeking technological innovation.

The momentum of technological change is leading us into worrisome new (artificial?) realities. Existing technologies—the automobile, ammonia-based fertilizers, herbicides and pesticides, plastics, and social media—have unintended consequences that continue to snowball. And new technologies in development, such as artificial intelligence (AI, which could make hundreds of millions of people "redundant"), "deepfake" video (which could generate innumerable conspiracy theories and further exacerbate political polarization), CRISP-R gene editing (which could have unforeseen biological consequences throughout the ecosphere), and 5G wireless technology (which requires ever more energy and broadcasts intense electromagnetic fields with possibly significant biological impacts) now threaten to take us further into dangerous, uncharted territory.[27]

While technological change may eventually decelerate for the inadvertent reasons mentioned above, technology theorists say it would be better to *deliberately and controllably* slow the pace of change, so that we can pick and choose which technologies to save and which to let go. In order to gain control of the speed of the conveyor belt bringing us new tools and techniques, society should investigate the possible unintended consequences of technologies before permitting widespread adoption. This notion, known as the Precautionary Principle, has been endorsed in the 1982 World Charter for Nature and other international agreements.[28] However, its adoption and application by national governments has been sporadic, inconsistent, and weak.

Management of technological change for the benefit of nature and culture would be one of the critical signs of a maturing Superorganism. If we're going to move in that direction, economic gain on the part of manufacturers cannot continue to be sufficient reason for the adoption of any significantly new process or tool. We will need new democratic deliberative bodies tasked with assessing proposed technologies, and we will need to provide them with mandate and authority sufficient for them to actually do their job.[29]

You needn't wait for such bodies to emerge. Examine the technologies you use and depend on. Which ones truly serve your needs, and which ones are imposed by society at large but make you less autonomous and happy? Formulate a plan to free yourself from the latter.[30]

Learning to Live with Less Energy and Stuff

Changing our relationship with technology isn't simply a matter of getting rid of harmful gadgets. We must alter the *direction* of technological research and development so that it is relevant to a society whose rate of energy usage is declining, not growing.

As the fossil-fuel era winds down (whether as a result of government climate action or oil, coal, and gas depletion), we will almost certainly have less energy available for human activities. How will we adjust? What are the implications for food, transport, manufacturing, and the economy? Our future lies with renewable sources of energy—whether firewood (again), or water, solar, and wind. But just as important as the source of our energy is the quantity of it that we use. Theoretically, we could still overpower natural systems using electricity from solar panels and wind turbines—though that's unlikely, because at the current rate of investment we simply won't build enough of them to be able to maintain present levels of energy usage as coal, oil, and natural gas deplete. There's been considerable research regarding alternative energy sources in recent years, but not nearly as much on how to get by with less. One way or another, we will have to adapt to energy decline, and technology will necessarily play a significant role in how we do that.

We should start by asking: How much energy does it actually take to enable us to lead happy lives filled with opportunities for aesthetic enjoyment? Fortunately, global happiness surveys show that, once basic human needs for food and shelter are met, higher levels of energy usage don't translate to higher levels of self-reported happiness.[31] Those surveys suggest that reducing energy usage need not make us miserable; if we do it well, it could actually do the opposite.

The general direction of the research that's required is clear: we need fewer tools classifiable as high tech (ones that use more energy and that have globalized processes embedded in their manufacture and distribution) and more tools that we might consider low tech (ones that use less energy and that depend on local skills and supply chains). The technology of the future need not necessarily entail a recapitulation of technologies from the agrarian era—though many of those were ingenious and deserve to be preserved or revived. In many cases, we should be able to find even more elegant solutions than those, using scientific knowledge developed during the past century. For example, people have been using wind and water power for a range of purposes for a very long time (as we saw in Chapter 4), but tools for energy capture were typically inefficient; today we have the benefit of decades of research into optimal designs for sails, paddles, and blades, enabling the production of sailboats, windmills, and watermills that can outperform traditional designs.[32]

In general, the technological strategies we should be investigating are ones that tend to substitute human labor for machine power; ones that de-mechanize services of all sorts (which translates to more face-to-face interactions among people in daily life, less pushing of buttons); and ones that balance performance with conviviality. Fortunately, a few smart people are already thinking along these lines, and they have produced a small but growing literature. Typically, these pathfinders are not people who hate technology, but engineers who understand the current human ecological predicament and who happen to love finding beautiful ways to solve problems.

Susan Krumdieck taught for many years at the Department of Mechanical Engineering, University of Canterbury, New Zealand, where she was the director of the Advanced Energy and Material Systems Lab (she's recently moved to the Orkney Islands of Scotland, where she runs an energy transition project). No stranger to high technology, Susan authored a PhD thesis titled "Experimental characterization and modeling for the growth rate of oxide coatings from liquid solutions of metalorganic precursors by ultrasonic pulsed injection in a cold wall low pressure reactor." However, since learning about the global sustainability crisis, she has focused her research and teaching on what she calls *Transition Engineering*—the title of her recent book, which explores engineering strategies in response to the mega-issues of global climate change, decline in world energy, scarcity of key industrial minerals, and local environmental constraints.[33] Through her years of teaching, Susan has developed a global network of engineers working on a wide range of technical problems entailed in society's shift away from unsustainable activities.

Low-Tech Magazine is an online publication written and edited by Kris De Decker, who lives in Barcelona, Spain.[34] De Decker began his career as a freelance tech and science journalist. "After ten years or so," he told me in an email exchange,

> ...it became clear to me that the focus on technological solutions is doing more harm than good. In a nutshell, that's how *Low-Tech Magazine* was born. As a consequence, I also started looking more critically at my own lifestyle, and so I stopped flying (which I never did often anyway), stopped driving cars, tried to power my apartment with solar energy, refused to switch to a smartphone, and so on. I didn't want to be a hypocrite and I discovered that "practice what you preach" is a very interesting research position. Another ten years later this approach has radically changed my life, and for the better. Lowering your ecological footprint also saves a lot of money, and this has given me the freedom to do the work that I like.

De Decker questions our society's blind belief in technological progress, and writes about the potential of past and often forgotten knowledge and technologies to aid in our transition to a sustainable society. Many of his experiments involve combining old technology with new knowledge and new materials, or applying old concepts and traditional knowledge to modern technology. Recent article titles include "How to Run the Economy on the Weather," "Reinventing the Greenhouse," and "Ditch the Batteries—Off-Grid Compressed Air Energy Storage."

In Australia, the Simplicity Institute, an education and research center, seeks to "envision and defend a 'simpler way' of life at a time when the old myths of progress, techno-optimism, and affluence are failing us." Samuel Alexander, co-director of the Institute, is a lecturer and researcher at the University of Melbourne; he and fellow Aussie Ted Trainer have pioneered a path they call "The Simpler Way," aimed at improving human life as we descend the scale of energy usage and societal complexity.[35]

Frenchman Philippe Bihouix worked as a construction engineer in the building industry and as an engineering analyst in various industrial sectors (energy, chemicals, transport, telecommunications, and aeronautics) before learning about climate change and other sustainability crises. His recent book, *The Age of Low Tech*, is more than a manifesto; he explores the principles of simple technologies, and imagines daily life in a world where simple technologies supply our needs.[36] With his far-ranging technical background, Bihouix is able to drill down into details about simpler communications systems, banks and finance, transportation, and food systems.

Speaking of food systems: nearly everyone who works and writes in the field of low tech agrees that the best place to start society's simplicity overhaul is with the ways we feed ourselves. The skills needed to get started retrofitting food growing are neatly bundled in permaculture—a set of design tools for living created back in the 1970s by ecologists David Holmgren and Bill Mollison, who understood then that industrial civilization would eventually reach its limits. A portmanteau of the words "permanent" and "agriculture," permaculture has sustainability baked into its very name. Usually applied in gardening and small-scale farming, permaculture promotes biodiversity, low energy usage, and water conservation.[37]

Recently, ecological agronomists have been experimenting with ways to sequester more atmospheric carbon into soil, as well as the

roots, wood, and leaves of plants. These efforts go by the name carbon farming; it has the potential to increase biological activity in soil,
aiding plant growth and increasing agricultural yield while improving soil water retention capacity and reducing fertilizer use (and the
accompanying greenhouse gas emissions). As of 2019, hundreds of
millions of hectares worldwide were being carbon farmed in some
fashion. Eric Toensmeier's book *The Carbon Farming Solution* is the
best overview and how-to manual available.[38]

While I've just mentioned a few power-down experts, the low-
tech efforts currently under way are in most cases being taken up by
people whose names we'll never hear or read. You can get in on the
fun. Spend more of your time working at energy-harvesting activities
such as gardening. Wean yourself from high tech. Mend your clothes
instead of buying new ones. Using simpler tools usually requires
more skill and attention, and therefore also tends to yield more satisfaction. Therefore, choose some simpler tools you want to get good at
using, learn from a master, and, as they say at music school, practice,
practice, practice. Meanwhile, take an occasional holiday from the
use of electricity and fossil fuels. For many years, my wife Janet and I
have done this as a way of marking the summer and winter solstices;
it's an excellent way to get back in sync with deeper cosmic, terrestrial, and biological rhythms.

SIDEBAR 28

Advice to Young People in the 21st Century

- Learn to grow food. Study permaculture.
- Learn to read people. You're going to need to know
  whether people in your vicinity are trustworthy.
- Be trustworthy. Otherwise smart and trustworthy people
  won't associate with you.
- Learn to express yourself clearly and persuasively.
- It is okay not to reproduce. There are already plenty of
  people in the world.

- Learn to make decisions by consensus and to work collaboratively. Be a person with whom others enjoy working.
- Learn to repair and use relatively simple technologies. Studying to be a computer programmer or hacker could pay off in the short run, but over the longer term you'll benefit more from learning to fix farming and construction tools and small engines. Learn to make spare parts from junk.
- Learn how energy works. Be able to identify the sources of energy in your environment and find ways to harness that energy to do useful work.
- Learn to defend yourself. Sadly, for the remainder of this century the world is likely to be a more violent place. Even if that turns out not to be the case, martial arts can still be useful paths of self-discipline.
- Learn to heal the human body via nutrition, herbs, and basic emergency care.
- Learn to recognize the subjective effects of sex hormones, dopamine, and other brain chemicals, and find ways to override those effects when they threaten to push you off course. Instead, channel their effects to help achieve goals.
- Learn about nature. Memorize the names of local plants, birds, and insects, and observe their habits. Learn to be comfortable in the wild.
- Learn how to produce beauty via art, music, or movement, and how to engage others in creative, celebratory activities.
- Learn to emotionally process trauma and grief, and to help others do so. Learn when and how to use humor to release tension.

No one can do it all, but do your best.

Lessening Inequality

The era of climate change presents a troubling new context for the millennia-old social contest over inequality. Some environmental activists, at least privately, say that dealing with social justice issues must take a back seat to saving the planet. After all, governments, trade unions, and advocacy groups have been fighting poverty for decades; and activists have long been fighting racism, sexism, and other forms of oppression—but all of these ills are still with us to degrees varying by country. Climate change is something new, global, and existentially threatening: if we don't minimize the damage, civilization might not survive and human mortality might be (as we have already discussed) unprecedented. Equality would be nice, but survival is paramount. As some put it, "there is no justice on a dead planet."

Some social justice advocates, for their part, say that climate change can never be successfully addressed until inequality is curbed, or perhaps until capitalism is overturned. Without buy-in from the world's masses of poor and underprivileged, and without political defeat of the powerful fossil fuel industry, climate change mitigation efforts will never gain traction.

My own view is a hybrid of those just stated. Adapting to the twilight of the fossil fuel era and the end of growth will entail sacrifice and change. If the sacrifice is highly unequal, the result will be social chaos, and the transition will likely fail. I do not believe that society must utterly defeat inequality before it can address climate change; but we can and must address climate change and other overshoot crises in ways that reduce existing inequities and that avoid imposing new ones.

As I've already suggested, having less available energy is likely to mean that society will need more manual labor. In the worst scenario, the end of the fossil fuel era could imply the return of low-paid drudgery for people in currently industrialized nations, or possibly even debt peonage or slavery. Recall from our discussion in Chapter 3 of the "Pathologies of Power" that advantaged people whose power is being challenged often don't respond with equanimity; instead, they tend to become defensive and may lash out. As the physical power that enables modern consumer culture dissipates, people who have spent their lives accustomed to high levels of amenity are likely to be surprised and indignant; they may respond with anger and look for scapegoats—and hence for less-powerful people to abuse. Groups that are historic victims of the powerful will likely be the first to be targeted. In short, humanity could be setting itself up for unimaginable abuses of power if we don't set processes in motion now to prevent such an outcome.

We may have a narrow window of time still available for acknowledging and overcoming historic power abuses, and for avoiding new ones. As we have seen, economic inequality arose through the destruction of the gift economy and the commons; it stands to reason, therefore, that greater equity could be achieved by reviving these features of prestate societies. Doing so would require economic structural changes that may seem impossible for sclerotic industrial democracies that are largely captured and controlled by corporate interests. But when growth ends, banks fail, and governments are scrambling for ways to keep crisis at bay, political leaders may be in a receptive mood for radical ideas. Or it may be that current global and national power holders lose legitimacy and there are the opportunities to implement radical ideas locally. Either way, it would be essential for equity activists to seize the moment, as it could be a fleeting one, considering the disintegrative trends discussed earlier in this chapter.

Ultimately, promoting economic equality will require expanding the commons once more. The ethical basis for doing so is clear: no human being made land, rivers, or deposits of iron ore, gold, or fossil fuels through their own ingenuity or labor. Why, then, should a person or corporation be entitled to extract wealth from natural resources for purely private benefit? An obvious alternative to private ownership is for natural resources to be declared public goods to be owned and protected in common—a commonwealth. A widely respected American economist in the late 19th century named Henry George proposed we do just this, and his ideas have been put into practice successfully, for example, in publicly owned utilities and public transit systems.[39] George was, in effect, arguing for the validity and necessity of the commons, and his work brought that ancient customary institution into a more modern context. Modern agricultural and conservation land trusts are examples of "Georgism" at work—though on a scale limited by society's overall commitment to exclusionary private ownership of natural resources. If that commitment were to be even partly overturned, the trend toward economic inequality could be reversed.

Creating more economic equality will also mean simplifying and regulating financial markets. Many proposals along these lines were put forward immediately after the 2008 global financial crisis—including, for example, proposals for a financial transactions tax. However, policy makers quickly lost their enthusiasm for financial regulation once the recovery began.

In his 2020 Democratic presidential primary campaign, Senator Bernie Sanders raised the question: *Should billionaires exist?*—and proposed a wealth tax to relieve the super-wealthy of their excess monetary baggage. Naturally, some billionaires think they should continue to have the opportunity to expand their income and wealth with no limit.[40] However, Sanders touched a nerve among a swath of the public that is raw with resentment over the lavish lifestyles of the rich in the face of hard times for the working poor.

In the United States, inequality has persistent racial overtones that cannot be expunged solely by implementing a new economic paradigm. The nation must fully acknowledge the implications of the genocide of native peoples and the enslavement of Africans who were the sources of intergenerational wealth for European immigrants and their descendants. A Truth and Reconciliation Commission or some similar effort toward moral redress, reparations, and a reversal of lingering legal inequities would be significant steps toward healing. Fully equal rights and wages for women are also overdue. At the same time, all of humanity must come to terms with the inequalities of power between ourselves and other species; legal establishment of the rights of nature and of species is already being debated. In 2008, Ecuador became the first nation to enshrine the rights of nature in its Constitution.[41]

It is possible that the general crisis we are approaching will present a forbiddingly challenging context in which to try to overcome historical power imbalances and abuses. If that is the case, our overall prospects are dim. But it is also possible that crisis will offer new opportunities to effect fundamental change in social and cultural systems. I'm encouraged, in this regard, by the example of the Kurdistan Workers' Party, or PKK. Inspired in part by the writings of American ecosocialist Murray Bookchin, the Party made significant efforts to address ecological and social problems in tandem. As combatants in a deadly conflict (the Syrian civil war), the PKK are surely no saints; American President Donald Trump, with typical pugnacious hyperbole, went so far as to call them "worse than ISIS." Nevertheless, every member of the community has an equal say in popular assemblies that address the issues of their neighborhoods and towns, and the region operates as a federated system of self-determining municipalities. All ethnic groups within the region are guaranteed the right to teach and be taught in their own languages, and public institutions are required to work toward the elimination of gender discrimination. The PKK Social Contract also promotes a philosophy of ecological stewardship that is supposed to guide all decisions about town planning and agriculture.[42] The fact that the PKK has been able to do so much along these lines in the horrific context of the Syrian civil war suggests that crisis can indeed bring opportunity.

If you're concerned about inequality, begin to explore how you can rebuild the commons in your region. Consider taking a diversity and inclusion training. Who are the least advantaged within your community? What are the structures that keep them that way? How could you make a difference? What activist groups, nonprofit corporations, and cooperatives could you join and support?[43]

Population: Lowering It and Keeping It Steady

As we have seen, while any serious collective effort to rein in our human power of population growth is bound to be contentious, such an effort is necessary if we wish to minimize the overshoot crisis. So far, our instinct to avoid contention has had the upper hand. At the 1994 population summit in Cairo, rich and poor nations reached a quiet accommodation: if poor nations refrained from forcing a conversation on consumption (which would be uncomfortable for rich nations), rich nations wouldn't force a conversation on population (which would discomfort many poorer countries).[44] The ensuing code of silence has held for three decades, limiting any substantive intergovernmental discussion on either overconsumption or overpopulation. That refusal to deal with reality has constrained humanity's current options; and the longer we wait to engage with population and consumption issues, the worse our prospects become.

The arithmetic is daunting—not in its complexity, but its implications. Joel Cohen of Rockefeller University, in his aptly titled and exhaustively researched book, *How Many People Can the Earth Support?*, concludes that there is no widely agreed answer to the question; there are simply too many variables to consider. However, the number may well be considerably smaller than our current global population.[45] Some of the scenarios Cohen cites for future energy production and agricultural output are almost certainly overly optimistic. Energy and available material resources are what enabled population expansion, and less energy and a degraded environment will almost inevitably translate to a smaller global carrying capacity. Another variable is, how much space shall we leave for the rest of nature? Putting a number on global human carrying capacity post-fossil fuels necessarily involves choices and guesswork, but analysts who've used realistic scenarios for energy, food, raw materials, and biodiversity have come up with numbers in the range of a few hundred million up to three billion—a huge range that reflects a wide variety of assumptions concerning the amount of climate warming that's likely and the methods of food production that might be possible under those circumstances.[46]

Even the high end of that range presents a formidable challenge. How could we realistically transition from our current eight billion to three billion without massive human rights violations?

China's one-child policy was effective but coercive. Successful noncoercive efforts to end population growth have typically involved raising the education levels and social status of women, so that they can choose their desired family size. Often, such efforts are hindered by fundamentalist religious social norms that promote fertility and subjugate women. However, even some highly religious nations, such as Iran, have had success in dramatically reducing population growth this way.

Some of the most promising results have been achieved by Population Media Center (PMC), which enlists creative artists in countries with high population growth rates (which are usually also among the world's poorest nations) to produce radio and television soap operas featuring strong female characters who successfully confront issues related to family planning. The biggest barriers to population reduction are cultural, so the PMC team decided early on that merely distributing condoms wouldn't help much if the people they were trying to reach believed that women were inferior beings whose main purpose in life is to birth babies, and that men can attain higher status by having many children. In comparative studies, the PMC strategy has been shown to be the most cost-effective and humane means available of reducing high birth rates.

Unfortunately, PMC's efforts are (at present) too slow to get us to, say, three billion by the end of the century. A recent book by Christopher Tucker, *A Planet of 3 Billion*, advises ramping up such efforts. Tucker, Chairman of the American Geographical Society, advises empowering women across the world and giving them access to family planning information and contraceptives. The book is tied to an advocacy campaign called P3B.[47] Could such programs, even if turbocharged by more money and publicity, be sufficient to avert human die-off later this century? The math is not encouraging. A 2017 study published in the *Proceedings of the National Academy of Sciences* modelled various population policies and found that "even a rapid transition to a worldwide one-child policy leads to a population similar to today's by 2100."[48]

Meanwhile some demographers worry about a very different population trend—declining fertility rates in many countries, leading to falling population.[49] The three causes of declining fertility are families choosing to have fewer children, women deferring pregnancy to later in life, and environmental toxins causing rapidly falling sperm counts and sperm quality in men. The numbers for 2019 are shocking: In Japan, only 864,000 births were recorded—fewer than any year since records began in 1899. In the United States, the number of births was the lowest for 32 years and the fertility rate set an all-time record low of 1.73 children per woman. In China, the baby cohort dropped to 14.6 million, the lowest in 70 years (excluding the 1961 famine); the fertility rate is below 1.6 children per woman. In South Korea the fertility rate fell below 1.0. Israel, at 3.1 children per woman, is now the only industrialized country with a fertility rate above the 2.1 replacement level.[50] As if all of this isn't worrisome enough, the evidence suggests that declining fertility is affecting other species as well—including (but probably not limited to) mammals, birds, amphibians, and insects. In these cases, it is impossible to attribute the trend to deliberate choice or to contraception; it appears entirely the result of human-made, hormone-mimicking environmental chemicals. The word *omnicide* is starting to creep into discussions among biologists studying the problem.

For nations concerned about slowing economic growth and aging populations, declining fertility is perceived as a challenge that must be met with birth incentives. Of course, in the context of the strong likelihood that planetary human carrying capacity is set to shrink in the decades ahead, the "problem" of declining fertility sounds like news to be welcomed. Still, the economic challenge of an aging demographic is real. It is an unintended consequence of decades of rapid population expansion that must eventually end. The best way to deal with that challenge may be *not* to encourage immigration and increased births, but simply to adapt by taking care of the aging population in ways that don't overly burden the young, as Japan is attempting to do, such as by providing affordable housing that offers the elderly opportunities to mingle with people of other ages so as to reduce feelings of isolation.[51]

However, the possibility that birth rates are collapsing largely due to the accumulation of endocrine-disrupting chemicals in the environment adds a worrisome new wrinkle to the population discussion, since it raises the possibility of rapid, uncontrolled, involuntary population decline. If we don't do something to reduce chemical loading, humanity could be on the downward ramp to extinction even without "help" from nuclear war or climate change.

Over the long run, if humanity were to take control of population growth and decline through humane means such as education and a dramatic reduction in chemical pollution, this could be seen as another way to promote the maturing and awakening of the Super-organism. At some point, the corrections required would be subtle. Unfortunately, in the current situation we have allowed such enormous imbalances to accumulate that humane corrective measures of adequate strength are hard even to identify, much less to implement.

Don't leave it entirely to others to deal with the human population dilemma. Have the conversation with those you love: what size family makes sense? Child-free marriages should not be stigmatized (my wife and I made that choice 30 years ago, and we'd decide the same today). From an ecological standpoint, family size is likely the most important decision of your lifetime.[52] If you do decide to have children, minimize chemical exposure by eating only organic foods for six months prior to pregnancy and avoid environmental petrochemicals to the extent that you can.

Fighting Power with Power

In the last six sections we've discussed what humanity would need to do collectively in order to limit its powers so as to avert or mitigate societal collapse. In reality, the likelihood of all humanity agreeing on sufficient self-limits is exceedingly remote, at least in the next few decades. What's far more likely is a power struggle among groups, as outlined in the section "All Against All"—that is, among nations, among economic and social classes, among age strata, and among ethnicities and interest groups. Some of these divisions of society will be seeking to maintain outsized social power; others will be seeking to cut the powerful down to size, or to seize more power for themselves. Meanwhile, many people, including some who don't fit neatly into any of these classifications, will be seeking to preserve ecosystems and nonhuman species in the face of climate change, pollution, resource depletion, and habitat destruction.

The key to minimizing suffering and environmental damage, and to improving the prospects of succeeding generations, will be for groups and individuals interested in long-term power (via the optimum power principle) and power sharing (i.e., horizontal power) to overcome groups committed to maximizing vertical social power and power over nature. The latter can be said to comprise the forces of catabolic capitalism (a self-cannibalizing system whose insatiable hunger for profit can be fed only by devouring the society that sustains it), and will also include all the people who can be convinced to ally themselves with those forces.

If those interested in long-term and horizontal power are to have any possibility of success, it is essential that they band together in a strong coalition, as Craig Collins suggests in his forthcoming book *Catabolic Capitalism & Green Resistance*.[53] The four main branches of such an alliance would be:

- *Groups and individuals working to save the planet by halting climate change, biodiversity loss, and pollution.* Most of these groups and individuals work to confront the depredations of corporations and governments; some work to promote environment-friendly governmental policies or commercial practices.
- *Social justice advocates.* This category would include advocates for the poor, and for oppressed minorities of all kinds, including Indigenous peoples (who make up the majority of the population in some countries).
- *Groups opposing violence, especially state violence.* These would include anti-war groups, as well as organizations seeking to reduce the domestic proliferation of guns, international arms sales, and the militarization of police, as well as groups promoting peaceful methods of conflict resolution.
- *Builders of the new culture.* This final category is currently less internally organized than the previous three; it consists of individuals and organizations seeking to model and promote a sustainable, post-fossil-fuel way of living. Some members would be

experts in the attitudes and habits of horizontal power. Others would be permaculturists, ecovillage pioneers, and renewable energy advocates. Still others would be creative artists of all kinds who are seeking to enlist the human imagination in building a green future.

Each of these by itself is at a power disadvantage when compared to the forces still pushing society toward capital accumulation and profit. However, added together, these groups and individuals comprise a huge constituency. Hence the vital importance of coalition and cooperation among them—rather than competition and conflict, with litmus tests of ideological purity as a requirement for inclusion.

In the coming struggle, success for the coalition depends on building social power. That means changing hearts and minds, winning elections, putting people in the streets, and succeeding at negotiations. In many instances, it means gaining control of institutions—and redesigning them so that they can work in a post-fossil-fuel, post-growth era of reduced vertical social power.

SIDEBAR 29

Power Analysis and Organizing for Activists

Generations of activists have pioneered and refined methods for social change, which nearly always require overcoming entrenched power relations within society at large. The most effective long-term tactic for activism is *organizing*—which has been defined as "the process of building power as a group and using this power to create positive change in people's lives."[54] An organizing campaign can be summarized in the following steps:

1. **Identify the problem you want to address.** What is the issue that you think is most critical in the world, your nation, or your community?

2. **Trace root causes.** What lies back of the problem? How did it arise?

3. **Focus on a specific policy change or action you demand from institutional power holders.** Ideally it should be measurable, attainable, realistic, and timely.

4. **Identify who can actually make the change you want.** It's useless to make demands of people who are incapable of making the particular decision, law, or policy change you desire.

5. **Determine who or what influences that decision maker.** Often, decision makers are merely stand-ins for the real power holders in a country or community. However, decision makers are typically answerable to a range of other groups as well. Among officials' constituents, which are the ones most likely to be influenced by your campaign? Which are the ones who need to be influenced if your campaign is to succeed?

6. **Map the power in this situation.** Diagram the current and potential interactions between the decision makers, power holders standing behind them, those who are impacted by the current power imbalance, and those in a position to work for change.

7. **Create a strategy to influence decision makers.** Identify resources, supporters, targets, and tactics. Connect with all the groups and individuals who are motivated to work together on a campaign—which could involve nonviolent direct action, petitions, social media memes, and conventional media outreach.

8. **Craft a message, focusing on shared values and stories that connect.** Messages that confront and accuse power holders are appropriate in many situations (for example, in campaigns to end polluting activities), but they should be framed in ways aimed to appeal to as many potential allies as possible, and should focus on the specific problem rather than demonizing opponents unnecessarily.

9. **Do the work.** Recruit supporters, develop leaders, assess results periodically, and repeat until the goal is achieved.

As I have stressed often in this book, money is a key technology of vertical power. One of the key goals of the anti-collapse coalition must be the re-organization of the global economy away from its current orientation toward profit, and instead toward human and environmental well-being. Money must become a less important feature of people's daily existence. Nevertheless, mounting a challenge to the current structures of power—i.e., fighting vertical power with horizontal power—will require funding. Fortunately, it won't need as much money as will efforts on the part of the powerful to suppress resistance. In many cases, people will contribute what they can afford toward their own immediate efforts. However, mounting large or sustained projects will require capturing and diverting some of the billions currently controlled by the 1%. Currently, philanthropy— Peter Buffett calls it the "charitable-industrial complex"—is the primary pathway by which some of the wealth of the rich is distributed to causes benefiting nature and the rest of society; however, the philanthropic model is deeply flawed, as it relies on economic growth (foundations typically disperse only interest and investment income) and reflects the whims and preferences of the wealthy.[55] At the very least, philanthropists must be persuaded to spend down their charitable funds much more quickly.

If money is necessary to the success of social movements, money nevertheless opens the door to corruption. That's just one of the contradictions likely to hinder the process of forming and maintaining an anti-collapse coalition, especially in times of increasing social and economic stress. Whose priorities are most important? What strategies shall be pursued? How to keep *agents provocateurs* from derailing meetings and actions? Such questions have plagued horizontal power movements for decades. It's impossible to forecast whether the coalition I am describing will emerge and prevail; I am merely pointing out that, from a theoretical perspective, such an alliance is our best hope of averting the worst outcomes.

As we've seen, power is at the heart of the problems confronting humanity. But those who wish to address those problems can't be effective without exercising power in some form, and without reducing the power of other people or organizations that now wield enormous amounts of it. Is it the inevitable fate of those who wrest power from the powerful to themselves succumb to the lures of power?

That question is implied in the perennial vexation of the prepandemic era: should climate activists fly in airplanes to go to climate conferences? If they do, they're hypocrites. If they don't, there'll be fewer climate conferences, and activists will have surrendered a possibly useful organizing tool.

Here's one more paradox: activism is a contest for public opinion and public policy. In that contest, some people's interests and opinions inevitably become marginalized. And, as we have seen, partisanship and disunity are hurdles to addressing our converging crises. Should we engage in activism, knowing that our efforts could be polarizing?

There's no way to avoid these sorts of contradictions if one wishes to make a difference in the fate of the Earth and of humanity. Perhaps the great classic of ancient Hindu literature, the *Bhagavad Gita*, has wisdom to offer in this regard. The *Gita* is a dialogue between prince Arjuna and his guide and charioteer Krishna, which occurs beside a battlefield during a war between Arjuna's kinsmen and another tribe. Arjuna is overwhelmed with moral dread about the violence and death his actions may contribute to, even though he believes his kinsmen are in the right, and wonders if he should renounce his title and duty. Krishna counsels Arjuna to fulfill his warrior obligation, but to act without thought of self or attachment to outcome.[56]

Similarly, those of us with awareness of the crises ahead must have the courage to act, knowing that action will inevitably have unintended as well as intended consequences. Like Arjuna, we find ourselves playing a role assigned in part by fate; it's up to us to play it as cleanly and selflessly—and as effectively—as possible. At the end of Chapter 5, I noted that many people who care about climate change feel powerless. Their (and our) only hope now is to build social power around the project of doing things that will make a difference—winning elections; staging demonstrations, walkouts, and strikes; formulating sound policies and campaigning for them.[57]

Of course, the sad and alarming fact is that, just at the historical moment when we are called to confront the overaccumulation of power in various forms and to do so as fairly as possible, the mechanisms of horizontal social power tied to the ideals and institutions of democracy, which have for decades enabled citizens peacefully to confront power abuses, are being contested. The current rise of illiberalism, tribalism, and nationalism echoes trends of the 1930s, when the world was hurtling toward the chasm of World War II. Thus, we are saddled with a double duty—to fight against the forces driving the world toward ecological catastrophe, while at the same time doing everything we can to build trust and buttress the existing institutions of horizontal power including democracy itself, as well as cooperatives, unions, the nonprofit sector, and anything that might be lumped into what's being called the sharing economy.

History has shown that horizontal social power is the only force (outside of war, pandemic, or collapse) capable of overcoming concentrated vertical social power. From union organizing in the 19th century to anti-colonialist struggles in India in the early 20th century, to the Black Lives Matter protests of the past few years, people have collaborated, often in ingenious and courageous ways, to confront the juggernaut of control of the many by the few and for the few. It's heartening to see more and more people from various backgrounds devoting their lives to the development of horizontal power. Here are just a few examples.

Margaret Klein Salamon was a young clinical psychologist in New York City when Hurricane Sandy hit. She started to educate herself on climate change, and soon left the field of psychotherapy to found The Climate Mobilization—an organization that advocates a WWII-scale transformation of our economy and society to protect humanity and the natural world.[58]

Cooperation Jackson is a network of worker cooperatives in Jackson, Mississippi, that aims to develop a series of independent but connected democratic institutions to empower workers and residents of Jackson—particularly to address the needs of poor, unemployed, black or Latino residents. The development of Cooperation Jackson was largely inspired by the Mondragon federation of cooperatives in Spain, and by historical cooperative movements described in the works of W. E. B. Du Bois. Kali Akuno, co-founder and co-director of Cooperation Jackson, says he was seeking to bring cooperative economics to the urban context, complementing existing rural agricultural and utility co-ops.[59] The organizers of Cooperation Jackson have had to struggle with working in an economically depressed city in the poorest state in the US, and with a politically hostile governor and state legislature.

A group of activists and scientists who met in a café in Bristol, England, in April 2018 founded Extinction Rebellion—an organization that aims to use widespread nonviolent civil disobedience to force governments to act on climate change and biodiversity loss.[60] "XR," as it is colloquially know, has a decentralized structure and has organized numerous prominent demonstrations, primarily in the UK, though activists in cities across the world have participated.

A permaculture teacher named Rob Hopkins realized in 2006 that shifting away from fossil fuels would require a near-complete redesign of society around smaller and more cooperative communities; he went on to co-found the Transition Initiatives.[61] There are now hundreds of Transition Towns all across the globe.

Still other climate organizations—including 350.org, the Sunrise Movement, and World War Zero (WWZ)—have sprung up in recent months and years, some more mainstream, some more radical in their means and objectives, but all pushing in generally the same direction.

What strategies and tactics make most sense? There's no general agreement on that question; you can take your pick. Extinction Rebellion aims to create horizontal power through citizens' assemblies. The Sunrise Movement lobbies elected officials for a Green New Deal.[62] WWZ is looking to gain bipartisan support for climate action, and therefore doesn't endorse any particular policy.[63]

In my conversations with climate activists, there's a perennial commonality: just about everyone sees the need for a new cultural vision that can help society through the post-carbon transition. Currently our cultural visions are oriented either toward continued growth and consumerism, or apocalypse and doom. We have few shared realistic but attractive images of what life after fossil fuels could look like. Without those, our only motive for action is fear. Rob Hopkins has probably done the best job of anyone at trying to fashion such a vision.[64]

Help is also needed in figuring out how society can work optimally as we make our way back down from the pinnacle of societal complexity and high levels of energy usage. We need more local experimentation with rationing, cooperatives, and sharing institutions (such as tool libraries) in order to determine what works and what doesn't.

Strategies must adapt to changing circumstances. We all would like to avert climate change and the catastrophic loss of biodiversity. But it's already too late to do that fully. And at this point the things we will have to do to minimize environmental collapse will themselves likely have severe economic and social impacts. In the last few years, a consistent theme has emerged in my private conversations with scientists and activists: a Great Unraveling looms, though there is still a range of possible outcomes depending on how much we do over the next couple of decades to lower carbon emissions, reduce the load of environmental toxins, plant trees, build soil, and protect habitat. Some of us, at least, should be turning our attention to the problem of managing that Unraveling (if it can, indeed, be managed), so as to minimize the human and nonhuman casualties. Humanity is approaching a moment when the unimaginable becomes inescapable. Who will preserve our cultural achievements (math, science, art, and music) from loss? Who will protect ecosystems from wildfires when there is no funding for fire departments and no fuel for firetrucks? Somehow, we must keep the bad from devolving to the worst. It's actually not difficult to identify "no-regrets" strategies—i.e., things that could be done that would be useful whether collapse is inevitable, or large-scale societies are able to bend and adapt. Building community, localizing economies, and developing low-tech ways of meeting human needs all make sense regardless of the severity of the impacts that are now locked in.

Have a conversation with those closest to you: How will you all adapt? Where should you live, if you wish to be of most service or if you wish to avoid the worst environmental impacts? *How* shall you live in order to minimize your own negative impacts and make the biggest positive difference? What are you willing to sacrifice for the sake of future generations? For example, are you willing to be arrested at a demonstration?[65]

Long-Term Power Through Beauty, Spirituality, and Happiness

Let's now explore the alternative solution to the Fermi Paradox—a society that voluntarily limits its power by developing its aesthetic creativity and spirituality, striving for happiness rather than dominance over nature. This would necessarily be a society different in both broad strokes and details from industrial nations of the early 21st century. I have no desire merely to propose a fanciful utopia. Rather, I'm interested in what societies have already shown themselves capable of achieving by channeling familiar human foibles, drives, and appetites toward sustainable, prosocial habits and institutions. Our ability to deliberately design and shape our own culture may be limited, since the evolution of culture is largely a response to circumstances. Nevertheless, what follows may be within the realm of possibility—if not in this century, then perhaps in centuries to come.

In imagining a possible future culture (ideally, a global network of locally-adapted cultures with some similar characteristics—including robust methods for sharing power and amicably resolving disputes), I propose two overarching goals: the most happiness for the most people (including future generations), and the integration of society with nature in such a way that human habitation and biological diversity are integrated, and persist for as long as possible.

Social inequality breeds unhappiness, and alienation from nature leads people to abuse ecosystems in ways that eventually produce ruin, so inequality and nature estrangement are to be avoided if possible. Looking back on cultural history, it's clear that the development of agriculture represented a fateful turn toward both destructive power over nature and vertical social power. Horticultural societies (based on gardening rather than field cropping) entailed far less inequality, and kept the entire populace in closer contact with nature's lessons and limits. In such a society, everyone who is physically able is involved part-time in food production. In short, I think the permaculturists have got it right. The closer we can come to being a society of gardeners, the happier and more durable our way of life is likely to be.

Still, it may be utopian to think that we can entirely do away with inequality while subsisting in any way other than by hunting and gathering. Even in a gardening society, some individuals will work harder than others, while also convincing their friends and relatives to work more, in order to build up surpluses. However, seasonal surpluses need not be a serious impediment to happiness or sustainability. Even in many agricultural societies, traditions eventually arose that bound those with social power to duties on behalf of society as a whole. Wealthy and influential members of the community were tasked with defending it from outside attackers, with funding the arts, and with supporting specialists in matters of the sacred. The chivalric tradition in medieval Europe and the samurai tradition in pre-industrial Japan are only the most familiar of many examples of class systems that sought to uphold justice, honor, and duty (though abuses of power in these contexts, needless to recount, occurred nevertheless). Typically, such systems attached shame to conspicuous consumption or the exercise of power purely for personal benefit. If specialization or inequality exists in future societies, the people who benefit the most must be bound by tradition to contribute the most.

Aesthetics and spirituality play key roles in all human societies; in the best instance, these can be engines of happiness and motivators of prosocial, ecologically sustainable behavior. In the possible societies I am describing, they would be key elements.[66]

In earlier chapters we saw that beauty is a powerful attractor—and not just for humans, but for most higher organisms. In some societies, aesthetic products seem largely designed to impress the common people with the power and wealth of the elite classes. However, the arts can also serve community solidarity, and can help bind people to nature. Arts—whether painting, architecture, music, poetry, literature, drama, or movement—offer intrinsic rewards, but they needn't deplete resources or diminish the status of non-artists; indeed, they often spiritually enrich the entire community.[67]

Gardening itself can be an art, in which nature is both the model and the canvas. The Zen gardens of Edo-period Japan offer serene examples, as do some English and Persian gardens.

Becoming a good artist requires learning to control one's senses, body, mind, and emotions. In a highly aesthetic culture, everything is an art—conversation, the preparation of food, the planting of seeds and shoots, and the harvest later on. Art and spirituality are interwoven throughout daily life.

As we have seen, spirituality is invariably present in every society, though it can be expressed very differently depending on the cultural context. It serves human needs that are universal and innate. For one thing, spirituality can take us beyond language and the mental categories that we habitually impose on the world. Language has given us the ability to do wondrous things; but, as we have seen, it often misleads as much as it empowers. Nearly every spiritual tradition teaches a form of meditation or contemplation that enables the quieting of what Buddhists call the "monkey mind"—i.e., our restless, uncontrollable internal linking of thoughts. The goal of such exercises is a direct perception of reality as it is, not as we conceptualize it with words. Meditation is also a pathway for confronting our own mortality without denial or dread, and for overcoming the lingering mental and emotional impacts of trauma. Through meditation, awareness of death can become a motive for maximizing beauty and happiness within our finite period of existence.[68]

Some quasi-spiritual traditions seek to develop control of the body and mind to supernormal degrees. Examples include martial arts such as Tai Chi and Qigong, and meditation systems such as yoga (which comprises several schools and goals, and a wide range of practices). Many such traditions teach control of the autonomic nervous system via the breath, and some proponents claim the ability to extend self-control even to cellular level.[69] Pursuits along these lines could occupy ambitious members of society.

Whether or not we will need Big Gods seems to depend on whether we continue to associate in large multicultural groups and in big cities, and whether we are able to (or wish to) maintain big governments that take care of the disadvantaged and punish cheaters. If we return to a village-based gardening society, it could come about that Big Gods were merely a passing phase in human history.

The arts and spirituality are mutually supportive: some of the greatest art, from the Aboriginal cave drawings in Australia to Bach's St. Matthew Passion, has been created in service to the spirit, while beatific states of consciousness seem predictably to be engendered by great art. Both the arts and spirituality nurture self-control: the artist or musician trains her brain, senses, and muscles, while the spiritual seeker trains attention, emotions, and cravings. Both point us toward something greater than ourselves and encourage an attitude of humility.

We are striving and competitive animals. Having spent tens of thousands of years developing extraordinary powers of communication and invention, we are driven to find ways to use these abilities to our advantage. However, as we have seen, building empires and fortunes tends to get us into trouble. How shall we harmlessly occupy our big brains and our extraordinary abilities? Innumerable cultures have come up with essentially the same answer: strive for beauty, serenity, and wisdom. The need for benign ways to channel outsized human capabilities is one of the reasons societies have devoted large portions of their hard-won material and labor surpluses toward building beautiful temples; it's also one of the reasons prominent families in traditional societies encouraged some of their sons and daughters to become monks and nuns.

Spirituality and the arts also fill basic human needs for community. Seasonal festivals, rife with concentrated aesthetic and spiritual experiences, make life fun for everyone by celebrating the cycles of time. David Fleming, author of *Surviving the Future*, was one of the few futurists who could see humans in three dimensions as complex beings with needs and drives. He wrote:

> Celebrations of music, dance, torchlight, mime, games, feast and folly have been central to the life of community for all times other than those when the pretensions of large-scale civilization descended like a frost on public joy. Carnival is a big word: it spans the buffoonery of the Feasts of Fools, the erotic Saturnalia of Rome, the holy holidays of the Church's calendar and the agricultural year, and local days of festival in which communities, for most of history, have put down their work and concentrated on enjoying themselves.[70]

Fleming believed that carnival must play a key role in any future culture that's worth our effort in building it. Play is essential to brain development and happiness, so it's important that we value it and make room for it.

Happiness and beauty are best shared with others—whether humans or nonhumans. We all take spiritual and aesthetic pleasure from interaction with other kinds of animals, and with nature in general. In today's urban environment, human spiritual and aesthetic needs are fulfilled to a certain extent by dog and cat ownership. But this is thin gruel compared with the simmering interspecies gumbo in which our ancestors stewed, strove, and played. Our arts and spirituality will be immeasurably enriched if they flow from a sense of community that extends to all living beings.

Influence based on bribes and threats, no matter how nicely it is dressed up, is a pathway only to vertical power. And, whether it is deployed in the context of a family or in global geopolitics, vertical power creates servility and resentment on one hand, and a sense of entitlement on the other. The only influences that can reliably build horizontal power are those of inspiration and moral example. Many priests and gurus have learned to feign moral example in order to build and exercise subtle forms of vertical power, but this eventually just immunizes genuine spiritual seekers against priests and gurus. Religious zealots likewise claim moral example, but do so by defining morality in sectarian terms and decrying those who do not adhere to those terms (and often also by hiding their own corruption). Real moral example does not call attention to itself or insist on comparisons. A healthy culture encourages moral example, but not false piety.

Through spirituality and the arts, it might eventually be possible for humanity to develop horizontal power to a far higher degree than has been the case at any time since the start of the agrarian revolution several millennia ago. This power could even take the form of a matured collective intelligence that is finely attuned to the ecosphere from which it emerged. However, in saying this I may be veering dangerously close to utopianism.

Can we get there from here? Obviously, I'm not just advocating increased funding for the arts or more church attendance. I'm envisioning a complete reorientation of society's structures and aspirations away from profit and toward the goals of happiness and sustainability. Doing so would require us to change our relationship with power, from focusing on control of the environment and other people to control of ourselves. This would, of course have ramifications for just about everything we do—no more luxury vacations, no more digital devices, no more retail therapy. Life would be lived far more simply and closer to the ground. But general satisfaction and the enjoyment of life's simple gifts would be maximized.

Again, this is not utopia (in many people's minds, a world without 24-hour sports networks and social media would hardly be utopia in any case!). We don't have to assume the existence of perfect humans with no conflicting drives in order to hope for such a future. Competition and violence will likely always play roles in human experience, and, even if we all return to gardening, there may still be some kinds and degrees of inequality among us. Societies will always experience cycles of growth and retreat. But it's clear that some societies do better than others at keeping their members happy without stealing from future generations. We can learn from them, and perhaps exceed their achievements. It's also undeniable that societies, and the

humans that inhabit them, are still evolving, so that what was true
of Paleolithic humans or early state societies is not necessarily true
of us today: over time, it seems that we are becoming more sociable
and peaceful. If we set our considerable powers to the task, we could
create something beautiful and durable together.

◆ ◆ ◆

As we've seen, power has a long history, originating with the universe
itself. On our planet, it has worked its way up through bacteria, larger
eukaryote cells with mitochondria, and multicelled life. Guided by
the maximum power principle and by fascination with beauty, species proliferated, competed, and gained remarkable abilities of movement, cognition, and cooperation. Finally, one species was able to
outdo even the fastest and strongest of its competitors by specializing
in communication and tool-making. Most recently, fossil fuels made
that species the undisputed master of the planet. It's been a long road
to the top. But we humans, the winners and beneficiaries, find ourselves on a precarious pinnacle.

If, as William H. Calvin has argued, human intelligence evolved
partly in response to dozens of instances of rapid climate change
during the past 2.5 million years, then the current instance of human-
caused extreme climate change may have equally profound evolutionary consequences for the survivors. It may be foolish for us now,
in current conditions, and with our current mindset, to try to predict
the results. But we would be equally foolish to assume that existing
institutions will persist in anything like their familiar forms.

There can be no perfect, stable society. Imbalance and imperma-
nence are baked into biological existence. But we are in a particularly
explosive moment now. History shows that overconcentrations of
physical, economic, military, and political power create instability,
and, in the past few decades, humanity has found ways to build and
concentrate these kinds of power as never before. The strong likelihood is that we are headed toward what economists glibly call a
"correction," though not just in stock market values but also in popu-

lation and consumption levels. If we hope to minimize the shock and casualties, we will need to mobilize cooperation and behavior change, aiming to limit our own collective power at a speed and scale that are unprecedented.

Fortunately (or unfortunately, depending on how you look at it), cultural evolution is now happening faster than ever. There's certainly no guarantee that it will work to our advantage: the internet and social media could easily create opportunities for extraordinary levels of cooperation, but along competing lines, thereby defeating any effort to build a unified coalition of humanity willing to check its power now so that it can sustain itself and the biosphere over a much longer period.

Nevertheless, the possibility now exists for rapid shifts in human understanding and behavior—and such shifts are essential if we are to create future societies that live happily within natural limits. As I have argued repeatedly in these pages, our only way out of our current predicament is to tamp down various forms of power, often to significant degrees. We humans are well acquainted with the problem of overaccumulation of power, and cultural evolution has supplied plenty of ways of solving it—from the ancient Australian Aboriginal tradition of avoiding hunting the red kangaroo in its mating season, to trade unions and democracy, environmental regulations, and modern billionaires like Tom Steyer who say, "Please tax me."[71] As I am writing, today's local newspaper here in Santa Rosa, California, features a story about crab fishermen on the Sonoma coast who are voluntarily delaying their crabbing season (thereby incurring a substantial financial loss) in order to protect migrating whales.

For rhetorical purposes, it is difficult to altogether avoid an either/or framing of the choices and outcomes before us. But, of course, reality will be complicated. It is pointless to imagine a future in which power self-limitation is entirely absent from human society, because such a condition has never before existed. It is just as unrealistic to paint an imaginary picture of a world in which all human power excesses have been quickly, sufficiently, and amicably checked.

However, we can be fairly confident that, one way or another, human power *will* be reined in through *some combination* of collective moral struggle on one hand, and, on the other, social/ecological unraveling triggered by climate change, biodiversity loss, resource depletion, economic collapse, political polarization, famine, population decline, pandemic, social fragmentation, and war. The actual trajectory of future events will be determined by *how much* collective self-limitation humanity can muster—what quantity of carbon emissions we are able to forgo, and how many nuclear weapons we dismantle. It may also be shaped by a power struggle between the persistent forces of capital accumulation and an anti-collapse coalition. Can campaigners forge durable alliances? Can they communicate effectively with the public and take strategic advantage of opportunities? Or will self-consuming capitalism win the day? In the best instance, we humans will learn collectively and rapidly to live equitably and peacefully within limits to a much greater degree than we do now; in the worst, society will uncontrollably descend the ladder of cultural evolution back to a condition that *can* be sustained with whatever resources are left.

There would be obvious differences in the two routes and outcomes—cooperation and self-limitation on a significantly increased scale in one case, mayhem and ecological devastation in the other. But, in the long run, there might also be some similarities. In the aftermath of a Great Unraveling, our few surviving descendants—learning from hard experience—might eventually adopt cultural narratives similar to ones that Indigenous peoples used in order to protect biodiversity and to keep human population levels within the carrying capacity of the environment. These narratives might also be similar at least in some ways to those we would need quickly and intentionally to develop if we are to stave off utter collapse.

Those narratives would likely encode a deep cultural skepticism of power in all its forms, and a profound reinforcement for habits of self-restraint and self-control. We cannot do away with power, nor should we; it is necessitated by the fact that we are organisms—and especially since we are big-bodied, linguistic, tool-making mammals.

But if we wish to avoid outcomes that are awful to contemplate and far worse to experience, we can and must rein in the extreme powers that currently threaten our success and even our survival. If we're truly smart, we can do so in ways that are beautiful, and that make our descendants happy for a long time to come.

Notes

Chapter 1: Power in Nature

1. There are organisms living in the ocean depths that derive energy not from sunlight, but from Earth's internal heat as it is released from sea-floor vents. But these are rare exceptions to the rule.
2. For more about the maximum power principle, see Charles Hall, *Maximum Power: The Ideas and Applications of H. T. Odum*. Boulder, CO, University Press of Colorado, 1995.
3. Pallab Ghosh, "Earliest Evidence of Life on Earth 'Found'," *BBC News*, March 1, 2017. bbc.com/news/science-environment-39117523, accessed September 4, 2020.
4. See Nick Lane, *Power, Sex, Suicide: Mitochondria and the Meaning of Life*, Oxford: Oxford University Press, 2005, pp. 140 ff.
5. For more on proton pumping, Ibid., pp. 150–153.
6. Ibid., pp. 99–100.
7. Kat McAlpine, "Bacteria 'Factories' Churn Out Valuable Chemicals," *The Harvard Gazette*, December 24, 2014. news.harvard.edu/gazette/story/2014/12/bacteria-churn-out-valuable-chemicals/, accessed September 4, 2020.
8. It's actually problematic to speak of "species" when discussing archaea or bacteria, because they continually adapt to changing environments by altering their genetic material.
9. Sarah Zhang, "The 'Dark Matter' of the Microbial World," *The Atlantic*, March 7, 2017. theatlantic.com/science/archive/2017/03/archaea-sequencing-challenges/518535/, accessed September 4, 2020.
10. A very few types of eukaryote cells do not currently have mitochondria, but evidence suggests they once did.
11. Plants and fungi do have cell walls, but they're very different from those of bacteria.
12. *Power, Sex, Suicide*, p. 161.
13. In a popular figure of speech, anything quickly gaining popularity is said to be "going viral." "Going bacterial" is impressive enough.
14. Most ocean ecosystems are characterized by an inverted food pyramid, in which consumers outweigh producers; this happens because aquatic

producers have a very rapid turnover of biomass, on the order of days, while consumer biomass turns over much more slowly—a few years in the case of many fish species.

15. *Power, Sex, Suicide*, p. 234.

16. John Stoughton, "The Human Brain vs. Supercomputers…Which One Wins?" *Science ABC*, October 14, 2019. scienceabc.com/humans/the -human-brain-vs-supercomputers-which-one-wins, accessed October 9, 2020.

17. Some creatures do clone themselves: for example, segmented worms and many echinoderms such as sea stars reproduce asexually by frag-mentation, developing new, genetically identical individuals from each segment.

18. The reason for this difference has to do with the elimination of possible malfunctions of the mitochondria if their genes were derived from two different populations. For a more detailed explanation, see *Power, Sex, Suicide*, pp. 337 ff.

19. Some animals (such as the sea anemone) that live in the flux of ocean currents can let their food come to them and therefore become, in effect, rooted like plants.

20. Kat McGowan, "How Plants Secretly Talk to Each Other," *Wired*, December 20, 2013. wired.com/2013/12/secret-language-of-plants/, accessed September 4, 2020.

21. Ibid.

22. See Frans de Waal, *Are We Smart Enough to Know How Smart Animals Are?*, New York: W. W. Norton, 2016.

23. Anthony Trewavas, "Green Plants as Intelligent Organisms," *Trends in Plant Science*, Vol 10, No. 9, September 1, 2005. cell.com/trends/plant -science/comments/S1360-1385(05)00171-8, accessed September 4, 2020.

24. See Derrick Jensen, *The Myth of Human Supremacy*, New York: Seven Stories Press, 2016, pp. 78–9. And then there's the fascinating possibility that plants can use sound to communicate: see livescience.com/27802 -plants-trees-talk-with-sound.html, accessed September 4, 2020.

25. The *competitive exclusion principle*, sometimes referred to as Gause's Law, states that two species competing for the same resource cannot coexist at constant population values, if other ecological factors remain con-stant. When one species has any advantage over another, then the one with the advantage will dominate in the long term, leading to either the extinction of its competitor or an evolutionary or behavioral shift toward a different ecological niche. The principle has been paraphrased in the maxim, "complete competitors cannot coexist."

26. National Research Council (US) Subcommittee on Laboratory Animal Nutrition, *Nutrient Requirements of Laboratory Animals: Fourth Revised*

Edition, Washington, DC: National Academies Press, 1995. ncbi.nlm.nih
.gov/books/NBK231918/, accessed September 4, 2020.

27. "Feed Requirements of Horses," agriculture.vic.gov.au/agriculture/live
stock/horses/feed-requirements-of-horses, accessed December 11, 2019.
"Nutrient Requirements for Horses," July 31, 2019 horses.extension.org
/nutrient-requirements-for-horses/.

28. "Energy Expenditure," February 25, 2015. asianelephantnutrition.word
press.com/2015/02/25/big-body-lots-of-energy-maintenance/, accessed
September 4, 2020.

29. Staverie Boundouris, "Power of a Space Shuttle," The Physics Factbook,
2001. hypertextbook.com/facts/2001/StaverieBoundouris.shtml, accessed September 4, 2020.

30. "Utility-Scale Wind Energy," windexchange.energy.gov/markets/utility
-scale, accessed December 11, 2019.

31. Ben Zientara, "How Much Electricity Does a Solar Panel Produce?" *Solar
Power Rocks* (blog), November 6, 2019. solarpowerrocks.com/solar-basics
/how-much-electricity-does-a-solar-panel-produce/, accessed September 4, 2020.

32. Andrew Armstrong, "New Orleans and the Mississippi River," Hydro
International, January 1, 2008, hydro-international.com/content/article
/new-orleans-and-the-mississippi-river. U.S. National Park Service,
"Mississippi River Facts," nps.gov/miss/riverfacts.htm, accessed December 11, 2019.

33. "2016 U.S. Gazetteer Files," United States Census Bureau. (www2.census
.gov/geo/docs/mapsdata/data/gazetteer/2016_Gazetteer/2016_gaz_place
_06.txt, retrieved December 11, 2019. NOAA, "Storm Events Database—
Event Details," NOAA National Centers for Environmental Information.
ncdc.noaa.gov/stormevents/eventdetails.jsp?id=675433, accessed December 11, 2019.

34. Chris Landsea, "Frequently Asked Questions," NOAA Hurricane Research
Division, Atlantic Oceanographic and Meteorological Laboratory. aoml
.noaa.gov/hrd/tcfaq/D7.html, accessed December 11, 2019.

35. Herman Pontzer et al., "Energy Expenditure and Activity among Hadza
Hunter-Gatherers: Hazda Energetics and Activity," *American Journal of
Human Biology* 27, no. 5 (September 10, 2015): 628–37 (doi.org/10.1002
/ajhb.22711).

36. GS 361 Energy and Resources in Perspective, Western Oregon University, "Historical Perspectives of Energy Consumption." people.wou.edu
/~courtna/GS361/electricity%20generation/HistoricalPerspectives.htm,
accessed December 11, 2019. Paolo Malanima, "Energy in History," in
The Basic Environmental History, edited by Mauro Agnoletti and Simone
Neri Serneri, 4:1–29, Cham: Springer International Publishing, 2014,

doi.org/10.1007/978-3-319-09180-8_1. Mid-Atlantic Masonry Heat, "Amount of Energy in Firewood," November 9, 2019. midatlanticmasonryheat.com/blog-entry/amount-energy-firewood. "Energy Expenditure and Activity," pp. 628–37. Mike Schira, "How Much Heat Energy Is in Firewood?" MSU Extension, March 3, 2014. canr.msu.edu/news/how_much_heat_energy_is_in_firewood accessed December 11, 2019.

37. "Historical Perspectives of Energy Consumption." Kees Klein Goldewijk, "Cattle per Capita," Clio Infra. clio-infra.eu/Indicators/CattleperCapita.html, accessed December 11, 2019. Malanima, "Energy in History." Mid-Atlantic Masonry Heat, "Amount of Energy in Firewood." David Pimentel and Marcia Pimentel, *Food, Energy, and Society*, third edition, Boca Raton, FL: CRC Press, 2008. Pontzer et al., "Energy Expenditure and Activity." Schira, "How Much Heat Energy Is in Firewood?"

38. Joyce Tyldesley, "The Private Lives of the Pyramid-Builders," BBC, Ancient History in depth, February 17, 2011. bbc.co.uk/history/ancient/egyptians/pyramid_builders_01.shtml, accessed September 4, 2020.

39. World Bank, "GDP per Capita (Current US$) | Data," data.worldbank.org/indicator/NY.GDP.PCAP.CD?most_recent_value_desc=false, accessed December 11, 2019. World Bank, "Energy Intensity Level of Primary Energy (MJ/$2011 PPP GDP) | Data," data.worldbank.org/indicator/EG.EGY.PRIM.PP.KD?view=chart, accessed December 11, 2019, World Bank, "Energy Use (Kg of Oil Equivalent per Capita) | Data," data.worldbank.org/indicator/EG.USE.PCAP.KG.OE, accessed December 11, 2019.

40. World Bank, "GDP per Capita (Current US$) | Data." World Bank, "Energy Intensity Level of Primary Energy" World Bank. "Energy Use."

41. BP. "Statistical Review of World Energy," 2019, bp.com/content/dam/bp/business-sites/en/global/corporate/pdfs/energy-economics/statistical-review/bp-stats-review-2019-full-report.pdf, accessed December 11, 2019.

42. Richard Prum, *The Evolution of Beauty*, New York, Anchor Books, 2017, p. 27.

43. Even a single species of bird may be more specialized or generalized in what it eats depending upon where it happens to live and the diversity of other birds in the community. This has been studied among island groups, and is a classic set of research on niche partitioning based on community interactions.

44. In his book *Becoming Wild: How Animal Cultures Raise Families, Create Beauty, and Achieve Peace* (2020), ecologist Carl Safina explores cultural evolution among nonhuman animals in fascinating detail.

Chapter 2: Power in the Pleistocene

1. See William Calvin, *The Ascent of Mind: Ice Age Climates and the Evolution of Intelligence*. New York: Bantam Books, 1990.

2. A recent study suggests that climate change may also have been a sig-

nificant factor. Pasquale Raia et al., "Past Extinctions of Homo Species Coincided with Increased Vulnerability to Climate Change," *One Earth*, Vol. 3, No. 4, October 23, 2020. cell.com/one-earth/fulltext/S2590-3322 (20)30476-0, accessed November 14, 2020.

3. Richard Wrangham, *Catching Fire: How Cooking Made Us Human*, New York: Basic Books, 2009, p. 57.

4. Ibid., p. 81.

5. Ibid., p. 87.

6. Ibid., p. 121.

7. Kris De Decker, "Too Much Combustion, Too Little Fire," *Resilience*, January 23, 2020. resilience.org/stories/2020-01-23/too-much-combustion -too-little-fire/, accessed September 4, 2020.

8. Becky Little, "Neanderthals Knew How to Start a Fire," *History*, August 31, 2018. history.com/news/neanderthals-fire-evidence-archaeology, accessed September 4, 2020.

9. William Catton, *Overshoot: The Ecological Basis of Revolutionary Change*. Chicago, University of Chicago Press, 1982.

10. Jen Viegas, "Humans First Wore Clothing 170,000 Years Ago," *Seeker*, January 6, 2011. seeker.com/humans-first-wore-clothing-170000-years -ago-1765156178.html, accessed September 4, 2020.

11. Reed et al. (2004). "Genetic Analysis of Lice Supports Direct Contact between Modern and Archaic Humans," *PLoS Biology*, Vol. 2, No. 11. November, 2004.

12. Wrangham is far from being alone in this view. Many other evolutionary scientists, reaching back to Darwin himself, have come to similar conclusions regarding human self-domestication. I'm focusing on Wrangham here because his writings are recent and he expresses his ideas clearly and forcefully; if the reader wishes to explore this subject further, Wrangham's book is the best place to start.

13. Our least aggressive primate relative, the bonobo, also shows signs of self-domestication, at least in Wrangham's view.

14. Ben James, "A Sneaky Theory of Where Language Came From." *The Atlantic*, June 10, 2018. getpocket.com/explore/item/a-sneaky-theory -of-where-language-came-from?utm_source=pocket-newtab, accessed September 4, 2020. See also Oren Kolodny and Shimon Edelman, "The Evolution of the Capacity for Language: The Ecological Context and Adaptive Value of Cognitive Hijacking," *Philosophical Transactions of the Royal Society B*, February 12, 2018. royalsocietypublishing.org/doi /full/10.1098/rstb.2017.0052, accessed September 4, 2020. Noam Chomsky and a few other linguists resist this evolutionary view of the development of language.

15. John Shea, *Stone Tools in Human Evolution*. Cambridge, MA, Cambridge University Press, 2017, pp. 84–101.

16. Paleontologists have searched for evidence of the evolution of the physical apparatus of human speech—the descended larynx and the hyoid
bone—but such evidence is hard to find in ancient skeletal remains and
is so far inconclusive. Cultural anthropologist Quentin Atkinson, in a
2011 paper, proposed treating phonemes (basic vocal sounds) like genes
in order to trace language origins. African languages have more phonemes than other languages, and the further away from Africa you go,
the fewer one encounters. This suggests the basic elements of language
existed in *sapiens* before its exit from Africa 70,000 years ago. But
Atkinson's conclusions were later criticized. Why focus on phonemes,
but not other elements of language like the passive voice or subordinate
clauses? If you choose either of the latter, you arrive at a different site for
the origin of language, and a different date as well.

17. Cultural evolution is observable outside human society in about half of
bird species, in which individual birds learn their songs from their parents, leading to varying local song "dialects."

18. For more on this subject, see, for example, Richard Lippa, *Gender, Nature
and Nurture*, London, Routledge, 2005.

19. See Susan Carol Rogers, "Female Forms of Power and the Myth of Male
Dominance," *American Ethnologist* Vol. 2, No. 3, "Sex Roles in Cross-
Cultural Perspective" (Nov., 1975), pp. 727–756.

20. Tia Ghose, "Male Aggression: What Chimps Can Reveal about People,"
Live Science, November 13, 2014. livescience.com/48743-aggressive
-chimps-reproduce-more.html, accessed September 4, 2020.

21. Frans De Waal, "Bonobo Sex and Society," *Scientific American*, June 1,
2006. scientificamerican.com/article/bonobo-sex-and-society-2006-06/,
accessed September 4, 2020.

22. "Sex Differences in Crime." Wikipedia. en.wikipedia.org/wiki/Sex_dif
ferences_in_crime, accessed September 4, 2020.

23. Richard Wrangham and Dale Peterson, *Demonic Males: Apes and the
Origins of Human Violence*, New York, Houghton Mifflin, 1996, p. 113.

24. "List of Countries by Intentional Homicide Rate." Wikipedia, en.wiki
pedia.org/wiki/List_of_countries_by_intentional_homicide_rate#By
_country,_region_or_dependant_territory, accessed September 4, 2020.

25. Alice Dreger, "Where Masturbation and Homosexuality Do Not Exist,"
The Atlantic, December 4, 2012. theatlantic.com/health/archive/2012/12
/where-masturbation-and-homosexuality-do-not-exist/265849/, accessed September 4, 2020.

Chapter 3: Power in the Holocene

1. New evidence suggests humans may have reached the Americas as
early as 33,000 years ago. See Lorena Becerra-Valdivia and Thomas
Higham, "The Timing and Effect of the Earliest Human Arrivals in North

America," *Nature*, July 22, 2020. nature.com/articles/s41586-020-2491-6, accessed September 4, 2020.

2. See Marvin Harris, *Cultural Materialism: The Search for a Science of Culture*, 2nd edition, New York, Viking, 1980; and Marvin Harris and Orna Johnson, *Cultural Anthropology*, 7th edition, Boston, Pearson, 2007.

3. Vaclav Smil, *Energy in World History*, Cambridge, MA, MIT Press, 2017, p. 21. For an up-to-date and highly readable account of early plant domestication and state formation, see James C. Scott, *Against the Grain: A Deep History of the Earliest States*, New Haven, Yale University Press, 2017.

4. Harris and Johnson, Ibid., p. 172; see also Peter Turchin, *Ultra Society: How 10,000 Years of War Made Humans the Greatest Cooperators on Earth*, Chaplin, CT, Beresta Books, 2016, p. 11.

5. Historian Ian Morris agrees that war has been a significant factor in cultural evolution. In his book *War! What Is It Good For?: Conflict and the Progress of Civilization from Primates to Robots* (2014), he argues that the history of violence across many thousands of years shows that war has made the world safer and richer by creating larger and more internally pacified societies. The lesson of the last 10,000 years of military history, he argues, is that we can learn to manage war, but cannot realistically wish it away.

6. Quoted in Turchin, *Ultra Society*, p. 149.

7. It's tempting, but misleading, to see the ancient shifts from hunting and gathering to simple horticulture, to complex horticulture, to agrarian statehood as an inevitable progression. However, as noted in the text, many societies stopped off at one or another of these levels of food production and social organization and simply stayed there. For example, pre-Columbian California had complex cultures that never adopted agriculture, pottery, architecture, or writing—as other Native American groups had done—because there was no environmental requirement to do so. The land was so abundantly productive that tribelets found it more to their advantage to develop advanced horticultural practices. Instead of making pottery, they concentrated on basketry, and made some of the finest baskets in the world. Benign environmental conditions were spread over large areas, so there was relatively little organized violent competition for control of specific resources. Had they wished to develop chiefdoms, the Californians certainly could have done so, but evidently the need never arose.

8. "Indians 101: Disease and Indians in the 16th Century," *Daily Kos*, February 19, 2019. dailykos.com/stories/2019/2/19/1835704/-Indians-101 -Disease-and-Indians-in-the-16th-Century, accessed September 4, 2020.

9. It's only a metaphor, and metaphors can mislead us if we adhere to them too strongly (I'll use quotation marks to highlight metaphoric uses of the

terms predation, predator, prey, and predatory). But metaphors can also sometimes aid understanding. The idea of human-on-human "predation" is hardly original: we're all accustomed to speaking of "predatory" lending practices and sexual behavior. The fact that literal human-on-human predation, in the form of cannibalism, was surprisingly common among both pre-state and archaic state societies lends validity to an exploration of metaphoric human-on-human "predation" in its many forms.

10. I wrote this section of the book before encountering James C. Scott's *Against the Grain: A Deep History of the Earliest States*, which propounds essentially the same ideas regarding human-on-human "domestication." It's reassuring to see a more eminent scholar reaching similar conclusions when confronting the same collection of evidence. New Haven, Yale University Press, 2017.

11. See, for example, Andrew Lawler and Jerry Adler, "How the Chicken Conquered the World," *Smithsonian Magazine*, June 2012. smithsonian mag.com/history/how-the-chicken-conquered-the-world-87583657/, accessed October 22, 2020.

12. Cattle were domesticated about 10,500 years ago, goats 10,000 years ago, sheep between 11,000 and 9,000 years ago. The first unmistakable signs of slavery only appear in the archaeological record about 5,500 years ago. Thus, it is extremely unlikely that human slavery served as a model or inspiration for animal domestication; however, the reverse seems likely.

13. Today, the genetic bifurcation of humanity into "improved" and "unimproved" species through the engineering of designer babies by the wealthy is a realistic possibility.

14. Tim Ingold, *The Perception of the Environment*, New York, Routledge, 2011.

15. Jacob Mikanowski, "Wild Thing: How and Why did Humans Domesticate Animals—and What Might This Tell Us about the Future of Our Own Species?" *Aeon*, November 28, 2016. aeon.co/essays/how-domestication -changes-species-including-the-human, accessed September 4, 2020.

16. "Atlantic Slave Trade." Wikipedia, accessed September 2, 2020.

17. See Edward E. Baptist, *The Half Has Never Been Told: Slavery and the Making of American Capitalism*, New York, Basic Books, 2016.

18. Quoted in Andrew Nikiforuk, *The Energy of Slaves: Oil and the New Servitude*, Vancouver, Greystone, 2012, p. 15. See also, James Oakes, *The Ruling Race: A History of American Slaveholders*, New York, W. W. Norton, 1998.

19. Quoted in Nikiforuk, Ibid., p. 17.

20. Walter Rodney, *How Europe Underdeveloped Africa*, Howard University Press, 1981 (reprint), p. 137.

21. See Jill Bolte Taylor, *My Stroke of Insight: A Brain Scientist's Personal Journey*. New York, Plume, 2006.

22. See Burton Mack, *Who Wrote the New Testament?: The Making of the Christian Myth*, San Francisco, HarperSanFrancisco, 1995, pp. 20–21.

23. *Ultra Society*, pp. 193 ff. For an alternative view, see Eric Cline, 1177 BC: *The Year Civilization Collapsed*, Princeton, NJ, Princeton University Press, 2014.

24. See Ara Norenzayan, *Big Gods: How Religion Transformed Cooperation and Conflict*, Princeton, NJ, Princeton University Press, 2013.

25. While Buddhism doesn't feature a Big God in the same way Christianity and Islam do, the Buddha is described in scriptures as the all-seeing "eye of the world," and sometimes depicted as such in iconography. Further, Buddhists are advised to police their own desires and cravings in the quest for enlightenment.

26. Joe Hill, "The Preacher and the Slave."

27. *Big Gods*, p. 133.

28. Karl Jaspers, *The Origin and Goal of History* (1949), New York, Routledge Revivals, 2011. In their recent book *Seshat History of the Axial Age* (2019), Daniel Hoyer, Jenny Reddish, and co-authors use an extensive historical database to explore evidence for the Axial Age; they conclude that there was actually no single "Axial Age" in human history. Instead, egalitarian ideals and constraints on political authority co-evolved together with greater sociopolitical complexity (i.e., the development of empires) in several places and times.

29. *Ultra Society*, p. 181.

30. These advantages are discussed in Jared Diamond's *Guns, Germs, and Steel*.

31. "The Doctrine of Discovery, 1493," *History Resources*, The Gilder Lehrman Institute of American History, gilderlehrman.org/history-resources /spotlight-primary-source/doctrine-discovery-1493, accessed September 4, 2020.

32. This section owes an enormous debt to the work of communication theorist Marshall McLuhan (1911–1980). I was fortunate to meet McLuhan in his office at the University of Toronto in 1978. Here I have sought to apply his way of thinking to communication technologies that have emerged since his passing.

33. Gary Snyder, *The Practice of the Wild* (1990), New York, Counterpoint, 2010 (reprint).

34. A parenthetical note: over all, chapters 2, 3, and 4 of this book present a chronology of power. However, in this chapter on the origins of social power, it seemed best to discuss the evolution of communication technologies and money all the way up to the present, rather than saving recent developments in these subject areas for the next chapter, which focuses especially on the influence of fossil fuels on events of the last couple of centuries.

35. The Chinese had actually invented printing much earlier, perhaps around the year 650, employing carved wood blocks. Moveable metal type was adopted in the 12th century, still long before Gutenberg.

36. David Graeber, *Debt: The First 5,000 Years*, Brooklyn, Melville House, 2011.

37. Ibid., p. 400, note 47.

38. Seshat Global History Databank, seshatdatabank.info/, accessed September 4, 2020.

39. Niall Ferguson, *The House of Rothschild*. New York, Penguin, 1999.

40. See Nitzan and Bichler, *Capital as Power*, London, Routledge, 2009.

41. William D. Nordhaus and James Tobin, "Is Economic Growth Obsolete?" in William Moss, ed., *The Measurement of Economic and Social Performance*, National Bureau of Economic Research, 1973.

42. Robert M. Solow, "The Economics of Resources or the Resources of Economics," *The American Economic Review*, Vol. 64, No. 2, Papers and Proceedings of the Eighty-sixth Annual Meeting of the American Economic Association, May 1974, pp. 1–14.

43. Nordhaus, "Reflections on the Economics of Climate Change," *Journal of Economic Perspectives*, Vol. 7, No. 4, Fall 1993, pp. 11–25.

44. This sidebar is loosely based on Blair Fix's essay, "Can the World Get Along Without Natural Resources?" *Economics from the Top Down*, June 18, 2020, economicsfromthetopdown.com/2020/06/18/can-the-world-get-along-without-natural-resources/, accessed September 4, 2020.

45. For a good overview, see Ana Guinote and Theresa Vescio, eds., *The Social Psychology of Power*, New York, Guilford Press, 2010.

46. *Social Psychology of Power*, p. 5.

47. Jennifer Overbeck, "Concepts and Historical Perspectives on Power," in *Social Psychology*, p. 33.

48. S. E. Asch, "Effects of Group Pressure on the Modification and Distortion of Judgments," in H. Guetzkow, ed., *Groups, Leadership, and Men*, Pittsburgh, Carnegie Press, 1953, pp. 177–190.

49. Stanley Milgram, *Obedience to Authority*, New York, Harper & Row, 1969.

50. Philip Zimbardo, The Lucifer Effect: Understanding How Good People Turn Evil. New York, Random House, 2007.

51. Kali Trzesniewski et al., "Low Self-Esteem During Adolescence Predicts Poor Health, Criminal Behavior, and Limited Economic Prospects During Adulthood," *Developmental Psychology*, Vol. 42, No. 2, pp. 381–390.

52. Jerry Unseem, "Power Causes Brian Damage." *The Atlantic*, June 18, 2017. getpocket.com/explore/item/power-causes-brain-damage?utm _source=pocket-newtab, accessed September 4, 2020.

53. Dacher Keltner, *The Power Paradox: How We Gain and Lose Influence*, New York, Penguin, 2016. See also, Dacher Keltner, "Don't Let Power Corrupt You," *Harvard Business Review*, October 2016. hbr.org/2016/10/dont-let -power-corrupt-you, accessed September 4, 2020.

54. Paul Hawken, *Blessed Unrest: How the Largest Social Movement in History Is Restoring Grace, Justice, and Beauty to the World*, New York, Penguin, 2008.

55. For a more complete list with extensive discussion, see Dirk Moses and Donald Bloxam, *Oxford Handbook of Genocide Studies*, Oxford, Oxford University Press, 2010.

56. Jacob M. Rabbie, "The Effects of Intragroup Cooperation and Intergroup Competition on In-Group Cohesion and Out-Group Hostility," in *Coalitions and Alliances in Humans and Other Animals*, Alexander Harcourt and Frans de Waal, eds., Oxford, Oxford University Press (1992), pp. 175–205.

57. See Richard Wrangham and Dale Peterson, *Demonic Males*, pp. 197–8.

58. Jared Diamond unpacks the reasons for European world dominance in his book, *Guns, Germs, and Steel: The Fates of Human Societies*, New York, W. W. Norton, 1998.

59. Daphne Blunt Bugental, "Paradoxical Power Manifestations: Power Assertion by the Subjectively Powerless," in *Social Psychology*, pp. 209–229.

60. Though he rarely uses the word "power," a good overview of status in human relations is presented in Robert Sapolsky, *Behave: The Biology of Humans at Our Best and Worst*, pp. 425–477.

Chapter 4: Power in the Anthropocene

1. I borrow the phrase "Great Acceleration" from J. R. McNeill and Peter Engelke, who coined it in their 2016 book, *The Great Acceleration: An Environmental History of the Anthropocene Since 1945*, Cambridge, MA, Harvard University Press, 2014. While McNeill and Engelke date the start of the Great Acceleration to 1945, and there is a good argument for doing so, the entire fossil fuel interval has been characterized by a general acceleration of trends such as population growth, consumption growth, and resource depletion. I would argue, therefore, that assuming a somewhat earlier start date (say, 1820) for the Great Acceleration aids explanation and understanding.

2. Vaclav Smil, *Energy and Civilization*, Cambridge, MA, MIT Press, 2017, p. 37.

3. Richard Lee, "Kung Bushmen Subsistence: An Input–Output Analysis," in A. Vayda, ed., *Environment and Cultural Behavior: Ecological Studies in Cultural Anthropology*, Garden City, NY, Natural History Press, Published for American Museum of Natural History, 1969, pp. 47–79.

4. E. Galan et al., "Widening the Analysis of Energy Return on Investment (EROI) in Agro-Ecosystems: Socio-Ecological Transitions to Industrialized Farm Systems (the Vallès County, Catalonia, c.1860 and 1999)," *Ecological Modeling* 2016 v. 336, pp. 13–25.

5. *Energy and Civilization*, p. 79.

6. Ibid., p. 81.

7. Zia Haq, "Biomass for Electricity Generation," *BioCycle*, Vol. 43, No. 11, October 2011.

8. "Historical Perspectives of Energy Consumption," wou.edu/las/physci/GS361/electricity%20generation/HistoricalPerspectives.htm, accessed September 4, 2020.

9. Quoted in Barbara Freese, *Coal: A Human History*, London, Arrow Books, 2006, pp. 203–4.

10. John Long, "Coal's Formation Is a Window on an Ancient World," *The Conversation*, June 2016, http://theconversation.com/coals-formation-is-a-window-on-an-ancient-world-54333, accessed September 4, 2020.

11. David Stradling and Peter Thorsheim, "The Smoke of Great Cities: British and American Efforts to Control Air Pollution, 1860–1914," *Environmental History*, Vol. 4, No. 1, January 1999, pp. 6–31.

12. Historian Ian Morris, in *Foragers, Farmers, and Fossil Fuels*, argues that technological innovation was a response to high labor costs. See p. 100.

13. Joseph Needham's masterwork, *Science and Civilization in China* (seven volumes, the first published in 1954), details the many innovations in science and technology that occurred in China long before their appearance in Europe and North America.

14. For a more extended discussion of China's early industrial revolution, see William McNeill, *The Pursuit of Power*, Chicago, University of Chicago Press, 1982, pp. 24–62.

15. Andrew Nikiforuk, *The Energy of Slaves: Oil and the New Servitude*, Vancouver, Greystone Books, 2012, p. 28.

16. Lewis Mumford, *The Myth of the Machine, Vol. II: The Pentagon of Power*, New York, Harcourt Brace Jovanovich, 1964, p. 147. On smaller scales, wage labor can be traced back to Mesopotamia in the third millennium BCE; see Ian Morris, *Foragers, Farmers, and Fossil Fuels*, Princeton, NJ, Princeton University Press, 2015, p. 63.

17. Timothy Mitchell, *Carbon Democracy: Political Power in the Age of Oil*, London, Verso, 2011, p. 12.

18. *Carbon Democracy*.

19. See Richard Heinberg, *Blackout: Coal, Climate and the Last Energy Crisis*, Gabriola Island, BC, New Society, 2009.

20. "Air Traffic by the Numbers," *Federal Aviation Administration*, faa.gov/air_traffic/by_the_numbers/, accessed September 4, 2020.

21. Darrin Qualman, "Another Trillion Tonnes: 250 Years of Global Material Use Data," April 9, 2019. darrinqualman.com/global-material-use/, accessed September 4, 2020.

22. For detailed descriptions of the rise of consumerism and advertising, see: Stuart Ewen, *Captains of Consciousness: Advertising and the Social Roots of the Consumer Culture* (New York: McGraw-Hill, 1976); and Kerryn Higgs, *Collision Course: Endless Growth on a Finite Planet* (Cambridge, MA: MIT Press, 2016).

23. Hannah Ritchie and Max Roser, "Urbanization," *Our World in Data*, November, 2019. ourworldindata.org/urbanization, accessed September 4, 2020.

24. careerplanner.com/DOTindex.cfm, accessed September 4, 2020.

25. A good summary of the process can be found in John Perkins, *Confessions of an Economic Hit Man*, New York, Penguin, 2004.

26. Michael Mann, *The Sources of Social Power. Volume 3: Global Empires and Revolution, 1890–1945*, Cambridge, MA, Cambridge University Press, 2012, p. 153.

27. *Sources of Social Power*, p. 160.

28. Walter Scheidel, *The Great Leveler: Violence and the History of Inequality from the Stone Age to the Twenty-First Century*, Princeton, NJ, Princeton University Press, 2017.

29. "World Nuclear Industry Status Report 2019." worldnuclearreport.org/, accessed September 4, 2020.

30. I'm grateful to Nate Hagens for alerting me to the concept of the human Superorganism; see his "Economics for the Future: Beyond the Superorganism," *Ecological Economics*, Vol. 169, March 2020.

31. Kevin Kelly: *Out of Control: The New Biology of Machines, Social Systems, and the Economic World*, New York, Basic Books, 1995 (reprint).

32. Roger LeBaron Hooke, "Humans May Surpass Other Natural Forces as Earth Movers," *Science Daily*, July 9, 2004, sciencedaily.com/releases /2004/07/040709083319.htm, accessed September 2, 2020. The study cited here suggests humans moved 40 billion tons of rock and soil in 1994. Since world consumption of most resources has more than doubled in the intervening quarter-century, a doubling of that figure seems likely.

Chapter 5: Overpowered

1. Adapted from Onno de Jong, *For a Future*, Apple Books, 2018. forafuture. com, accessed September 3, 2020.

2. *Next 10*, "California's Green Innovation Index." next10.org/publications /2019-gii, accessed September 2, 2020.

3. Richard Heinberg and David Fridley, *Our Renewable Future: Laying the Path for 100 Percent Clean Energy*, Washington, DC, Island Press, 2016. Full text available at ourrenewablefuture.org, accessed September 2, 2020.

4. Other researchers have come to similar conclusions. For example, Tim Morgan (former head of research at Tullett Prebon) argues that it is surplus energy—the energy left over once energy required for energy-producing activities has been subtracted—that has driven economic expansion, and that a transition to renewables will necessarily result in declining surplus energy (see Tim Morgan, *Surplus Energy Economics*

website surplusenergyeconomics.wordpress.com, accessed September 2, 2020). In a recent paper, Carey King of the Energy Institute at the University of Texas, Austin, shows the inadequacy of current growth-based economic modeling of the renewable energy transition and proposes a new model that incorporates data-derived relationships between energy use, resource extraction, and economic growth. His conclusion is that the renewable energy transition will entail trade-offs with consumption, population, and wages; these trade-offs will depend on the path taken (whether high or low rate of investment). Carey King, "An Integrated Biophysical and Economic Modeling Framework for Long-Term Sustainability Analysis: The HARMONY Model," *Ecological Economics*, Vol. 169, March 2020. doi.org/10.1016/j.ecolecon.2019.106464, accessed September 2, 2020.

5. *Our Renewable Future*, p. 140.
6. Kevin Anderson and Alice Bows-Larkin, "Avoiding Dangerous Climate Change Demands De-Growth Strategies from Wealthier Nations," *KevinAnderson.Info*, November 2013. kevinanderson.info/blog/avoiding -dangerous-climate-change-demands-de-growth-strategies-from -wealthier-nations, accessed September 2, 2020. See also Patrick Moriarty and Damon Honnery, "Can Renewable Energy Power the Future?" *Energy Policy*, Vol. 93, June 2016, pp. 3–7. sciencedirect.com /science/article/pii/S030142151630088X, accessed September 2, 2020.
7. Rachel Kaufman, "The Risks, Rewards and Possible Ramifications of Geoengineering Earth's Climate," *Smithsonian*, March 11, 2019. smithson ianmag.com/science-nature/risks-rewards-possible-ramifications-geo engineering-earths-climate-180971666, accessed September 3, 2020.
8. Christopher Flavelle, "Climate Change Threatens the World's Food Supply, United Nations Warns," *New York Times*, August 8, 2019. nytimes .com/2019/08/08/climate/climate-change-food-supply.html, accessed September 3, 2020.
9. Matt Richtel and Andrew Jacobs. "A *Mysterious Infection, Spanning the Globe in a Climate of Secrecy*," New York Times (April 6, 2019). forhuman liberation.blogspot.com/2019/04/3217-mysterious-infection-spanning .html, accessed September 3, 2020.
10. "Living Planet Report 2020," WorldWildlife.org. livingplanet.panda.org /en-us, accessed September 3, 2020.
11. NatureNeedsHalf.org, Accessed September 3, 2020.
12. In January 2021, 50 countries announced their commitment to protect at least 30 percent of the globe's land and ocean by 2030. See *Campaign for Nature*, "50 Countries Announce Bold Commitment..." campaign fornature.org/50-countries-announce-bold-commitment-to-protect-at -least-30-of-the-worlds-land-and-ocean-by-2030, accessed February 2, 2021.

13. Douglas J. McCauley et al., "A Mammoth Undertaking: Harnessing Insight from Functional Ecology to Shape De-Extinction Priority Setting," *Functional Ecology* 31, no. 5, 2017.

14. See Sergey Zimov, "Chapter Fourteen," in *Woolly: The True Story of the Quest to Revive One of History's Most Iconic Extinct Creatures*, ed. Ben Mezrich, New York, Atria Books, 2017.

15. See David Shultz, "Should We Bring Extinct Species Back from the Dead?" *Science Magazine*, September 2016. sciencemag.org/news/2016 /09/should-we-bring-extinct-species-back-dead, accessed September 2, 2020.

16. Author Eileen Crist argues that terms like "resources" and "natural capital" frame nature from the perspective of human exploitation and finance, and that healing our relationship with nature requires us to rethink our choice and usage of words. While I agree in principle, in instances like this it is a challenge to find alternative terms that aren't awkward. Since we're on the topic of word choice, I should also mention that some people find the terms "produce" and "production" objectionable when used in connection with the activities of the fossil fuel industry. Of course, oil companies do not make or manufacture crude oil, as the word "produce" implies; they merely extract it from the Earth's crust. While I agree with this objection, I occasionally use the familiar terms anyway so as to avoid overuse of the few available alternatives.

17. The degree to which resource depletion contributed to the collapse of past societies is debated. For example, in the case of Easter Island, Jared Diamond has argued that deforestation played a significant role in population decline during the period immediately prior to European contact (see Jared Diamond, *Collapse: How Societies Choose to Fail or Succeed*, New York, Viking, 2005, pp. 79–118). Archaeologists Terry Hunt and Carl Lipo have disputed this interpretation of the evidence (see Robert DiNapoli et al., "A Model-Based Approach to the Tempo of 'Collapse': The Case of Rapa Nui [Easter Island]," *Journal of Archaeological Science*, Vol. 116, April 2020).

18. It is possible, in principle, to make nitrogen fertilizers by using electricity from renewable sources to produce hydrogen from water, then using the hydrogen to make ammonia via the Fischer-Tropsch process. However, see the previous discussion regarding practical near-term limits to the deployment of renewable energy technologies.

19. Calculations based on data in "Overconsumption?: Our Use of the World's Natural Resources." *Friends of the Earth.UK*, September 14, 2012. cdn.friendsoftheearth.uk/sites/default/files/downloads/overconsump tion.pdf, accessed September 3, 2020.

20. See Ugo Bardi, *Extracted: How the Quest for Mineral Wealth Is Plundering the Planet*, White River Junction, VT, Chelsea Green, 2013. Christopher

Clugston, *Blip: Humanity's 300 Year Self-Terminating Experiment with Industrialism*, St. Petersburg, FL, Book Locker, 2019.

21. Peter Turchin, *Ages of Discord: A Structural-Demographic Analysis of American History*, Chaplin, CT, Beresta Books, 2016.

22. *Ages of Discord*, p. 204.

23. See "Democracy Index," Wikipedia. en.wikipedia.org/wiki/Democracy_Index, accessed September 3, 2020.

24. Chuck Collins, Dedrick Asante-Muhammed, Josh Hoxie, and Sabrina Terry, "Dreams Deferred: How Enriching the 1% Widens the Racial Wealth Divide." *Institute for Policy Studies*. ips-dc.org/wp-content/up loads/2019/01/IPS_RWD-Report_FINAL-1.15.19.pdf, accessed November 5, 2020.

25. See charts here: "US Economic and Social Inequality," *New York Times*, July 2, 2020. nytimes.com/interactive/2020/07/02/opinion/politics/us-economic-social-inequality.html, accessed September 3, 2020.

26. Martin Gilens and Benjamin Page, "Testing Theories of American Politics: Elites, Interest Groups, and Average Citizens," *Perspectives on Politics*, Vol. 12, Issue 3, September 2014, pp. 564–581. cambridge.org/core/journals/perspectives-on-politics/article/testing-theories-of-american-politics-elites-interest-groups-and-average-citizens/62327F513959D0A304D4893B382B992B, accessed September 2, 2020.

27. "An Economy for the 99%: It's Time to Build a Human Economy that Benefits Everyone, Not Just the Privileged Few." *Oxfam*, January 16, 2017. policy-practice.oxfam.org.uk/publications/an-economy-for-the-99-its-time-to-build-a-human-economy-that-benefits-everyone-620170, accessed October 28, 2020. See also Greg Sargent, "The Massive Triumph of the Rich, Illustrated by Stunning New Data," *Washington Post*, December 9, 2019.

28. Douglas Main, "How the World's Most Widely Used Insecticide Causes Fish Declines," *National Geographic*, November 2019. nationalgeographic.com/animals/2019/11/neonicotinoid-insecticides-cause-fish-declines-japan/#close, accessed September 3, 2020.

29. "Neonicotinoid Pesticides Are Slowly Killing Bees." *PBS Newshour*, June 29, 2017. pbs.org/newshour/science/neonicotinoid-pesticides-slowly-killing-bees, accessed September 3, 2020.

30. See Rebecca Harrington, "By 2050, The Oceans Could Have More Plastic than Fish," *Business Insider*, January 26, 2017. businessinsider.com/plastic-in-ocean-outweighs-fish-evidence-report-2017-1, accessed September 3, 2020.

31. "PFAS Chemicals and You," *Science Friday*, November 1, 2019. sciencefriday.com/segments/pfas-dupont-lawsuit-robert-bilott, accessed September 3, 2020.

32. Julian Cribb, *Poisoned Planet: How Constant Exposure to Man-Made*

Chemicals is Putting Your Life at Risk, New York, Allen and Unwin, 2015, p. 149.

33. Aya Norenzayan, *Big Gods: How Religion Transformed Cooperation and Conflict,* Princeton, NJ, Princeton University Press, 2013, p. 152.

34. Ibid., p. 153.

35. Thomas Malthus, *An Essay on the Principle of Population.* London, 1798. gutenberg.org/ebooks/4239, accessed September 3, 2020.

36. Allen Good and Perrin Beatty, "Fertilizing Nature: A Tragedy of Excess in the Commons," *PLoS Biology,* August 16, 2011. journals.plos.org/plos biology/article?id=10.1371/journal.pbio.1001124, accessed September 3, 2020.

37. David Wuepper et al., "Countries and the Global Rate of Soil Erosion," *Nature Sustainability,* Vol. 3, December 2, 2020. nature.com/articles /s41893-019-0438-4, accessed September 3, 2020.

38. George Monbiot, "Lab-Grown Food Will Soon Destroy Farming—and Save the Planet," *The Guardian,* January 8, 2020. theguardian.com /commentisfree/2020/jan/08/lab-grown-food-destroy-farming-save -planet, accessed September 3, 2020.

39. Elizabeth Dunbar, "Climate Curious: How Much Does Population Growth Contribute to Climate Change?" *MPR News,* December 11, 2019. mprnews .org/story/2019/12/11/climate-curious-how-much-does-population -growth-contribute-to-climate-change; and churchandstate.org.uk/2019 /06/e-o-wilson-runaway-population-growth-at-epicenter-of-environ mental-problems, accessed September 3, 2020.

40. "Democracy Index 2019," *The Economist Intelligence Unit,* March 2020. eiu.com/topic/democracy-index, accessed September 3, 2020.

41. T. Parrique et al., "Decoupling Debunked: Evidence and Arguments Against Green Growth as a Sole Strategy for Sustainability," European Environmental Bureau, 2019. mk0eeborgicuypctuf7e.kinstacdn.com /wp-content/uploads/2019/07/Decoupling-Debunked.pdf, accessed September 2, 2020.

42. See Carmen Reinhart and Kenneth Rogoff, *This Time Is Different: Eight Centuries of Financial Folly,* Princeton, NJ, Princeton University Press, 2009.

43. See *This Time Is Different*; and Steve Keen, *Debunking Economics: The Naked Emperor of the Social Sciences,* New York, Palgrave, 2001.

44. *This Time is Different.*

45. Enda Curran, "The Way Out for a World Economy Hooked on Debt? More Debt," *Bloomberg,* Dec. 1, 2019. bloomberg.com/news/articles/2019-12 -01/the-way-out-for-a-world-economy-hooked-on-debt-yet-more-debt, accessed September 2, 2020. See also Geopolitical Futures staff, "Rising Global Debt." *Geopolitical Futures,* January 3, 2020. geopoliticalfutures .com//pdfs/rising-global-debt-geopoliticalfutures-com.pdf?utm_source =newsletter&utm_medium=email&utm_term=https%3A%2F%2Fgeo

politicalfutures.com%2F%2Fpdfs%2Frising-global-debt-geopolitical
futures-com.pdf&utm_content&utm_campaign=PAID+-+Everything+as
+it%27s+published, accessed September 2, 2020.

46. Peter Cook, "Is the U.S. Economy Really Growing? (Spoiler Alert:
No!)" *Zerohedge*, March 16, 2018. zerohedge.com/news/2018-03-16/us
-economy-really-growing, accessed September 2, 2020.

47. Ben King, "What Is Quantitative Easing and How Will It Affect You?"
BBC News, June 18, 2020. bbc.com/news/business-15198789, accessed
September 3, 2020.

48. Stephen Williams and Samuel Alexander, "MMT, Post-Growth Economics, and Avoiding Collapse," in Haydn Washington, ed., *Ecological
Economics: Solutions for the Future,* Chapter 7. In press.

49. Will Bedingfield, "Universal Basic Income, Explained," *Wired UK*, August
25, 2019. wired.co.uk/article/universal-basic-income-explained, accessed September 3, 2020.

50. See "MMT, Post-Growth Economics."

51. "List of Incidents Involving Ricin." Wikipedia. en.wikipedia.org/wiki
/List_of_incidents_involving_ricin, accessed September 3, 2020.

52. See: airwars.org, accessed September 3, 2020.

53. Letta Taylor, "The Truth about the United States Drone Program," Human
Rights Watch, March 24, 2014. hrw.org/news/2014/03/24/truth-about
-united-states-drone-program, accessed September 3, 2020.

54. See "Nuclear Weapons: Who Has What," Arms Control Association. arms
control.org/factsheets/Nuclearweaponswhohaswhat, accessed September 3, 2020.

55. "Nuclear Weapon Yield," Wikipedia. en.wikipedia.org/wiki/Nuclear
_weapon_yield, accessed September 3, 2020.

56. Seth Baum, "The Risk of Nuclear Winter," *Federation of American Scientists,* May 29, 2015. fas.org/pir-pubs/risk-nuclear-winter, accessed September 3, 2020.

57. Andrew Osborn, "Russia Says It Has Deployed First Hypersonic Nuclear-
Capable Missiles," *Reuters*, December 27, 2019. reuters.com/article/us
-russia-nuclear-missiles/russia-says-it-has-deployed-first-hypersonic
-nuclear-capable-missiles-idUSKBN1YV1M1, accessed September 3, 2020.

Chapter 6: Optimum Power

1. Michio Kaku, *The Future of Humanity: Our Destiny in the Universe*, New
York, Anchor, 2019.

2. Decades of UFO reports notwithstanding, there are other possible explanations for these. See, for example, John Michael Greer, *The UFO Chronicles: How Science Fiction, Shamanic Experiences, and Secret Air Force Projects
Created the UFO Myth*, Aeon Books, 2021.

3. Scientists have deliberately sent out such signals in the unmanned Voyager spacecraft, and via the Search for Extraterrestrial Intelligence (SETI) program. Theoretically, ordinary Earth-based radio and television broadcasts could be detected by a sufficiently advanced society located at a distance of several light years.

4. There are two more solutions to the Fermi Paradox that I consider credible. It is possible that, while the evolution of bacteria and other prokaryotes can get started relatively easily, the evolution of eukaryotes, and therefore of all multicelled organisms, is really hard to ignite. Nick Lane makes this argument in his book *Power, Sex, Suicide*. Even if that's not the case, it may be that, as Ajit Varki and Danny Brower argue in *Denial*, the evolution of high intelligence—and therefore the awareness of mortality—must lead to a pervasive state of anxiety and cautiousness, and therefore a decisive reduction in evolutionary fitness, unless it is accompanied by the highly unlikely simultaneous evolution of the ability to consciously deny reality. These are not mutually exclusive arguments; both could be true. The upshot is that either high intelligence, or multi-celled life in general, or both, are likely to be extremely rare in the universe.

5. The actual mechanics of ageing and death are a topic of ongoing research. The leakage of free radicals from mitochondria appears to be a key to the ageing process. For a discussion of this research, see Nick Lane, *Power, Sex, Suicide: Mitochondria and the Meaning of Life*, Oxford, Oxford University Press, 2005, pp. 405–446.

6. See, for example, Lindsey Harvell and Gwendelyn Nisbett, eds., *Denying Death: An Interdisciplinary Approach to Terror Management Theory*, New York, Routledge, 2016.

7. For information on antibiotic immunity, see "Antimicrobial Resistance," World Health Organization, July 31, 2020. who.int/news-room/fact -sheets/detail/antimicrobial-resistance, accessed September 4, 2020.

8. This conclusion is also supported by the work of Earth system and environmental scientists, led by Johan Rockström from the Stockholm Resilience Centre, and Will Steffen from the Australian National University, who have identified nine "planetary boundaries" that circumscribe a safe space within which humanity can continue to thrive for many generations to come. Transgressing even one boundary could lead to catastrophic risk; as of 2009, two boundaries had already been crossed, while others were in imminent danger of being crossed. See "The Nine Planetary Boundaries," Stockholm Resilience Centre, stockholmresilience .org/research/planetary-boundaries/planetary-boundaries/about-the -research/the-nine-planetary-boundaries.html, accessed November 3, 2020.

9. For more on the adaptive cycle, see "Adaptive Cycle," Resilience Alliance, n.d., resalliance.org/adaptive-cycle, accessed September 4, 2020.

10. Joseph Tainter, *The Collapse of Complex Societies,* Cambridge, MA, Cambridge University Press, 1988.

11. Peter Turchin, *Historical Dynamics: Why States Rise and Fall,* Princeton, NJ, Princeton University Press, 2003.

12. I use the term "balancing" in this context with some trepidation. It's useful, in that it points our thinking in the right general direction; but it can also mislead: in nature, there is rarely if ever a condition of stable "balance"—only dynamic imbalance.

13. See "2000-Watt Society." 2000watt.swiss/english.html, accessed September 3, 2020.

14. Glenda Yenni, "Self-Limitation as an Explanation for Species' Relative Abundances and the Long-Term Persistence of Rare Species," Dissertation, Utah State University, 2013. See also: G. Yenni et al., "Strong Self-Limitation Promotes the Persistence of Rare Species," *Ecology,* Vol. 93, No. 3, 212, pp. 456–461; and G. Yenni et al., "Do Persistent Rare Species Experience Stronger Negative Frequency Dependence than Common Species?" *Global Ecology and Biogeography,* Vol. 26 (2017), pp. 513–523.

15. G. Yenni, Dissertation.

16. G. Yenni, personal communication.

17. G. Yenni, personal communication. Unfortunately, the pika's habitat is becoming even rarer as a result of climate change, which could threaten its ability to persist at low abundance. See Lucas Moyer-Horner et al., "Predictors of Current and Longer-Term Patterns of Abundance of American Pikas (*Ochotona Princeps*) across a Leading-Edge Protected Area," *PLoS One,* November 30, 2016. journals.plos.org/plosone/article?id=10.1371/journal.pone.0167051, accessed September 3, 2020.

18. Lee, R. B. "Reflections on Primitive Communism," in T. Ingold, D. Riches, and J. Woodburn (eds.), *Hunters and Gatherers Vol. 1,* Oxford, Berg, 1988, pp. 252–268.

19. "Tribal Conservationists in the Congo Basin," *Survival,* n.d. survivalinternational.org/articles/3473-conservationistscongobasin, accessed September 3, 2020.

20. Colding, J. and C. Folke, "Social taboos: 'Invisible' systems of local resource management and biological conservation," *Ecosystems* 4, 2001, pp. 85–104.

21. See also Jim Robbins, "Native Knowledge: What Ecologists Are Learning from Indigenous People," *Yale 360,* April 26, 2018. e360.yale.edu/features/native-knowledge-what-ecologists-are-learning-from-indigenous-people, accessed September 3, 2020.

22. Clark Monson, "Indigenous Resource Taboos: A Practical Approach

Towards the Conservation of Commercialized Species," Dissertation, University of Hawaii, August 2004. scholarspace.manoa.hawaii.edu /bitstream/handle/10125/11606/uhm_phd_4488_r.pdf?sequence=2, accessed September 3, 2020.

23. *Science*, 13 Dec 1968: Vol. 162, Issue 3859, pp. 1243–124. science.science mag.org/content/162/3859/1243, accessed September 3, 2020.

24. "Elinor Ostrom's 8 Principles for Managing a Commons," *Commons Magazine*, October 2, 2011. onthecommons.org/magazine/elinor-ostroms-8 -principles-managing-commmons, accessed September 3, 2020.

25. All quotes are from the anthology *Less Is More*, by Goldian Vandenbroeck. Rochester, VT, Inner Traditions, 1991.

26. My concern here is not with the question of whether God or gods really exist, but solely with the social and psychological impacts of religious belief and practice.

27. Today, religious principles guide many people toward activism on behalf of climate change mitigation and other environmental issues. Pope Francis has even proposed adding "ecological sin" to church teachings. Jon Queally, "While Warning of Nazi-Like Fascism and Corporate Crimes, Pope Francis Proposes Adding 'Ecological Sin' to Church Teachings," *Common Dreams*, November 16, 2019, commondreams.org /news/2019/11/16/while-warning-nazi-fascism-and-corporate-crimes -pope-francis-proposes-adding, accessed September 3, 2020.

28. For Marx and Engels, working-class democracy was essential for any "dictatorship" over the former ruling class. Lenin and his successors subverted this revolutionary doctrine by making the vanguard party a political substitute for actual working-class democracy.

29. See Thomas Piketty, *Capital in the Twenty-First Century*, Cambridge, MA, Belknap, 2017.

30. Jason Hickel, "The Dark Side of the Nordic Model," Al Jazeera, Dec. 6, 2019. aljazeera.com/indepth/opinion/dark-side-nordic-model-19120510 2101208.html, accessed September 3, 2020.

31. See Adam Hochschild, *Bury the Chains: Prophets and Rebels in the Fight to Free an Empire's Slaves*, New York, Houghton Mifflin, 2005. See also Andrew Nikiforuk, *The Energy of Slaves*, Vancouver, Greystone, 2012, pp. 22–26.

32. This is the founding principle of the Breakthrough Institute. See also David Roberts, "John Kerry and the Climate Kids: A Tale of 2 New Strategies to Fight Climate Change," *Vox*, December 10, 2019. vox.com/energy -and-environment/2019/12/10/20996651/climate-change-john-kerry -world-war-zero-sunrise-movement, accessed September 3, 2020.

33. This latter view is propounded, for example, by Tim Jackson in *Prosperity Without Growth: Economics for a Finite Planet*, London, Earthscan, 2009.

34. See Roger Hallam, *Common Sense for the 21st Century*, White River Junction, VT, Chelsea Green, 2019.

35. It is important to note, however, that some environmentalists now advocate the expansion of nuclear power as a way of cutting carbon emissions.

36. Stan Cox, *Any Way You Slice It: The Past, Present and Future of Rationing*, New York, The New Press, 2013, p. 24.

37. Ibid., p. 25.

38. Ara Norenzayan, *Big Gods: How Religion Transformed Cooperation and Conflict*, Princeton, NJ, Princeton University Press, 2013, pp. 170–190.

39. Don Ross, "Game Theory," *Stanford Encyclopedia of Philosophy*, 1997 (revised 2019). plato.stanford.edu/entries/game-theory, accessed September 3, 2020. See also Max Willner Giwerc, "Game Theory and Disarmament: Thinking Beyond the Table." E-International Relations. e-ir.info/2018/12/18/game-theory-and-disarmament-thinking-beyond -the-table, accessed September 3, 2020.

40. Deng and Conitzer, "Disarmament Games," users.cs.duke.edu/~conitzer /disarmament_full_version.pdf, accessed September 3, 2020.

41. Peter John Wood, "Climate Change and Game Theory," papers.ssrn.com /sol3/papers.cfm?abstract_id=1883944, accessed September 3, 2020.

42. *Prosperity without Growth*, pp. 143–157.

43. Coalition theory, a subset of game theory, is generally better at accounting for the real-world prospects for cooperation. Coalitions arise where at least three actors (individuals, groups, or countries) are in contention, and no single actor acting unilaterally can achieve an optimal outcome. Coalition theories attempt to explain why alliances emerge, why they take the forms they do, how they persist, and why they terminate.

44. Varki and Brower, *Denial*, New York, Twelve Books, 2013, pp. 110–134.

45. See Thomas Moynihan, *X-Risk: How Humanity Discovered Its Own Extinction*, Boston, MIT Press, 2020.

46. Tali Sharot, *The Optimism Bias: A Tour of the Irrationally Positive Brain*, New York, Vintage Reprint, 2012.

47. Yuval Noah Hariri, *Sapiens: A Brief History of Humankind*, New York, Harper, 2015, p. 32.

48. See, for example, Herman Daly, *Beyond Growth: The Economics of Sustainable Development*, Boston, Beacon, 1996; and Rob Dietz and Dan O'Neill, *Enough Is Enough: Building a Sustainable Economy in a World of Finite Resources*, San Francisco, Barrett-Koehler, 2011.

49. See Martin LaMonica, "Greta Thunberg's Radical Climate Change Fairy Tale Is Exactly the Story We Need," *The Conversation*, September 27, 2019. theconversation.com/greta-thunbergs-radical-climate-change-fairy-tale -is-exactly-the-story-we-need-124252, accessed September 3, 2020. See also Varki and Brower, *Denial*, p. 161.

50. Studies show that the satisfaction associated with status signals is related not to the absolute size of the signal (i.e., the size of one's car or house), but its size relative to others in the vicinity.

51. See Melissa Dahl, "A Classic Psychology Study on Why Winning the Lottery Won't Make You Happier," *The Cut*, January 13, 2016. http://nymag .com/scienceofus/2016/01/classic-study-on-happiness-and-the-lottery .html, accessed September 3, 2020.

Chapter 7: The Future of Power

1. Most prominent futurists forecast exactly this outcome. They do so, again in my view, by ignoring or arbitrarily minimizing the trends discussed in Chapter 5, or by proposing solutions that are unworkable, unscalable, or unaffordable.

2. See Thomas Carothers and Andrew O'Donahue, eds., *Democracies Divided: The Global Challenge of Political Polarization*, Washington, DC, Brookings Institution Press, 2019.

3. I'm grateful to Craig Collins for this framing of the three phases of capitalism's adaptation to the evolution of industrial societies' energy base. See "Catabolism: Capitalism's Frightening Future," *Counterpunch*, December 3, 2019. counterpunch.org/2018/11/01/catabolism-capitalisms -frightening-future, accessed September 6, 2020.

4. See R. R. Reno, *Return of the Strong Gods: Nationalism, Populism, and the Future of the West*, Washington, DC, Gateway Editions, 2019.

5. The story is told by Matthieu Auzanneau in *Oil, Power and War*, White River Junction, VT, Chelsea Green, 2018, pp. 389–412.

6. See, for example, Edward Lucas, *The New Cold War: How the Kremlin Menaces Both Russia and the West*, London, Bloomsbury, 2008.

7. Evan Osnos, "Fight Fight, Talk Talk," *The New Yorker*, Jan. 13, 2020.

8. Russian military analyst Andrei Martyanov offers an overview of military capability in the 21st century. claritypress.com/product/the-real -revolution-in-military-affairs, accessed September 4, 2020.

9. See Mark Fischetti, "Climate Change Hastened Syria's Civil War," *Scientific American*, March 2, 1915.

10. See John Michael Greer, *The Long Descent: A User's Guide to the End of the Industrial Age*, Gabriola Island, BC, New Society, 2008.

11. John Gowdy, "Our Hunter-Gatherer Future: Climate Change, Agriculture, and Uncivilization," *Futures*, Vol. 115, January 2020.

12. See Christopher Tucker, *A Planet of 3 Billion: Mapping Humanity's Long History of Ecological Destruction and Finding Our Way to a Resilient Future, A Global Citizen's Guide to Saving the Planet*, Atlas Observatory Press, 2019.

13. See Albert Bates and Kathleen Draper, *Burn: Igniting a New Carbon Drawdown Economy to End the Climate Crisis*, White River Junction, VT, Chelsea

Green, 2019. Bates and Draper discuss captured carbon as a performance booster in products.

14. See footprintnetwork.org, accessed September 4, 2020.

15. Thomas Wiedmann et al., "The Material Footprint of Nations." *Proceedings of the National Academy of Sciences of the United States of America*, May 19, 2015 112 (20) 6271–6276; first published September 3, 2013 doi.org/10 .1073/pnas.1220362110, accessed September 4, 2020.

16. T. Vaden et al., "Raising the Bar: On the Type, Size and Timeline of a 'Successful' Decoupling," *Environmental Politics*, 24 June, 2020. See also T. Vaden et al., "Decoupling for Ecological Sustainability: A Categorization and Review of Research Literature," *Environmental Science & Policy*, Vol. 112, October 2020, pp. 236–244.

17. Adapted from Ed Simon, "What Viktor Frankl's logotherapy can offer in the Anthropocene," *Aeon*, February 11, 2020. aeon.co/ideas/what-viktor -frankls-logotherapy-can-offer-in-the-anthropocene, accessed September 4, 2020.

18. See Alice Friedemann, *When Trucks Stop Running: Energy and the Future of Transportation*, New York, Springer, 2016.

19. Jason Bradford, "The Future is Rural: Food System Adaptations to the Great Simplification," *Post Carbon Institute*, 2019. postcarbon.org /publications/the-future-is-rural, accessed September 4, 2020.

20. For a compelling recent restatement of the "small is beautiful" argument, see Helena Norberg-Hodge, *Local Is Our Future: Steps to an Economics of Happiness*, Local Futures, 2019.

21. See David Korowicz, "Brexit: Systemic Risk and a Warning," *Geneva Global Initiative*, Sept. 25, 2019. genevaglobalinitiative.org/brexit -systemic-risk-and-a-warning, accessed September 4, 2020.

22. See, for example, David Pilling, "5 Ways GDP Gets It Totally Wrong as a Measure of Our Success," January 17, 2018, *World Economic Forum*. weforum.org/agenda/2018/01/gdp-frog-matchbox-david-pilling-growth -delusion, accessed November 11, 2020.

23. For more discussion about GDP, see Richard Heinberg, *The End of Growth*, Gabriola Island, BC, New Society, 2011, pp. 231–259.

24. Marvin Harris, *Cultural Materialism: The Struggle for a Science of Culture*, New York, Vintage Books, 1980.

25. *Foragers, Farmers, and Fossil Fuels*, p 204.

26. See Mats Larsson, *The Limits of Business Development and Economic Growth*, New York, Palgrave Macmillan, 2004.

27. See Environmental Health Network, ehtrust.org, accessed September 4, 2020.

28. See D. Kriebel et al., "The Precautionary Principle in Environmental Science," *Environmental Health Perspectives*, Sept, 2001; Vol. 109, No. 9, pp. 871–876.

29. The Wikipedia entry on the Office of Technology Assessment tells its story fairly and succinctly. en.wikipedia.org/wiki/Office_of_Technology _Assessment, accessed September 4, 2020.

30. See Dmitry Orlov, *Shrinking the Technosphere: Getting a Grip on Technologies that Limit Our Autonomy, Self-Sufficiency and Freedom*, Gabriola Island, BC, New Society, 2016.

31. "World Happiness Report," worldhappiness.report, accessed September 4, 2020.

32. An excellent example is the organization Farm Hack, farmhack.org, accessed November 12, 2020. The people engaged in this community rely on modern communication technology to share ideas about how to improve farm equipment, with a focus on low-energy, easy-to-repair designs.

33. Susan Krumdieck, *Transition Engineering: Building a Sustainable Future*, Boca Raton, CRC Press, 2020.

34. *Low Tech Magazine*, lowtechmagazine.com, accessed September 4, 2020.

35. Ted Trainer, *The Simpler Way: Collected Writings of Ted Trainer*, edited by Samuel Alexander and Jonathan Rutherford, Australia, Simplicity Institute, 2020.

36. Philippe Bihouix, *The Age of Low Tech: Towards a Technologically Sustainable Civilisation*, Bristol, Bristol University Press, 2020.

37. Many books and courses on permaculture are available. Start with "What Is Permaculture?" *Permaculture Principles*, Permacultureprinciples.com, accessed September 4, 2020.

38. Eric Toensmeier, *The Carbon Farming Solution*, White River Junction, VT, Chelsea Green, 2016.

39. For information on Henry George's writings and "Georgist" economic ideas, see the Henry George Institute, henrygeorge.org, accessed September 2, 2020.

40. Others disagree. See Sheelah Kolhatkar, "The Ultra-Wealthy Who Argue that They Should Be Paying Higher Taxes," *The New Yorker*, December 30, 2019. newyorker.com/magazine/2020/01/06/the-ultra-wealthy-who -argue-that-they-should-be-paying-higher-taxes, accessed September 3, 2020.

41. "Ecuador Constitution in English." *Political Database of the Americas*, January 31, 2011. pdba.georgetown.edu/Constitutions/Ecuador/english08 .html, accessed September 4, 2020.

42. Debbie Bookchin, "How My Father's Ideas Helped the Kurds Create a New Democracy," *New York Review of Books*, June 15, 2018. nybooks.com /daily/2018/06/15/how-my-fathers-ideas-helped-the-kurds-create-a-new -democracy, accessed September 4, 2020.

43. See Hannah Salwen and Kevin Salwen, *The Power of Half: One Family's Decision to Stop Taking and Start Giving Back*, New York, Mariner, 2011.

44. Karin Kuhlemann, "The Elephant in the Room: The Role of Interest Groups in Creating and Sustaining the Population Taboo," in *Climate Change Denial and Public Relations*, Almiron and Xifra, eds., London, Routledge, 2019.

45. Joel Cohen, *How Many People Can the Earth Support?* New York, W. W. Norton, 1996.

46. See, for example, calculations by Chris Rhodes (Chris Rhodes, "How Many People Can the Earth Support…Really?" *Energy Balance*, December 29, 2008. ergobalance.blogspot.com/2008/12/how-many-people-can-earth-support.html Accessed September 4, 2020) and Folke Gunther ("The Carrying Capacity for Humans Without Fossil Fuels." *Holon*, n.d. holon.se/folke/kurs/logexp/carrying.shtml, accessed September 4, 2020).

47. Christopher Tucker, *A Planet of 3 Billion*. Alexandria, VA, Atlas Observatory Press, 2019. Website: planet3billion.com/index.html, accessed September 4, 2020.

48. Corey Bradshaw and Barry Brook, "Human Population Reduction Is Not a Quick Fix for Environmental Problems," *PNAS*, Vol. 111, No. 46, November 18, 2014. ncbi.nlm.nih.gov/pmc/articles/PMC4246304, accessed September 4, 2020.

49. James Gallagher, "'Remarkable' Decline in Fertility Rates." BBC News, November 9, 2018. bbc.com/news/health-46118103, accessed September 4, 2020.

50. Jeremy Grantham, "Chemical Toxicity and the Baby Bust," *GMO*, February 6, 2020. gmo.com/americas/research-library/chemical-toxicity-and-the-baby-bust, accessed September 4, 2020.

51. Alex Martin, "The Gray Wave: Japan Attempts to Deal with Its Increasingly Elderly Population," *Japan Times*, November 16, 2019. japantimes.co.jp/news/2019/11/16/national/social-issues/gray-wave-japan-attempts-deal-increasingly-elderly-population/#.XiiR8heIYlI, accessed September 4, 2020.

52. Bill McKibben, *Maybe One: A Case for Smaller Families*, New York, Plume, 1999.

53. Collins' framing of the coming power struggle against catabolic capitalism (including the four constituencies of the horizontal power coalition) is also set forth in two articles, "Meet Cannibalistic Capitalism: Globalization's Evil Twin," *Truthout* July 30, 2012 truthout.org/articles/meet-catabolic-capitalism-globalizations-evil-twin, accessed September 11, 2020; and "Cannibalistic Capitalism and Green Resistance," *Truthout* August 31, 2012 truthout.org/articles/cannibalistic-capitalism-and-green-resistance, accessed September 11, 2020. These articles and others related to the subject are archived at Collins' site, catabolic-capitalism.com/.

54. See "Youth Activist Toolkit," *Advocates for Youth*, from which this sidebar

has been adapted. advocatesforyouth.org/wp-content/uploads/2019/04
/Youth-Activist-Toolkit.pdf, accessed November 11, 2020. Other resources
include: "Power Mapping," *Beautiful Rising*, beautifulrising.org/tool
/power-mapping, accessed November 11, 2020. "Power Analysis," *Racial
Equity Tools*. racialequitytools.org/module/power-analysis, accessed
November 11, 2020.

55. Peter Buffett, "The Charitable-Industrial Complex," *New York Times*, July
26, 2013. nytimes.com/2013/07/27/opinion/the-charitable-industrial
-complex.html, accessed November 13, 2020.

56. George Monbiot makes this point well in his essay, "If Defending Life
on Earth Is Extremist, We Must Own That Label," *The Guardian*, January
22, 2020. theguardian.com/commentisfree/2020/jan/22/defending-life
-earth-extremist-police-extinction-rebellion, accessed September 4,
2020.

57. See Eitan Hersh, *Politics Is for Power: How to Move Beyond Political Hobby-
ism, Take Action, and Make Real Change*, New York, Scribner, 2020.

58. Margaret Klein Salamon with Molly Gage, *Facing the Climate Emergency:
How to Transform Yourself with Climate Truth*, Gabriola Island, BC, New
Society, 2020.

59. See "Revolutionary Black Nationalism for the Twenty-First Century:
Interview with Kali Akuno," *LeftRoots*, leftroots.net/revolutionary-black
-nationalism-for-the-twenty-first-century-interview-with-kali-akuno,
accessed November 13, 2020.

60. See Roger Hallam, *Common Sense for the 21st Century*, White River
Junction, VT, Chelsea Green, 2019.

61. Rob Hopkins, *The Transition Handbook: From Oil Dependency to Local Resilience*, Cambridge, MA, UIT Cambridge, 2014. Website: transitionnetwork
.org, accessed September 4, 2020.

62. Website: sunrisemovement.org, accessed September 4, 2020.

63. Website: worldwarzero.com, accessed September 4, 2020.

64. Rob Hopkins, *From What Is to What If: Unleashing the Power of Imagination
to Create the Future We Want*, White River Junction, VT, Chelsea Green,
2020.

65. A perennial classic guide for activism is Saul Alinsky's *Rules for Radicals:
A Practical Primer for Realistic Radicals*, New York, Vintage, 1989.

66. The British economist John Maynard Keynes came to a similar conclusion. He thought that when economic science was finally perfected,
everyone would be able to enjoy a high enough standard of living that
they could devote themselves largely to the arts, either as producers or
consumers.

67. Chris Jordan discusses beauty as a motivator for change in his TEDx talk.
ted.com/talks/chris_jordan_can_beauty_save_our_planet, accessed September 4, 2020.

68. Psychedelic drugs, used in many traditional religious or spiritual settings, can likewise take the mind beyond the realm of words; they are also being used in clinical settings to reduce hospice patients' fear of death. See Michael Pollan, *How to Change Your Mind: What the New Science of Psychedelics Teaches Us about Consciousness, Dying, Addiction, Depression, and Transcendence*, New York, Penguin, 2018.

69. See James Nestor, *Breath: The New Science of a Lost Art*, New York, Riverhead Books, 2020.

70. David Fleming, *Surviving the Future: Culture, Carnival and Capital in the Aftermath of the Market Economy*, White River Junction, VT, Chelsea Green, 2016, p. 53.

71. See Chuck Collins, *Born on First Base: A One Percenter Makes the Case for Tackling Inequality, Bringing Wealth Home, and Committing to the Common Good*, White River Junction, VT, Chelsea Green, 2016.

Index

A

abolitionists, 283
activism, 283–285, 343–344, 346
adaptive cycle, 264–265, 293, 301, 307
Adler, Alfred, 156
aesthetic decadence, 51, 93–94
aesthetics, 91–96, 350–356
The Age of Low Tech (Bihouix), 331
agricultural states
    adaptive cycle in, 266
    characteristics of, 106–112
    defense, 123–124
    disease in, 112–113
    domestication, 114–121
    energy use, 163–165
    gender roles, 84
    religion, 123, 125–128
    self-limiting behavior, 274–275
    transition to, 102
    values, 324
agriculture, 183, 238–239, 314
Akhenaten, 126
Akuno, Kali, 348
Alexander, Samuel, 331
algae, 29
American pika (*Ochotona princeps*), 272
anarchism, 280
animals
    behaviors, 41–46
    domestication, 78, 114–115, 120
    effect of size, 34–36
    emotions, 40
    energy use, 163
    extinction, 216–222, 260–266, 305
    as labor, 164, 166
    motion, 38–39
    physical power, 47

    proto-human powers, 49–55
    senses, 38
    warm-blooded advantage, 39–40
Anthropocene epoch, 160
anti-collapse coalition, 342–350
ants, 30, 45, 46, 53
Any Way You Slice It (Cox), 286
aquatic ape theory, 64–65
archaea, 23, 25, 26, 28
archaic states, 128
art, 91–96, 350–356
artists, power of, 12
Asch, Solomon, 148
Ashoka, 128–129
atheists, 131
Atlantic slave trade, 117–118
ATP (adenosine triphosphate), 25–27
Atwood, Margaret, 324–325
Australia, 70, 284
autism, 297
automobiles, 178–179, 208, 214
Axial Age, 128, 275–278

B

Babbage, Charles, 175
bacteria
    cooperation, 53
    disease, 112–113
    energy generation, 23–25, 26, 27
    food, 32
    genetics, 28–29
    photosynthesis, 31
    reproduction, 29–30
Bakunin, Mikhail, 280
banking, 143
beauty
    attraction and, 49–51

clothing evolution, 74–75
  influence of, 5, 8
  pursuit of art, 91–96
  striving for, 350–356
behaviors, 40–46
Bhagavad Gita, 346
Bichler, Shimshon, 145–146
Big God religions, 126–127, 238, 275–278, 287
Big Gods (Norenzayan), 128, 238
Big Men, 102–105, 274
Bihouix, Philippe, 331
biological evolution, 88–89
biological weapons, 249–250
bipedalism, 58–60
bonobos, 84–85, 88
The Botany of Desire (Pollan), 120
bowerbirds, 50
brain, 37, 59–60, 71–72, 75
Britain, 170–171, 177, 281, 284, 286
Brower, Danny, 293
Buddhism, 126, 131

C
California, 209–210
Calvin, William H., 61, 356
capital, 146
capitalism, 143–145, 161, 300, 307
Carbon Democracy (Mitchell), 176, 282
carbon dioxide, 206–208
carbon emissions, 268
carbon farming, 331–332
The Carbon Farming Solution (Toensmeier), 332
carbon sequestration, 209, 212
cars, 178–179, 208, 214
Carson, Rachel, 235
Catabolic Capitalism & Green Resistance (Collins), 342
Catching Fire (Wrangham), 69–72, 78
Catton, William, 73
cave art, 80–81, 91–92
cells. *See* archaea; bacteria
chemical industry, 172
chemical weapons, 249
chemotrophs, 24
Cherokee, 104–105
chiefdom societies, 104–105, 111–112

children, 119–120, 273, 277
chimpanzees
  gender hierarchy, 84–85, 88
  language, 44
  tools, 53, 66
  warfare, 45, 46
China, 173–174, 177, 236, 310
Christianity, 126, 129–131
civilizations, 110, 258–259, 265–266
class systems, 107–109, 114, 120, 351
climate change
  attempts to resolve, 207–210
  causes, 206–207
  comet impact and, 62–63
  denial of, 293–297
  in denial scenario, 305–312
  economic growth and, 321–322
  economists' view on, 146
  effect on human development, 61
  environmental movement, 284–285
  minimizing collapse, 349
  self-reinforcing feedbacks, 268
  transition from fossil fuel, 210–216
The Climate Mobilization, 347
clothing, 73–75
Clube, Victor, 62
club-winged manakins, 51
coal, 170–177, 195
coalition for change, 342–350
Cohen, Joel, 338
Cold War, 192
cold-blooded animals, 39–40
collapse, 10–11, 260–266
Collins, Craig, 342
colonialism, 130–131
commons, 232, 274–275, 290, 335–336
communication, 43–44, 75–83
communication technologies, 6, 131–138
communism, 187, 280
communities, development of, 100–102
complex societies, 265–266
computers, 137–138
consumerism, 138, 181, 299
consumption, 241–242, 312–316, 328–332
Cook, Earl, 175
cooperation
  effect of wars, 316–317
  global, 290–292, 322

in nature, 52–53
in societies, 198–201
studies in, 288–290
trust and, 306
Cooperation Jackson, 347–348
coronavirus pandemic, 216, 217–218, 310
The Cosmic Winter (Clube, Napier), 62
Costa Rica, 291
Cox, Stan, 286
credit, 139–143
crises, 1–2, 268
crops, 165
cultural evolution
    effect of communication technologies,
        131–138
    gender characteristics and, 89
    quest for beauty, 92, 94–96
    rate of, 54, 82–83, 357
    religion and, 125–129
cyanobacteria, 31–32
cyberweapons, 250

D
Daly, Herman, 145
Darwin, Charles, 49–50, 51
DDT, 235
De Decker, Kris, 330
death, 260–266
Debs, Eugene, 280
debt, 139–143, 242–248
Debt (Graeber), 139
deception, 42–43
decoupling, 315
degrowth, 291–292
democracy, 280
democratization, 175–177
Demonic Males (Wrangham,
    Peterson), 91
denial, 293–297, 305–312
Denial (Varki, Brower), 293
despair, 317–318
diesel, 180
disarmament, 290–291
disease, 112–113, 249–250
DNA, 28–29
domestication, 78–79, 114–121
dopamine, 36–37, 298
Dugin, Aleksandr, 190

E
Earth, 204–205, 207–208
eating, 32–34
ecological footprint, 241
economics, 139–147, 244–245
economists, 145–146
economy
    assumption of growth, 181–182
    debt, 242–248
    decoupling from consumption, 315
    effect of COVID-19, 217–218
    effect of wars, 190–191
    efficiency and, 313–315
    as entity, 181
    inequality, 187, 228–233
    oil and, 183–184
    self-reinforcing feedbacks, 307
ectotherms, 39–40
Edelman, Shimon, 80, 81
Edison, Thomas, 194
efficiency, 313–316
electricity, 193–198, 211–212
emotions, 40–41
endotherms, 39–40
energy
    advantage of, 5
    consumption decrease, 269–270,
        328–332
    effect on human values, 323–324
    measurement of, 21–22
    in origin of life, 20
    power and, 3–4
    pre-fossil fuel use of, 162–169
    renewable sources, 210–211, 315
    transfer in food, 33–34
    transfer of, 13
    transitions, 212–216
    See also electricity; fossil fuels
energy efficiency, 215
environmental impacts, in agricultural
    states, 110
    See also climate change
environmental movement, 284–285
EROEI (Energy Returned on Energy
    Invested), 162–164
Escherichia coli, 28
An Essay on the Principle of Population
    (Malthus), 239

eukaryotes, 25–27, 28, 32, 52
evolution
    cell development, 23–25
    control of power, 4
    gender and, 84–85
    group level, 31
    origin of life, 19–21
    use of power and, 17–18
The Evolution of Beauty (Prum), 50, 89
The Evolution of God (Wright), 127
exclusionary power, 12, 46
exploding ants (*Colobopsis
        saundersi*), 30
extinction, 216–222, 260–266, 294, 305
Extinction Rebellion, 285, 348

F
family planning, 339
fascism, 187
Fermi Paradox, 258–259, 301, 304
fertility rates, 339–341
fertilizers, 165, 182–183, 236
fighting, 44–46
financial regulation, 336
financialization, 232–233
fire, 55, 69–73
Fleming, David, 354
food
    cooking, 69–70, 71–72
    effect on culture, 102
    energy from, 32–34
    production, 164–165
    rationing, 285–287
Foragers, Farmers, and Fossil Fuels
        (Morris), 323–324
fossil fuels
    coal, 170–177, 195
    effect of, 9–10, 161–162, 293
    factors of production and, 146
    gasoline, 172, 178–179, 208
    natural gas, 182–183
    oil, 177–178, 180, 183–193
    power increase and, 4–5
    resource depletion, 222–228, 307
    role in alternate energy sources,
        210–211
    transition from, 212–216, 314
    wars, 185–193

Four Horsemen of leveling, 228–229
Frankl, Viktor, 317–318
freight shipment, 180
Fridley, David, 212–213
future
    denial scenario, 305–312
    faith in, 317–318
    power struggles, 341–350
    self-restraint scenario, 312–317
    shifts in power, 356–359
future risk, 299

G
game theory, 288–292, 316–317
Gandhi, Mahatma, 98, 283
gardening, 102–103
gasoline, 172, 178–179, 208
gay rights, 284
GDP (Gross Domestic Product), 320–321
gender
    characteristics, 87–89
    effect of industrialization, 181
    power and, 83–85, 154
    roles, 90–91, 119–120
    violence and, 86–87
generalist species, 51–52
genes, 28–29
genocide, 113, 152–154, 249, 253–254
geopolitics, 183–185, 188–190, 309–311
George, Henry, 335–336
Germany, 185–187, 188
gift economy, 139, 140–141
Gilens, Martin, 231–232
Gini index, 231
global cooperation, 290–292, 322
global economy
    debt, 242–248
    in denial scenario, 308–309
    future of, 319–320, 321–323, 345
    oil and, 183–184
global financial system, 191
The Goodness Paradox (Wrangham), 78
gorillas, 53
governments, 281–282, 287–288, 308
Graeber, David, 139
grains, 109
Great Acceleration, 9, 160–161, 210, 243,
        269, 301

The Great Leveler (Scheidel), 190, 228–229, 282
Great Unraveling, 349, 358
Green America, 151
greenhouse gas emissions, 207–210
Greenspan, Alan, 295
group selection, 88, 105–106
guns, 253–254

H
Half Earth (Wilson), 220
happiness, striving for, 350–356
Harari, Yuval Noah, 296
Hardin, Garrett, 274
Hardy, Alister, 64
Harris, Marvin, 101–102
Haushofer, Karl, 190
Hawken, Paul, 151
health, impacts of pollution, 237
Heidegger, Martin, 156
Hill, Joe, 127
Holmgren, David, 331
Holocene epoch, 98
hominins, 58–61, 63–64, 66–69
Homo erectus, 60, 67, 71, 72, 75
Homo habilis, 60, 67
Homo neanderthalis, 60, 67, 72, 75, 81
Homo sapiens
    adaptation by, 100
    adoption of clothing, 73–75
    aquatic ape theory, 64–65
    cooperation, 53, 198–201
    development of power, 8–9
    domestication, 115–116
    energy use, 163
    language development, 75–83
    migration of, 61, 63–64
    physical developments, 58–60
    physical power, 48–49
    population ecology, 263–264
    reproduction, 30
    tools, 66–69, 96
    use of fire, 69–73
Homo species, 60
Hopkins, Rob, 348, 349
horizontal social power
    definition, 12
    religions and, 275–278
    shift from, 100
    social scale and, 147–148
    strategies for, 342–350
    through spirituality and art, 355
horsepower, 22
horticultural societies
    development of, 100–105
    equality in, 119, 351
    persistence of, 111
    slavery, 118
    social power, 274
    spiritual practices, 122–123
How Many People Can the Earth Support (Cohen), 338
Human Scale (Sale), 151–152
humans. See *Homo sapiens*
hunter-gatherer societies
    energy use, 163
    gender roles, 90–91, 119
    resistance to change, 111
    self-limiting behavior, 273–274
    spiritual practices, 122, 128
    transition from, 100, 102
    values, 324
hydro power, 196

I
ideas, power of, 12
income, 230, 231
India, 236
Indigenous peoples
    colonialism, 129–131
    genocide, 113, 153, 249, 253–254
    resource management, 273–274, 275
industrial societies. See modern society
industrialization, 175–177
industry, 175, 180–181, 315
inequality, 6, 334–337, 350–351
information warfare, 310
Ingold, Tim, 116
institutional power, 150–152
intelligence, 41–42
International Panel of Climate Change (IPCC), 297
internet, 137
involuntary power limits, 260–266
irrigation, 164–165
Islam, 126, 131

J

Japan, 187–188, 190
Jaspers, Karl, 128
Jefferson, Thomas, 151
Jones, William, 125
Judaism, 126

K

Kaku, Michio, 258
Kallima butterfly, 42
Kaplan, Robert, 190
Kardashev, Nikolai, 258
Kayser, Manfred, 74
Keen, Steve, 145
Keltner, Dacher, 149
kerosene, 178, 180
kingdoms. *See* agricultural states
kings, 107, 123, 128
Kissinger, Henry, 190
Kittler, Ralf, 74
Kleiber's law, 34
Kolodny, Oren, 80, 81
Kropotkin, Peter, 277
Krumdieck, Susan, 329–330
Ku Klux Klan, 151
!Kung people, 163, 273
Kurdistan Workers' Party (PKK), 337

L

labor, 181, 184
labor unions, 175–176
land ownership, 109, 221
Lane, Nick, 34
language, 5, 75–83, 125
The Language Instinct (Pinker), 77
Lee, Richard, 163, 273
Leibniz, Gottfried, 17
life. *See* evolution
life's purpose, 317–318
lifespan, 34–35
Lincoln, Abraham, 281
Lloyd, William Forster, 274
Locke, John, 145
Lotka, Alfred, 17
lottery winner's syndrome,
      299–300
Low-Tech Magazine, 330
low-tech strategies, 329–333

M

Machiavelli, Nicolo, 155–156
MacKay, Charles, 295
Mackinder, Halford, 189–190
Mae Enga, 103–104
Magna Carta, 277
Mahon, Alfred Thayer, 189
Malthus, Thomas, 239
Man, Energy, Society (Cook), 175
man-made systems, 47
Mann, Michael, 157, 185–186
Man's Search for Meaning (Frankl), 317–318
Marx, Karl, 280
maximum power principle, 4, 17, 301
McGowan, Kat, 41
McLuhan, Marshall, 194
McNeill, William, 112
men. *See* gender
methane, 206
Middle East, 183–184, 193, 311
migrants, 308
Milgram, Stanley, 148
military, 6, 124, 192
Mitchell, Timothy, 176, 282
mitochondria, 26–27
Modern Monetary Theory, 247–248
modern society, 279–288, 160–162,
      170–173, 175–185, 324
Mollison, Bill, 331
money, 6, 139–147, 345
Monson, Clark, 274
Montgomery, Sy, 41
Morgan, Elaine, 64
Mormon Church, 238
Morris, Ian, 323–324
mortality, awareness of, 261, 293–294
movies, 136
multicelled organisms, 30–31, 34–38
      See also animals
Mumford, Lewis, 175
muscles, 36
music, 91–96
musical notation, 135–136
Mutual Aid (Kropotkin), 277

N

Napier, Bill, 62
nationalism, 300

Native Americans, 113, 153, 249, 253–254,
    130
natural gas, 182–183
natural resources, ownership, 335–336
natural selection, 88
natural systems, 47, 160–161
nature
    behaviors, 40–46
    disappearance of, 216–222
    imbalances, 293
    integration with, 350–352, 354
    power in, 8
    power limits, 260–266
    power systems in, 16–18, 31–34, 38–40,
        49–55
    reproduction, 37–38
    self-limitation, 30--31, 271–272
    specialization, 55–56
neoliberal economics, 232, 316
neonicotinoids, 235–236
neurons, 36–37
New Zealand, 284
Newcomen, Thomas, 171
Nietzsche, Friedrich, 156
nitrous oxide, 206
Nitzan, Jonathan, 145–146
nonhuman animals. *See* animals
nonrenewable resources, 222, 225–226
Nordhaus, William, 146
Norenzayan, Ara, 127–128, 238, 287
novelty, 298
nuclear power, 195–196, 236, 285
nuclear weapons, 192, 251–252
number systems, 139–140

O
Obhi, Sukhvinder, 149
oceans, 207
Odum, Howard, 17, 34
oil, 177–178, 180, 183–193
*On Economy and the Machine and Manufac-
    turers* (Babbage), 175
optimism bias, 295–296
The Optimism Bias (Sharot), 295
optimum power principle, 4, 260, 342–350
oral stories, 133
Ostrom, Elinor, 275
over-empowerment, 204–206

overpopulation. *See* population
Overshoot (Catton), 73
Overton Window, 296–297
ownership, 335–336

P
Page, Benjamin, 231
pandemics, 112–113
Papin, Denis, 171
patriotism, 300
permaculture, 331
pesticides, 235–236
Peterson, Dale, 91
PFAS (polyfluoroalkyl substances), 237
philanthropy, 345
photosynthesis, 31–32
physical power
    definition, 11
    examples, 47–49
    measures of, 21–22
    restraint, 279, 285
    social power and, 7, 154
Pinker, Steven, 77, 105
Pitt, William, 281
Plagues and Peoples (McNeill), 112
A Planet of 3 Billion (Tucker), 339
plants
    behaviors, 41–46
    domestication, 115, 120
    emotions, 40–41
    motion, 39
    perception, 38
    photosynthesis, 31–32
plastic pollution, 236–237
Pleistocene epoch, 98
political power
    destabilization, 309–310
    effect of communication technologies,
        138
    future changes, 311
    imbalance, 229–231
    industrialization and, 175–177
    leaders, 12–13
    power sharing theories, 280
Pollan, Michael, 120
pollution, 234–238
Polo, Marco, 170
population

cycles in nature, 262–263
in denial scenario, 305–312
growth rate, 30
increases in, 160, 238–241
management of levels, 316, 337–341
state development and, 106
trade-off with consumption, 312–314
Population Media Center (PMC), 339
positive feedbacks, 267–268
power
   as causative agent, 1–2
   definition, 3
   effect of accumulation, 7–8, 204–206
   energy reduction and, 215
   future shifts, 10–11, 356–359
   involuntary curbing, 260–266
   paradox of, 345–346
   pursuit of, 4
   relinquishing, 256
   tools of, 5
   types of, 11–13
   voluntary curbing, 266–279
powerlessness, 155
power-limiting behaviors. See self-
   limitation
Precautionary Principle, 327
predation, 114, 120–121, 262–263
primates, 58–59, 77, 84–85, 87–88, 199
printing press, 134–135
prisoner's dilemma game, 288–292
production, factors of, 146
property rights, 145–146
protests, 309
proto-humans. See hominins
proton pumps, 20, 23, 27
Prum, Richard, 50, 89, 91
Putin, Vladimir, 310

Q
quota rationing, 286–287

R
racism, 118, 336
radio, 136
rare species, 271–272
rationing, 285–287
Ratzel, Friedrich, 189
refugees, 308

regulations, 279–280
Reich, David, 124–125
Reinhart, Carmen, 245
religion
   development of, 82
   effect of communication technologies,
      132, 134, 135
   future of, 308, 353
   population levels and, 316
   self-limitation and, 287
   social power and, 99, 121–131, 275–278
renewable energy, 167–168, 210–212, 315
renewable resources, 222, 225–226, 314
reproduction, 29–30, 37–38
resources
   depletion, 222–228, 305–312
   management, 273–274, 275, 314
   ownership, 335–336
respiration, 25–26
rights of nature, 336
robotic weapons, 250–251
Rodney, Walter, 118
Rogoff, Kenneth, 245
Roman Empire, 117
Russia, 193

S
Salamon, Margaret Klein, 347
Sale, Kirkpatrick, 151–152
same-sex sexual behavior, 90–91, 284
Sanders, Bernie, 336
Sapiens (Harari), 296
Savery, Thomas, 171
Scheidel, Walter, 190, 228–229, 233, 282
science, effect of innovation, 160–161
scientific notation, 18–19
self-domestication, 78–79
self-limitation
   barriers to, 292–301
   energy use, 269–270
   importance of, 11
   in modern technology, 267–268
   in nature, 30–31, 266–267, 271–272
   throughout history, 273–279
self-reinforcing feedbacks, 267–269, 307
self-restraint scenario, 312–317
sense organs, 38
sexual attraction, 13

sexual relations, 83–85, 87–91, 284
sexual reproduction, 37–38, 50–51
sexual selection, 92, 95
Sharot, Tali, 295
Shea, John, 80
Silent Spring (Carson), 235
Silverman, David J., 253–254
Simplicity Institute, 331
single-celled organisms. *See* archaea;
    bacteria
slavery, 115–119, 141, 175, 283
Smil, Vaclav, 163
The Smoke of Great Cities (Stradling,
    Thorsheim), 172–173
Snyder, Gary, 133
social complexity, 5
social contact, 59–60
social justice, 334–337
social media, 193
social power
    definition, 12–13
    development of, 101–105, 205, 342–350
    measurement of, 98–99
    physical power and, 7
    pollution and, 234–235
    restraint, 279–288
    tools of, 6
    types of, 147–148
    See also horizontal social power; verti-
        cal social power
social programs, 281–282
social revolution, 175–177
socialism, 280
societies
    cooperation, 198–201
    effect of energy sources, 169
    future of, 319–320, 321–323
    imbalances, 293
    power restraint, 279–288
    pro-social behavior, 298–299
    realms of, 101–102
    self-limiting behavior, 273–279
    striving for happiness, 350–356
socio-political power, 175–177, 183–184
solar energy, 210–212
Solow, Robert, 146
The Soul of an Octopus (Montgomery), 41
sound recordings, 135–136

The Sources of Social Power (Mann), 157,
    185–186
South Africa, 291
specialist species, 51–52, 55–56
spiritual leaders, 12–13
spiritual practices. *See* religion
spirituality, 350–356
sports, 95–96
Stanford Prison Experiments, 149
state societies. *See* agricultural states
status, pursuit of, 297–298
steam engine, 171–172
Stephenson, George, 172
Stone Tools in Human Evolution (Shea), 80
Stoneking, Mark, 74
Stradling, David, 172–173
suffragists, 283–284
Sumeria, 131–132, 140
Summers, Graham, 246
Sun, 16, 20–21, 204
sunlight, 16–17
Superorganism, 198–201, 319–320,
    321–323
superorganisms, group-level evolution, 31
Surviving the Future (Fleming), 354

T
taboos, 273–274
taxation, 230, 231, 281
technologically advanced civilizations,
    258–259
technology
    assessment of, 325–328
    effect of innovation, 160–161
    extinct species revival, 221–222
    future strategies for, 328–332
    military innovations, 192
    solving climate change, 210–216
telegraph, 135
telephone, 135
television, 136–137
territorial power, 46
Terror Management Theory, 261
Tesla, Nikola, 194
Thorsheim, Peter, 172–173
Thunberg, Greta, 297
Thundersticks (Silverman), 253
Tiglath Pilesar, 129

Tjungurrayi, Jimmy, 133
Tobin, James, 146
Tocqueville, Alexis de, 118
Toensmeier, Eric, 332
tools, 5, 53–55, 66–69, 80
"The Tragedy of the Commons" (Hardin), 274
Trainer, Ted, 331
Transition Engineering (Krumdieck), 330
Transition Initiatives, 348
Trewavas, Anthony, 42
trust, 126, 288–292, 306–307, 308, 316–317
Tucker, Christopher, 339
Turchin, Peter, 106, 127, 229, 266
2,000-Watt Society, 269–270

U
United Nations, 207
United States
    coal use, 175, 177
    COVID-19 response, 310
    economic effects of war, 190–191
    economic recovery, 246–247
    global dominance, 191–193
    oil resources, 183
    rationing, 286
    social programs, 282
    taxation, 281
    voting rights, 284
    wealth inequality, 229–231
the universe, 16
uprisings, 309
USSR, 192, 280–281

V
values, 323–324
Varki, Ajit, 293
vertical social power
    definition, 12
    development of, 7, 9, 100, 105
    effect of, 147–150, 152, 154–155
    effect of religions, 275, 277–278
    overcoming, 347
violence, gender and, 86–87
violinists, 94–95
viruses, 19–20, 112–113
Volterra, Vito, 17

W
Waal, Frans de, 40
warm-blooded animals, 39–40
wars and warfare
    avoidance of, 316–317
    energy in, 185–193
    gender and, 86
    societal defense against, 100, 106, 124
    weapons development, 248–252
water power, 168
Watt, James, 21, 22, 171
watts, 21–22
wealth
    consumption decrease and, 316
    development of money, 139–147
    inequality, 6, 228–233, 281, 336
    lottery winner's syndrome, 299–300
    social power and, 98–99
wealth pump, 141
weapons, 6, 66–67, 253–254
weapons of mass destruction, 192, 248–252
Who We Are and How We Got Here (Reich), 124–125
wildfires, 204–205
wildlife. *See* nature
Wilson, Edward O., 78, 220
wind power, 167–168, 210–212
women, 181, 339, 283–284
Wood, Peter John, 291
wood, use of, 166–167
World War I, 185–186, 190
World War II, 187–188, 190–191, 286
World Wildlife Fund, 219
Wrangham, Richard, 69–72, 78, 81, 91
Wright, Robert, 127
writing, 131–134

Y
Yenni, Glenda, 271, 272
Younger Dryas, 62–63

Z
Zimbardo, Philip, 149

About the Author

RICHARD HEINBERG is the author of thirteen previous books including *The Party's Over*, *Powerdown*, *Peak Everything*, and *The End of Growth*. He is Senior Fellow of the Post Carbon Institute and is widely regarded as one of the world's most effective communicators of the urgent need to transition away from fossil fuels. Heinberg has given hundreds of lectures on our energy future to audiences around the world. He has been published in *Nature* and other journals and has been featured in many television and theatrical documentaries. He lives in Santa Rosa, CA.

ABOUT NEW SOCIETY PUBLISHERS

New Society Publishers is an activist, solutions-oriented publisher focused on publishing books for a world of change. Our books offer tips, tools, and insights from leading experts in sustainable building, homesteading, climate change, environment, conscientious commerce, renewable energy, and more—positive solutions for troubled times.

We're proud to hold to the highest environmental and social standards of any publisher in North America. When you buy New Society books, you are part of the solution!

- We print all our books in North America, never overseas

- All our books are printed on **100% post-consumer recycled paper**, processed chlorine-free, with low-VOC vegetable-based inks (since 2002)

- Our corporate structure is an innovative employee shareholder agreement, so we're one-third employee-owned (since 2015)

- We're carbon-neutral (since 2006)

- We're certified as a B Corporation (since 2016)

- We're Signatories to the UN's Sustainable Development Goals (SDG) Publishers Compact (2020–2030, the Decade of Action)

At New Society Publishers, we care deeply about *what* we publish—but also about *how* we do business.

To download our full catalog, please visit newsociety.com/pages/nsp-catalogue.

Sign up for New Society Publishers' newsletter for information on upcoming titles, special offers, and author events (https://signup.e2ma.net/signup/1425175/42152/).

ENVIRONMENTAL BENEFITS STATEMENT

New Society Publishers saved the following resources by printing the pages of this book on chlorine free paper made with 100% post-consumer waste.

TREES	WATER	ENERGY	SOLID WASTE	GREENHOUSE GASES
85	6,800	36	290	36,700
FULLY GROWN	GALLONS	MILLION BTUs	POUNDS	POUNDS

Environmental impact estimates were made using the Environmental Paper Network Paper Calculator 4.0. For more information visit www.papercalculator.org

Certified

B Corporation

FSC
www.fsc.org

MIX
Paper from responsible sources
FSC® C016245

new society
PUBLISHERS
www.newsociety.com